75

Fortschritte der Chemie organischer Naturstoffe

Progress in the Chemistry of Organic Natural Products

Founded by
L. Zechmeister

Edited by
W. Herz, H. Falk,
G. W. Kirby, R. E. Moore,
and Ch. Tamm

Authors:
L. A. Collett, M. T. Davies-Coleman,
D. C. Gournelis, G. G. Laskaris,
D. E. A. Rivett, R. Verpoorte

Springer-Verlag Wien GmbH

Prof. W. Herz, Department of Chemistry,
The Florida State University, Tallahassee, Florida, U.S.A.

Prof. Dr. H. Falk, Institut für Chemie,
Johannes-Kepler-Universität, Linz, Austria

Prof. G. W. Kirby, Chemistry Department,
The University of Glasgow, Glasgow, Scotland

Prof. R. E. Moore, Department of Chemistry,
University of Hawaii at Manoa, Honolulu, Hawaii, U.S.A.

Prof. Dr. Ch. Tamm, Institut für Organische Chemie der Universität Basel,
Basel, Switzerland

Originally published by Springer-Verlag Wien New York in 1998
Softcover reprint of the hardcover 1st edition 1998

Library of Congress Catalog Card Number AC 39-1015

Typesetting: Thomson Press (India) Ltd., New Delhi

Graphic design: Ecke Bonk
Printed on acid-free and chlorine-free bleached paper

SPIN: 10655417

With 26 Figures

ISSN 0071-7886
ISBN 978-3-7091-7340-4 ISBN 978-3-7091-6507-2 (eBook)
DOI 10.1007/978-3-7091-6507-2

Contents

List of Contributors

COLLETT, Miss L. A., Department of Chemistry, Rhodes University, P.O. Box 94, Grahamstown 6139, Republic of South Africa

DAVIES-COLEMAN, Prof. M. T., Department of Chemistry, Rhodes University, P.O. Box 94, Grahamstown 6139, Republic of South Africa

GOURNELIS, Prof. D. C., Laboratory of Pharmacognosy, Department of Pharmacy, Aristotelian University of Thessaloniki, 540 06 Thessaloniki, Greece

LASKARIS, Dr. G. G., Division of Pharmacognosy, Center for Drug Research, Leiden University/Vrije Universiteit Amsterdam, Einsteinweg 55, 2300 RA Leiden, The Netherlands

RIVETT, Prof. D. E. A., Department of Chemistry, Rhodes University, P.O. Box 94, Grahamstown 6139, Republic of South Africa

VERPOORTE, Prof. R., Division of Pharmacognosy, Center for Drug Research, Leiden University/Vrije Universiteit Amsterdam, Einsteinweg 55, 2300 RA Leiden, The Netherlands

Cyclopeptide Alkaloids

D. C. Gournelis[1], G. G. Laskaris[2], and R. Verpoorte[2]
[1] Laboratory of Pharmacognosy, Department of Pharmacy, Aristotelian University of Thessaloniki, Thessaloniki, Greece
[2] Division of Pharmacognosy, Leiden/Amsterdam Center for Drug Research, Leiden, The Netherlands

Contents

1. Introduction

Cyclopeptide alkaloids are defined as basic compounds embodying an ansa structure, in which a 10- or 12-membered peptide type bridge spans the 1,3 or 1,4 positions of a benzene ring (*1*). They are widely distributed among plants of the Rhamnaceae family, but their occurrence has also been confirmed in representatives of Asteraceae, Celastraceae, Euphorbiaceae, Menispermaceae, Pandaceae, Rubiaceae, Sterculiaceae and Urticaceae. These compounds are found in leaves, stem bark, root bark and seeds. They often occur in minute amounts and as complex mixtures. The total yield from dried plant material is between 0.01 and 1% and depends on many factors such as the region of growth, the season of collection, the maturity of the plant used as well as the method of isolation (*2–6*). The general structure of cyclopeptide alkaloids is designated in Fig. 1.

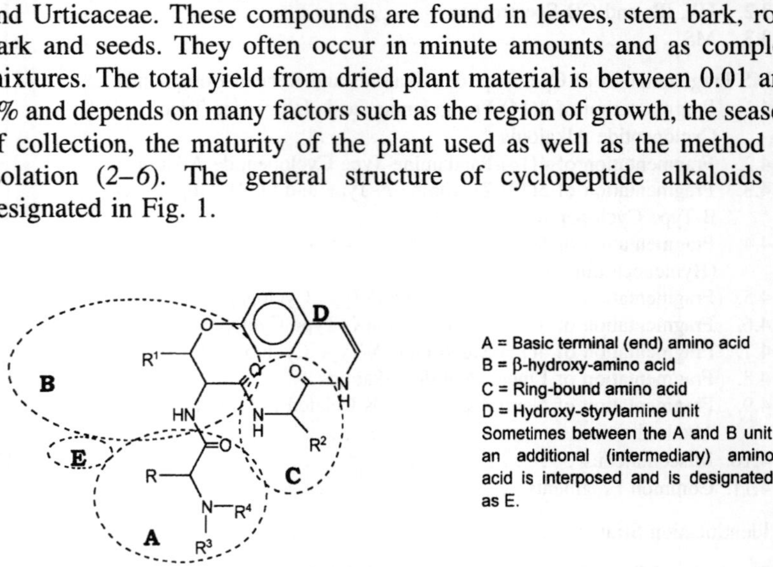

Fig. 1. General structure of cyclopeptide alkaloids

Although several reviews have been published covering 110 cyclopeptide alkaloids (*1–7*), there are no recent ones. In the present work we will review this field. The present article records all cyclopeptide alkaloids and related compounds reported in the literature from 1963 when the first representatives, adouetine-X, -Y and -Z (*9*) and zizyphine-A and -B (*10*), were isolated until the late 1995. In fact it starts from 1966 when the complete identification of pandamine was realized (*11*). Thus we will deal with 160 cyclopeptide alkaloids, including genuine cyclopeptide alkaloids (*sensu stricto*), some linear (open) peptide alkaloids as well as neutral compounds that do not exhibit basic properties and are not considered to be alkaloids. Linear peptides and neutral compounds are closely related structurally (and biogenetically) to genuine cyclopeptide alkaloids and have been isolated from the same sources as cyclopeptide alkaloids. A clear exception are the celenamides which have been isolated from sponges (*60, 61*). The linear peptides are considered to be the biogenetic precursors of the cyclic structures.

2. Classification

Different methods of classification have been proposed in previous review articles (*1, 2, 4–6*). In this review the compounds will be presented in a way that copes with the plethora of structures that have been isolated in the meantime. It was originally proposed by Joullié and Nutt (*6*) and comprises three classification steps:

First step:
1) Cyclopeptide alkaloids *sensu stricto* (151 compounds).
2) Linear peptide alkaloids (6 compounds).
3) Neutral compounds (3 compounds).

Second step:
This is confined only to cyclopeptide alkaloids *sensu stricto* and classifies them according to the size of the macrocycle: 13-, 14- or 15-membered rings. The 14-membered ring class has the most representatives.

Third step:
The 13-, 14- or 15-membered ring compounds are further divided according to the number of their units: 4 (having A, B, C and D units) and 5 (with A, B, C, D and E units). Consequently the cyclopeptide alkaloids *sensu stricto* are subdivided into groups with the following

Table 1. *Nomenclature of Cyclopeptide Alkaloids*

Number of units	13 atoms	14 atoms			15 atoms
		4(14)-compounds Nature of β-OH amino acid (unit B)			
4 units	4(13)-compounds (Nummularine-C-type) (Pro)	Frangulanine-type (Leu)	Integerrine-type (Phe)	Amphibine-F-type (Pro)	4(15)-compounds (Mucronine-A-type)
5 units	5(13)-compounds (Zizyphine-A-type) (Pro)	5(14)-compounds			
		Scutianine-A-type (Leu or Phe)		Amphibine-B-type (Pro)	

annotations: 4(13), 5(13), 4(14), 5(14) and 4(15). The 4(14) and 5(14) alkaloids are further subdivided according to the nature of the β-OH amino acid (B unit). The nomenclature is shown in Table 1.

Exceptions:
Within the 4(14)-compounds is the group of pandamine type compounds (8 compounds total) which contain a 2-alkoxy-2-(*p*-hydroxyphenyl)-ethylamine D unit, instead of styrylamine. Among the 5(14) compounds hymenocardine contains valine as the B unit and a 2-alkoxy-2-(*p*-hydroxyphenyl)ethylamine D unit, instead of styrylamine.

The nomenclature of every group, *e.g.* the frangulanine type, is based on the first representative to have been isolated and has been kept unaltered in this review by using the trivial names. This kind of classification allows easy correlation with or references to well known compounds and in this sense the prefixes *N*- and *O*-desmethyl or *N*- and *O*-methyl were retained.

We would like to stress that several cyclopeptide alkaloids have been recorded as new compounds without mention of or reference to previously isolated compounds possessing identical chemical structures. Some of the newly isolated cyclopeptide alkaloids must be identical with or stereoisomers of older ones. Therefore, daechuine-S10 (**15**) is superficially identical with nummularine-R (**14**), daechuine-S5 (**47**) with melonovine-A (**46**) or pubescine-A (**45**), lotusanine-A (**63**) with adouetine-Y′ (= myrianthine-B) (**62**), discarine-X (**73**) with nummularine-K (**74**), discarine-D (**97**) with crenatine-A (**96**) and AM-2 (**95**) with aralionine-B (**94**). Only for AM-2 (**95**) has the relationship with aralionine-B, although unresolved, been mentioned (*120*). Sanjoinine-G2 (**152**) has also been prepared from frangufoline (*63*).

3. Structure Elucidation – Stereochemistry

The main tools for structure elucidation nowadays are MS and NMR spectroscopy. MS is extensively used as an important tool for identification. Besides determining the molecular weight, it provides substantial information about the structure of the alkaloids since the various groups of cyclopeptide alkaloids fragment in a predicted and well described way. In recent years use of high field NMR spectroscopy has made structure elucidation of the cyclopeptide alkaloids considerably easier.

3.1. NMR Spectroscopy

It is being extensively used for structure elucidation of cyclopeptide alkaloids. In ^{1}H-NMR spectra the most eminent features are the styryl olefinic protons which resonate around δ 6.4 (β-H-Sty) and 6.6 (α-H-Sty) with a coupling constant of ca. 8 Hz. Aromatic proton signals occur at δ 6.9–7.8 usually as multiplets, while aromatic methoxyls resonate clearly around δ 3.8.

Broadbent and Paul presented ^{13}C-NMR spectra of a series of alkaloids, including 12 cyclopeptides (*8*). While MS is extensively used, NMR remains unique for determining the relative configuration of the β-OH amino acid. Thus in *erythro* compounds $J_{a,\beta} \cong 8$Hz, whereas in *threo* compounds $J_{a,\beta} \cong 2$Hz (*12–14*). The β-OH amino acid of the overwhelming majority of cyclopeptide alkaloids is in the L-*erythro* form. As an exception we mention scutianine-E (**78**) which contains D-*erythro*-β-OH-leucine. Furthermore while the other amino acids generally occur in the L-form, scutianine-E bears a D-*threo*-β-phenylserine as the ring-bound amino acid. ^{13}C-NMR spectroscopy is used for the elucidation of the absolute configuration of the β-OH amino acid: in compounds with the D-*erythro* configuration C-β resonates at δ 87, while in compounds with the L-*erythro* configuration C-β is found at δ 81.5; also C-α resonates at δ 53.8 for the D-*erythro* form and at δ 55 for the L-*erythro* form (*15*, *16*) (Table 2). ^{13}C-NMR spectra can be used to distinguish between *cis* or *trans* conformations when the β-OH amino acid is proline. The rule is: C-γ-Pro: *trans*, δ = 21.2–21.6 and *cis*, δ = 23.8–24.1

Table 2. *Assignments of the Configuration of the β-OH Amino Acid by NMR*

	C-α-shift (δ)	C-β-shift (δ)	$J_{\alpha,\beta}(^1H)(Hz)$
D-*erythro*	53.8	87	8
L-*erythro*	55	81.5	8
Threo			2

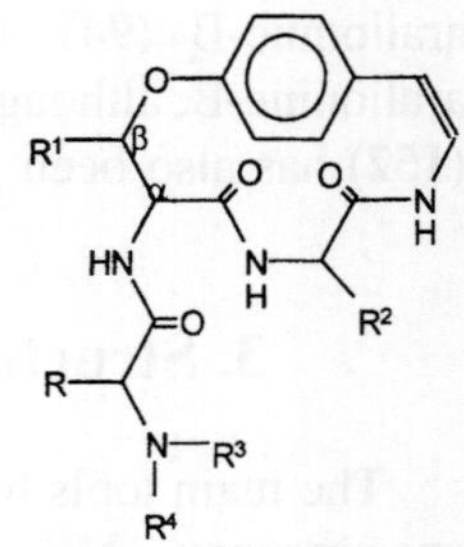

Table 3. *Assignments or* cis *or* trans *Conformations of Proline*

	C-γ shift (δ)
trans	21.2–21.6
cis	23.8–24.1

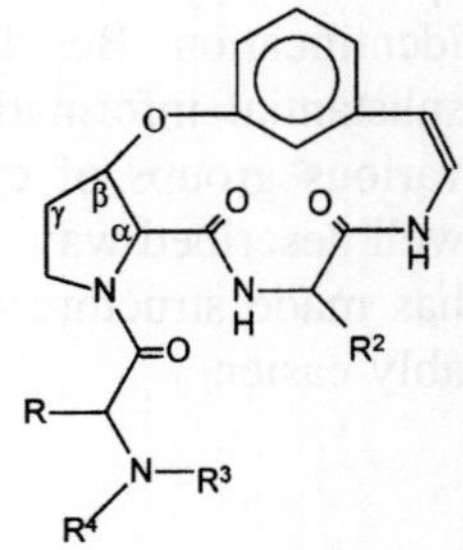

Table 4. *Cyclopeptide Alkaloids for which the Configuration of the β-Hydroxy-amino Acid has been determined by NMR Spectroscopy either by the Original Authors (denoted by "1") or from the Reported Data by us (denoted by "2")*

Compound	Configuration	Assignment	Reference
3	*trans*-β-OH-Proline	1	*71*
36	*trans*-β-OH-Proline	1	*71*
44	*erythro*-β-OH-Leucine	1	*98*
49	*erythro*-β-OH-Leucine	1	*47*
63	L-*erythro*-β-OH-Leucine	1	*118*
67	L-*erthryo*-β-OH-Leucine	2	*63*, *59*
95	N-Me-L-Phenylalanine	1	*120*
	erythro-β-OH-Phenylalanine	2	*120*
	L-isoleucine	1	*120*
158	*trans*-cinnamic acid	1	*63*, *59*
	threo or	2	*59*
	erythro-β-OH-Leucine	2	*63*
160	*trans*-cinnamic acid	1	*59*
	L-*erythro*-β-OH-Leucine	1	*59*

(*16*, *17*) (Table 3). Based on these rules the stereochemistry of some cyclopeptide alkaloids has been determined. Some examples can be found in Table 4.

The older use of chemical degradation as a means of structure elucidation is nowadays confined to identification of the amino acid composition. To gain information on the configuration, special procedures, like ozonolysis, should be applied so that the stereochemistry of the asymmetric centers (especially of the β-OH-amino acid) will be kept unaltered after hydrolysis (*12*, *18–20*). Enzymes have been used for this reason (L- or D-amino acid oxidases), but their selectivity is not absolute (*12*, *18*, *20*).

For some cyclopeptide alkaloids the absolute configuration of the molecule was deduced by X-ray crystallography (*21–23*).

3.2. UV, IR, and CD Spectroscopy

Interestingly, the 14-membered ring compounds most of the time do not show characteristic UV minima or maxima despite the presence of the styrylamine unit which, due to conjugation, should absorb intensively. The reason for this anomalous behavior is the strain of the ring system which hinders p-orbital overlap. This phenomenon sometimes is described as "end absorption". Some of the compounds belonging to this category show a weak absorption around 280 nm. When tryptophan is present, characteristic absorptions are observed at ca. 220, 270 and 290 nm. In 13-membered cyclopeptide alkaloids the strain is relieved and their UV spectra reveal absorptions around 270 (log $\varepsilon \sim 4.0$) and 320 nm (log $\varepsilon \sim 3.8$). 15-Membered cyclopeptide alkaloids exhibit a maximum around 275 nm (log $\varepsilon \sim 4.2$) and sometimes a shoulder at ca. 220 nm.

IR spectra are insensitive to ring strain and exhibit highly diagnostic peaks belonging to secondary amido groups (peptide bonds) at ca. 1680–1690 and 1630–1655 cm^{-1}. The occurrence of these peaks strongly suggests the existence of cyclopeptide alkaloids. Other bands are visible at: ~3300 (NH), 2780–2790 (*N*-Me), 1625 (styryl double bond), 1230–1240 (aryl-ether) and 2830 cm^{-1} (aromatic OCH_3).

The CD spectra of 13-membered cyclopeptide alkaloids reveal strong negative, Cotton effect bands at 324, 276, 254 and 219 nm and a small positive one at 232 nm. The spectra of 14-membered cyclopeptide alkaloids exhibit a weak positive band at 287 and an intense negative one at 239 nm. This behavior is due to the existence of the L-amino acid in position B or C. For the D-enantiomer the negative peak becomes intensely positive. This is illustrated by scutianine-D (**77**) which has L-β-OH-Leu and L-β-OH-Phe as the B and C units respectively and its isomer scutianine-E (**78**) having the corresponding D-forms.

3.3. MS

The following presentation of the mass spectral fragmentation of all cyclopeptide alkaloids and related compounds covers all mass spectra recorded in the literature cited. Several articles describe fragmentation mechanisms and the resulting fragments (*2*, *24–59*) and, without overlooking the danger of oversimplification, we can claim that there is a constant basic pattern of fragmentation in all cyclopeptide alkaloid groups except in the 4(15)-compounds. Based on these patterns, the MS fragmentation of cyclopeptide alkaloids is presented in such a way that fragments can be easily used as a tool in structure elucidation.

In every group of compounds (*e.g.* 5(14)-scutianine-A type) every letter that is used to denote a fragment is uniquely attributed to one fragment in this particular group.

4. Fragmentation of Cyclopeptide Alkaloids and Related Compounds

4.1 Fragmentation of 4(14)-Frangulanine- and -Integerrine-Type Cyclopeptide Alkaloids

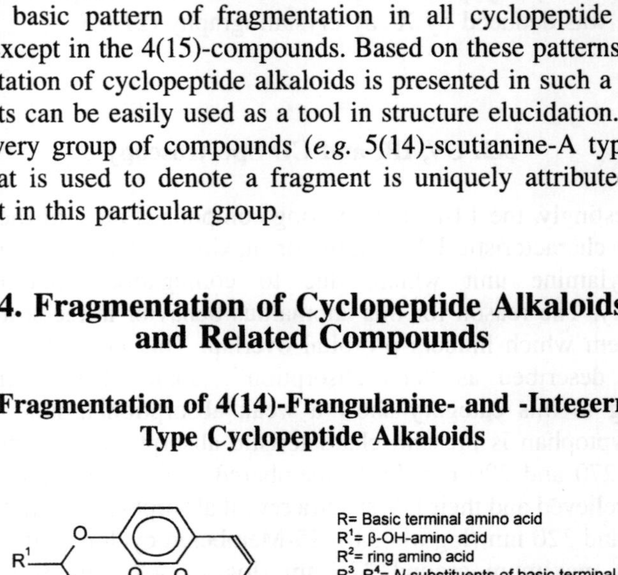

The main fragments of these 4(14) cyclopeptide alkaloids are shown below in Fig. 2 (*m/z* values of ions are in parentheses).

It has already been mentioned that in 4(14)-frangulanine-type cyclopeptide alkaloids R^1 is the β-OH-Leucine side chain, while in 4(14)-integerrine-type cyclopeptide alkaloids R^1 is the β-OH-Phenylalanine side chain. This leads to the generalization shown in Table 5.

Fragments **n-v** (Fig. 3) are of high diagnostic value. Like fragment **a** (base peak), fragments **o**, **p** and **n** denote the nature of the basic terminal amino acid. According to the literature, fragments **n** and **p** arise when R is a Val or Ileu side chain, with elimination of a C_2H_5 radical or neutral ethylene (C_2H_4), respectively, from the latter side chain. Fragment **o** is

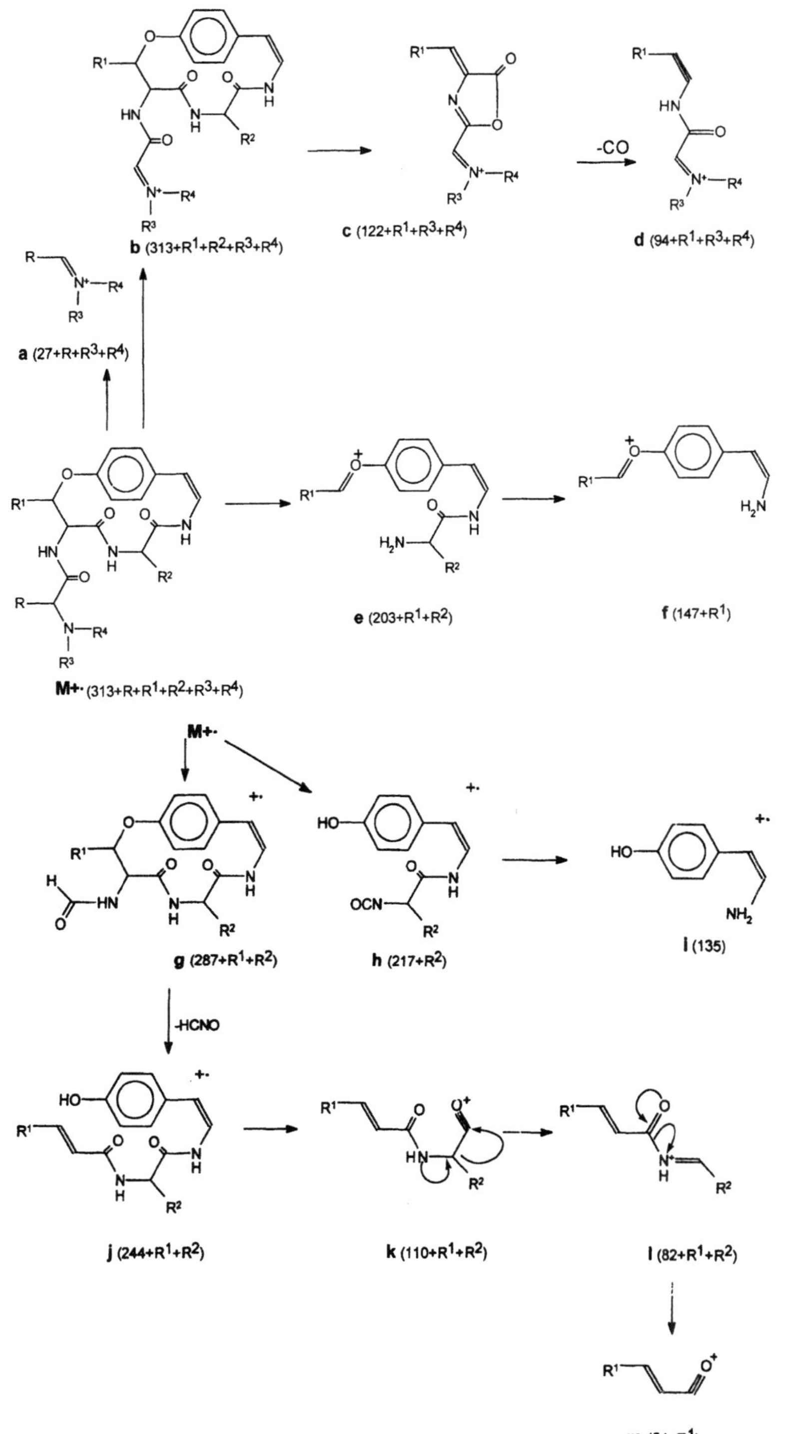

Fig. 2. Main fragments of 4(14)-frangulanine- and -integerrine-type cyclopeptide alkaloids

Table 5. *m/z Values of Fragments of 4(14)-Frangulanine- or -Integerrine-Type Cyclopeptide Alkaloids*

Fragment	R^1=β-OH-Leu side chain (Frangulanine-type)	R^1=β-OH-Phe side chain (Integerrine-type)
M+	$356+R+R^2+R^3+R^4$	$390+R+R^2+R^3+R^4$
b	$356+R^2+R^3+R^4$	$390+R^2+R^3+R^4$
c	$165+R^3+R^4$	$199+R^3+R^4$
d	$137+R^3+R^4$	$171+R^3+R^4$
e	$246+R^2$	$280+R^2$
f	190	224
g	$330+R^2$	$364+R^2$
j	$287+R^2$	$321+R^2$
k	$153+R^2$	$187+R^2$
l	$125+R^2$	$159+R^2$
m	97	131

formed from a Leu side chain, with elimination of neutral propene (C_3H_6). According to our survey the occurrence of other amino acids cannot be ruled out.

Fragment **q** affords information about the nature of ring-bound amino acid and is universal, while fragments **r** and **s** occur sporadically when R^2 is a Trp side chain. The *m/z* of **r** coincides with that of fragment **m**, when R^2 is a Leu or Ileu side chain.

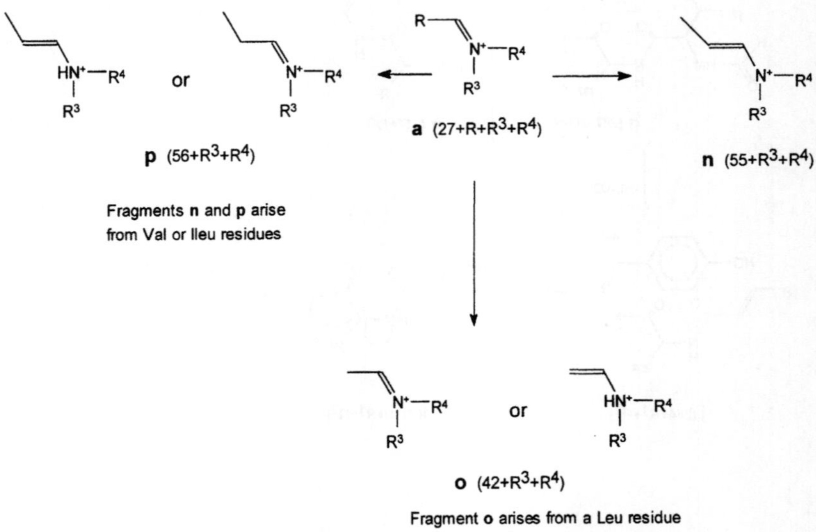

Fig. 3. Fragments in the mass spectra of 4(14)-frangulanine- and -integerrine type-cyclopeptide alkaloids

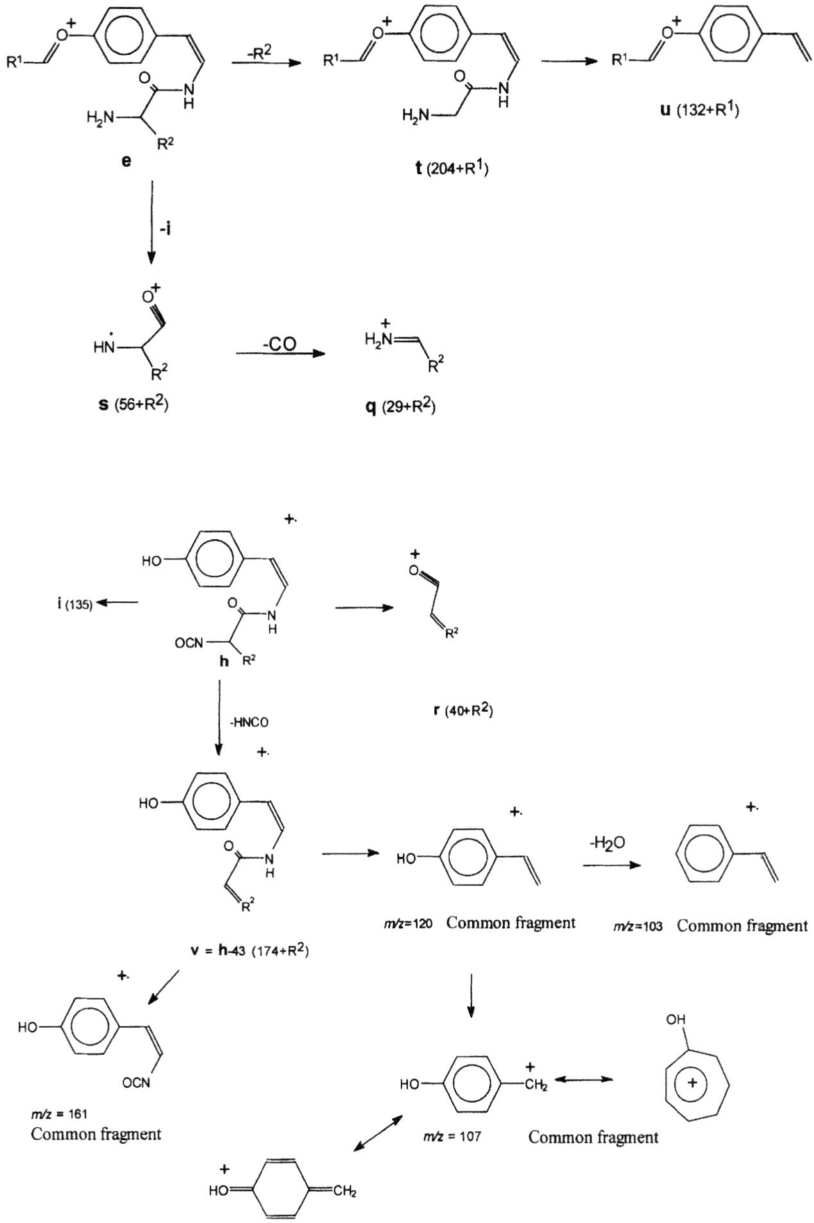
-R2
-i
-CO
e
t (204+R1)
u (132+R1)
s (56+R2)
q (29+R2)
i (135)
h
-HNCO
r (40+R2)
v = h-43 (174+R2)
-H2O
m/z=120 Common fragment
m/z=103 Common fragment
m/z = 161
Common fragment
m/z = 107
Common fragment

Fig. 3. (*Contd.*)

Fragments **t-v** not only reveal the nature of unit D of the cyclopeptide alkaloid (aminophenol moiety) but confirm the identity of R^1 (**t,u**) and R^2 (**v**).

Common fragments are fragments that give no information about R, R^1, R^2, R^3 and R^4 but confirm the nature of the aminophenol moiety. Furthermore their occurrence is indicative of 4(14) cyclopeptide alkaloids. Fragment **u** and common fragments with *m*/*z* 120, 107 and 103 are more frequent for the 4(14)-pandamine types.

Fragments **b**′ and **b**′-CO arise from fragment **b**, while fragments **w**, **w**-CO, **x**, **x**-CO, **y**, **y**-CO and **z** from the molecular ion, after elimination of the neutral molecule R^3R^4NH (Fig. 4).

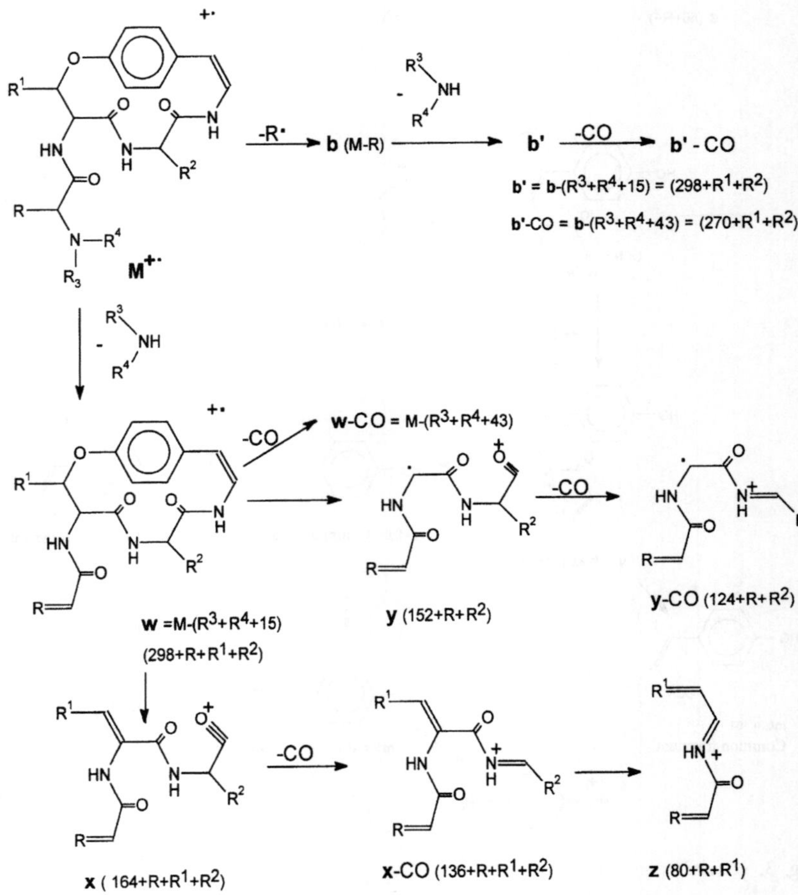

Fig. 4. Fragments in the mass spectra of 4(14)-frangulanine- and -integerrine-type cyclopeptide alkaloids

References, pp. 171–179

Ceanothines B, C and D which belong to the category of 4(14)-frangulanine-type cyclopeptide alkaloids, bear NMePro as basic terminal amino acid. In this case **a** = **a**′ (*m/z* = 84), while fragments **b**, **n**, **o** and **p** are absent.

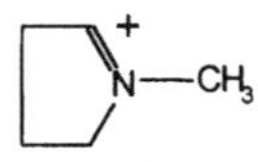

a' *m/z* = 84

4.2. Fragmentation of 4(14)-Pandamine-Type Cyclopeptide Alkaloids

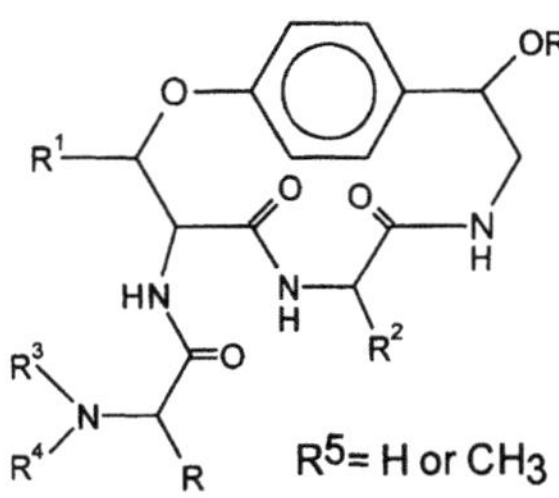

Mass spectra of 4(14)-Pandamine-type cyclopeptide alkaloids contain all the fragments previously described and in addition fragments **b**″, **f**′ and **i**′ (Fig. 5, Table 6). These last fragments are related to

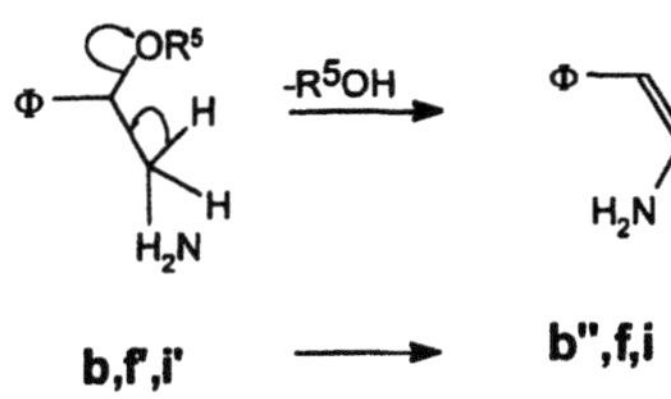

Fig. 5. Special fragments of 4(14)-pandamine-type alkaloids

Table 6. *m/z Fragments of 4(14)-Pandamine-Type Cyclopeptide Alkaloids*

Fragment	m/z	Fragment	m/z
M^+	$330+R+R^1+R^2+R^3+R^4+R^5$	f′	$164+R^1+R^5$
b	$M-R=330+R^1+R^2+R^3+R^4+R^5$	i	135
b″	$b-R^5OH=313+R^1+R^2+R^3+R^4$	i′	$151+R^5$
f	$147+R^1$		

fragments **b**, **f** and **i**. Fragment **b″** originates from **b** after elimination of the -OR^5 group of the styrylamine unit. When R = H then **b″** = **b**-18 and when R = Me then **b″** = **b**-32. Fragments **f′** and **i′** are like **f** and **i** but they contain the -OR_5 group.

Fragments **a**, **c-e**, **g-h** and **k-z** are the same as in 4(14)-frangulanine- and -integerrine-type spectra.

4.3. Fragmentation of 5(13)-Zizyphine-A-Type and 5(14)-Amphibine-B-Type Cyclopeptide Alkaloids

Fragments **a-w** are common to both categories (Figs. 7, 8 and 9). While fragments **a** and **b** are the same with previous groups, fragments **c-w** occur solely in these categories.

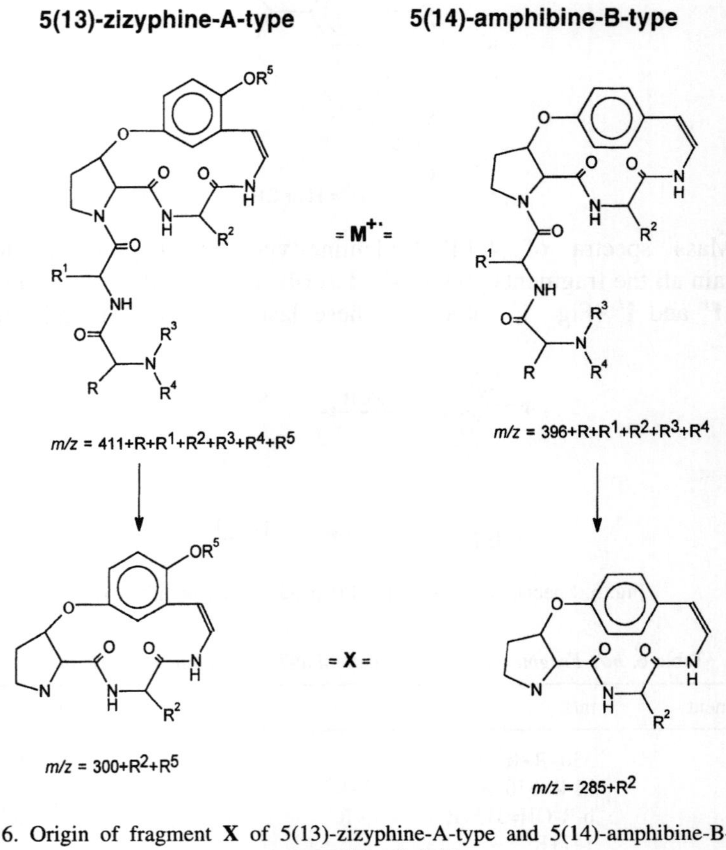

Fig. 6. Origin of fragment **X** of 5(13)-zizyphine-A-type and 5(14)-amphibine-B-type cyclopeptide alkaloids

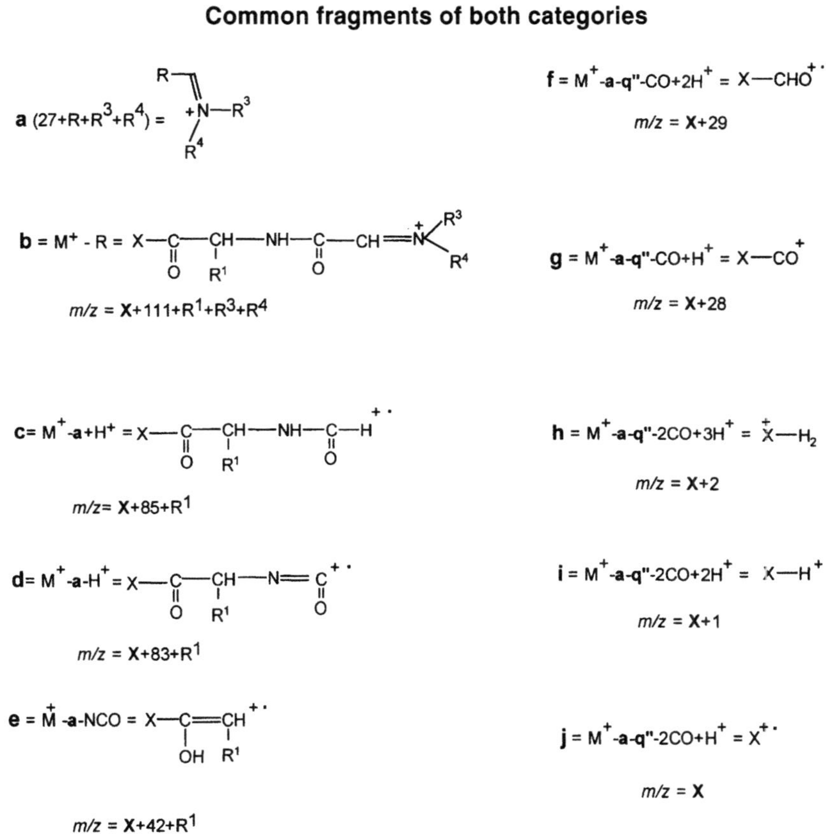

Fig. 7. Common fragments of 5(13)-zizyphine-A-type and 5(14)-amphibine-B-type cyclopeptide alkaloids

Furthermore, fragments **k**-**q**, **v** and **w** are exactly the same in both groups and are shown in Fig. 8.

Structures of fragments **r**-**u** are given in Fig. 9.

Fragments **q**′, **r**′ and **s**′ are homologues of **q**, **r** and **s** of 4(14)-type cyclopeptide alkaloids and are found in both categories (Fig. 10):

Fragments **q**″, **r**″ and **s**″ which are found in both categories are similar to **q**′, **r**′ and **s**′ but originate from the additional amino acid (molecule's fragment E) with the participation of R^1 (Fig. 11):

Fragment **i**′ (Fig. 12) from 5(14)-amphibine-B-type and fragment **x** from 5(13)-zizyphine-A-type alkaloids correspond to fragment **i** of

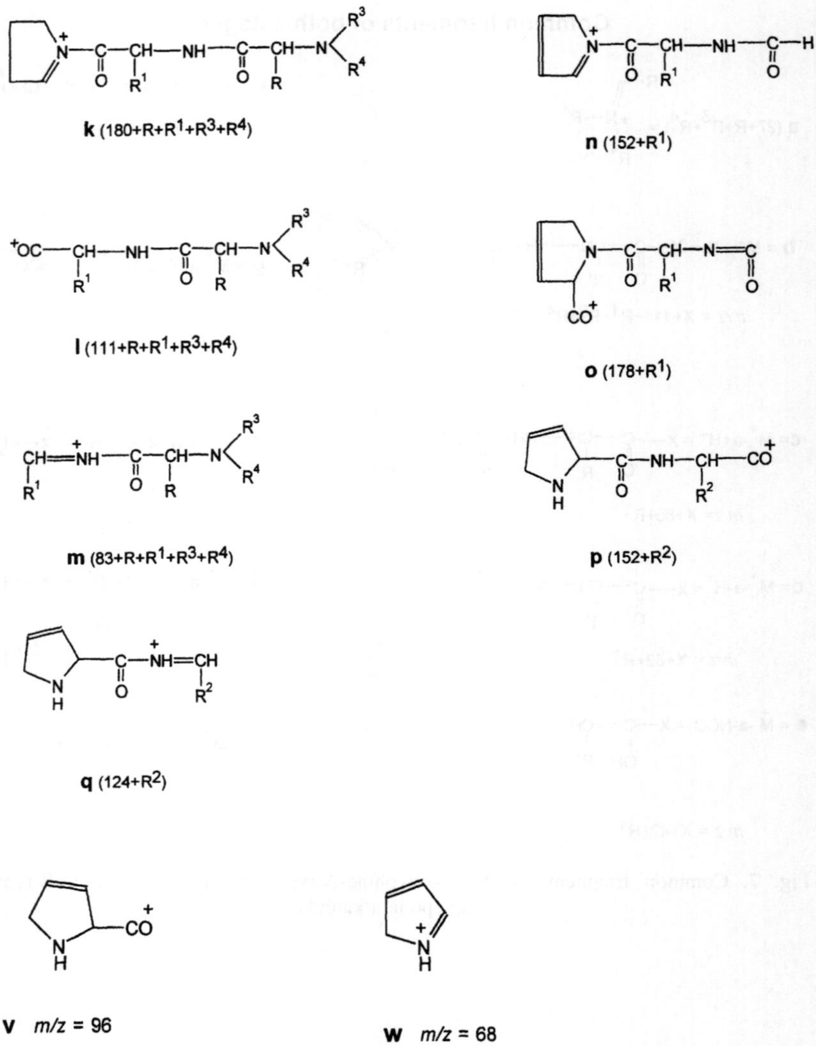

Fig. 8. Common fragments of 5(13)-zizyphine-A-type and 5(14)-amphibine-B-type cyclopeptide alkaloids

4(14)-frangulanine-, -integerrine- and -pandamine-type cyclopeptide alkaloids. In a few cases, 5(13)-zizyphine-A type alkaloids exhibit fragments **y** and **z**. We note that they are completely different from the **y** and **z** fragments of 4(14)-frangulanine-type cyclopeptide alkaloids.

References, pp. 171–179

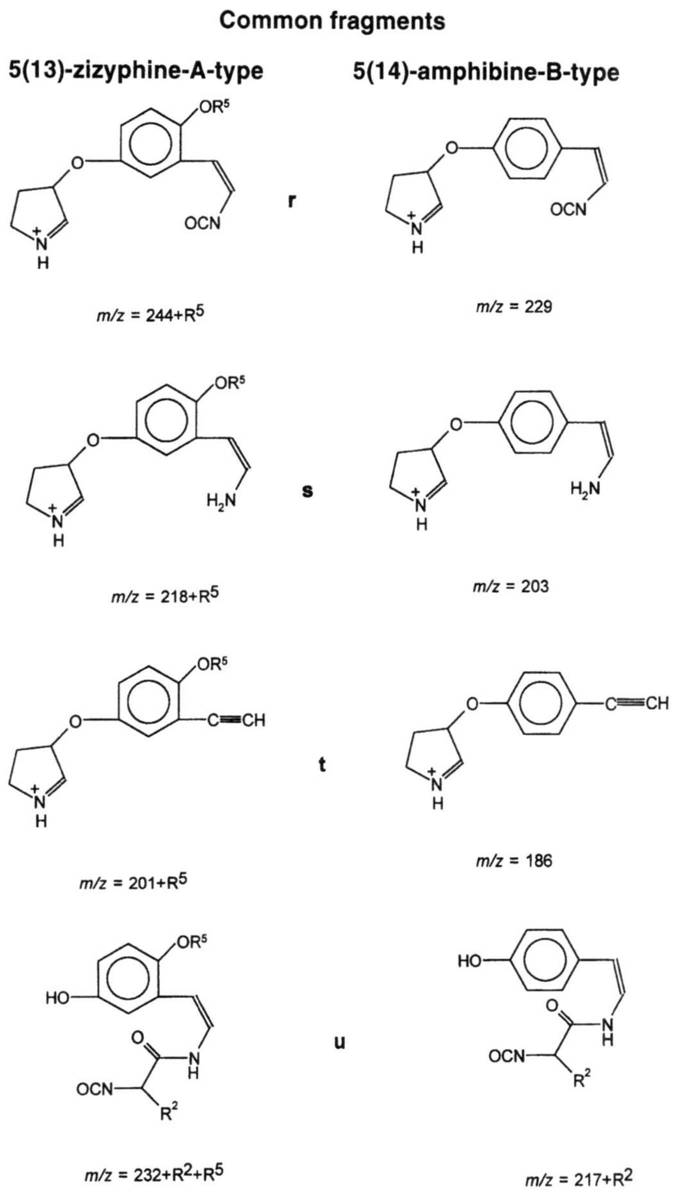

Fig. 9. Fragments **r**–**u** of 5(13)-zizyphine-A-type and 5(14)-amphibine-B-type cyclopeptide alkaloids

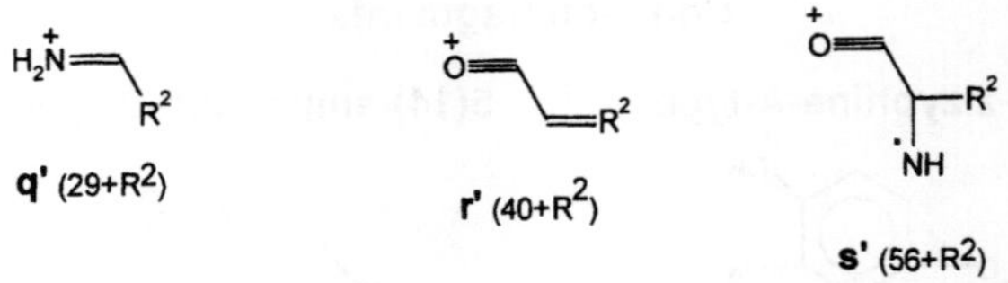

Fig. 10. Fragments **q′**, **r′**, **s′** of 5(13)-zizyphine-A-type and 5(14)-amphibine-B-type cyclopeptide alkaloids

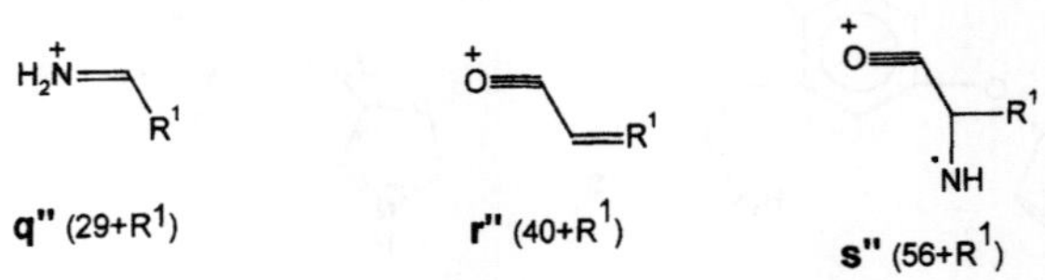

Fig. 11. Fragments **q″**, **r″**, **s″** of 5(13)-zizyphine-A-type and 5(14)-amphibine-B-type cyclopeptide alkaloids

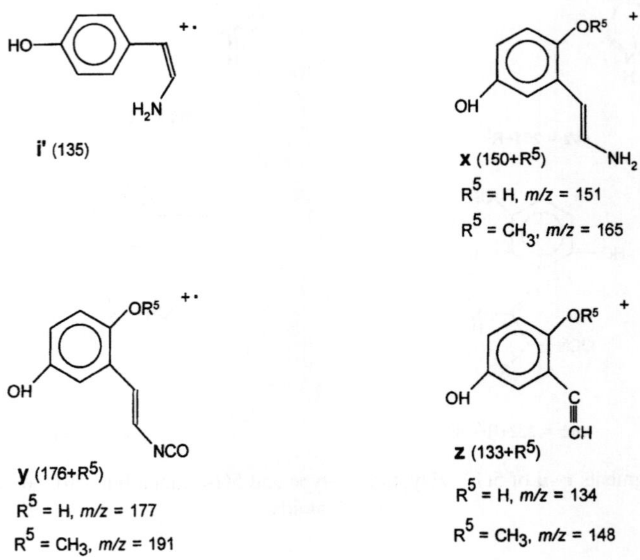

Fig. 12. Fragments **i′**, **x**, **y**, **z** of 5(13)-zizyphine-A-type and 5(14)-amphibine-B-type cyclopeptide alkaloids

Fragments **n**′, **o**′, **p**′, **e**′ = **e**-C_2H_5, **e**″ = **e**′-C_2H_5, **b**′ = **b**-R^3R^4NH, **b**′-CO, **w**′ = M^+-R^3R^4NH and **w**′-CO should also be mentioned. Fragments **n**′, **o**′, **p**′, **b**′ and **b**′-CO are homologues of **n**, **o**, **p**, **b**′ and **b**′-CO of 4(14)-cyclopeptide alkaloids, while **w**′ and **w**′-CO correspond to **w** and **w**-CO of the above mentioned group.

4.4. Fragmentation of 5(14)-Scutianine-A-Type Cyclopeptide Alkaloids (Hymenocardine included)

The main fragmentation pattern of this category follows that of the 4(14)-frangulanine-, -integerrine- and -pandamine-type cyclopeptide alkaloids and exhibits the same **a-m** and **q-v** fragments.

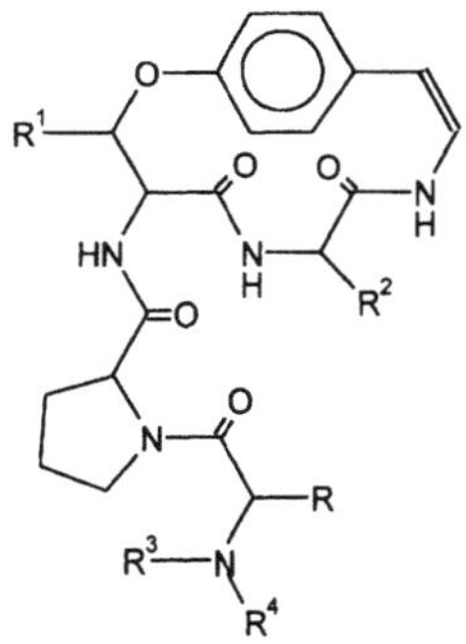

Fragments **c** and **d** have slightly modified structures due to the existence of Pro as additional amino acid. For the same reason the MS of all of these compounds contains fragment **q**′ which, in the case of hymenocardine, corresponds to **q**″ of 5(13)-zizyphine-A and 5(14)-amphibine-B cyclopeptide alkaloids.

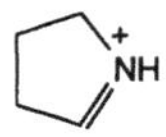

q' (70)

Because all 5(14)-scutianine-A type cyclopeptide alkaloids contain Phe as the basic terminal amino acid, fragments homologous to **n**, **o** and **p** of 4(14)-cyclopeptide alkaloids are absent. Fragments **w**, **x**, **y**, **z**, **c**′ and **d**′ are group specific, while **c**′ and **d**′ arise from the same fragmentation pattern giving **c** and **d** of 5(14)-amphibine-type cyclopeptide alkaloids (Fig. 13).

Hymenocardine is a special case because it contains Val as the β-OH-amino acid and an aminophenol moiety bearing a saturated acyl group (Fig. 14).

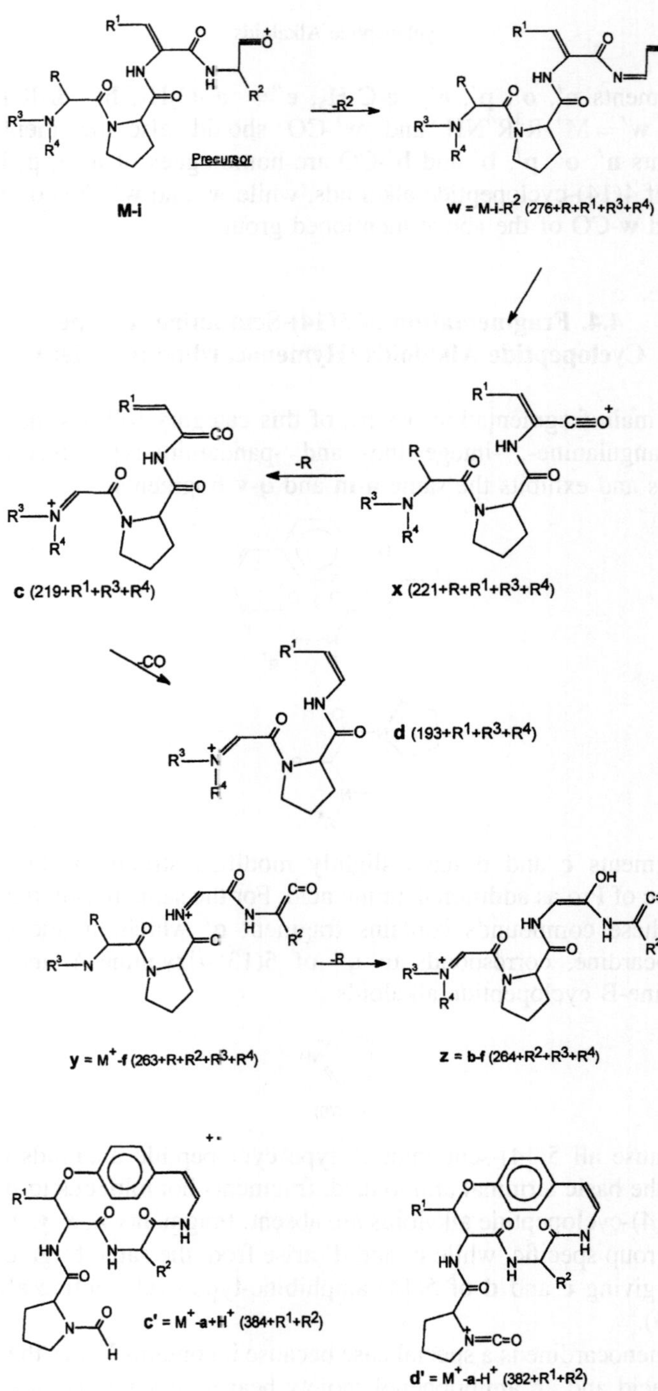

Fig. 13. Fragmentation of 5(14)-scutianine-A-type cyclopeptide alkaloids

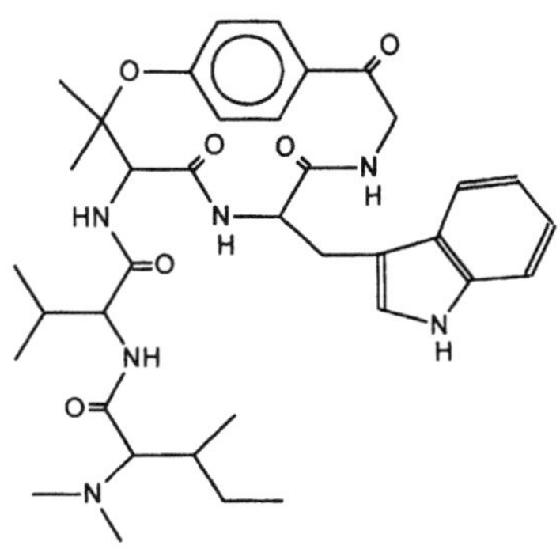

Fragments **a**, **b**, **d′**, **g**, **j**, **q**, **r**, **v**, **w** and **x** of hymenocardine follow the fragmentation pattern of 5(14)-scutianine-A-type cyclopeptide alkaloids. Due to the existence of Ile as the basic terminal amino acid it gives fragments **n** and **o** of 4(14)-cyclopeptide alkaloids. Furthermore two special fragments, namely **i** and **i′**, occur (Fig. 14).

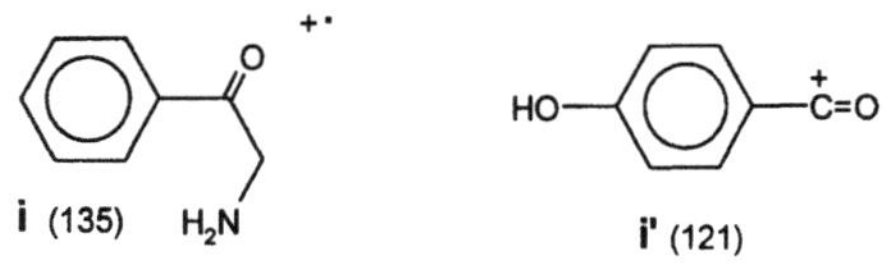

Fig. 14. Special fragments of hymenocardine

4.5. Fragmentation of 4(14)-Amphibine-F-Type Cyclopeptide Alkaloids

The fragmentation pattern of these alkaloids follows the lines of 5(14)-amphibine-B-type cyclopeptide alkaloids: fragments **a**, **b**, **r**, **s**, **t**, **v**,

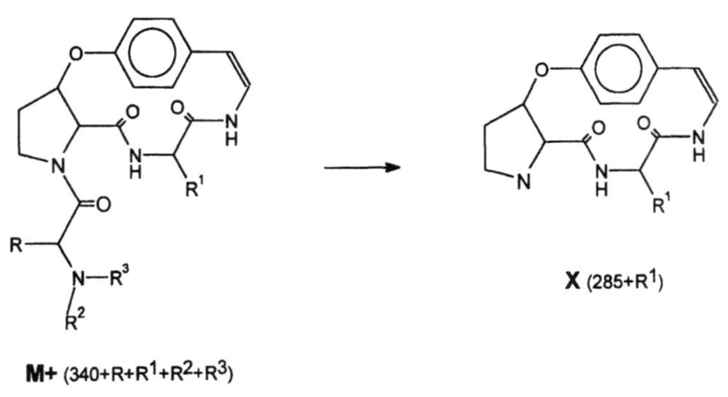

Fig. 15. Fragment **X** of 4(14)-amphibine-F-type cyclopeptide alkaloids

w, **i′**, **n′**, **o′**, **p′**, **q′**, **r′**, and **s′** are the same. Fragments **c-e** are absent. Fragments **f-j**, **p**, **q** and **u** are the same but the alkyl residue of the ring-bound amino acid is designated as R^1.

f = $X{-}CHO^{+\bullet}$ (X+29) = (314+R^1)

g = $X{-}\overset{+}{C}O$ (X+28) = (313+R^1)

j = $\overset{+}{X}$ (X+1) = (285+R^1)

h = $\overset{+}{X}{-}H_2$ (X+2) = (287+R^1)

i = $X{-}H^{+\bullet}$ (X+1) = (286+R^1

p (152+R^1)

q (124+R^1)

u (217+R^1)

Fragments **k-o** (Fig. 16) are homologous with **k-o** of 5(14)-amphibine-B-type cyclopeptide alkaloids without the additional amino acid. Furthermore **k** occurs in the spectra of all compounds of the group, unlike **l-o**.

Finally fragments **b′** and **b′**-CO are exactly the same, while **w′**, **w′**-CO, **y′**, **y′**-CO are homologous with **w**, **w**-CO, **y**, **y**-CO of 4(14)-cyclopeptide alkaloids.

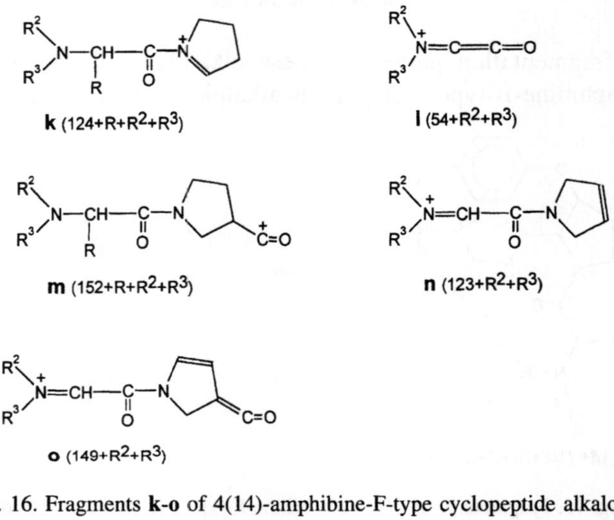

Fig. 16. Fragments **k-o** of 4(14)-amphibine-F-type cyclopeptide alkaloids

4.6. Fragmentation of 4(13)-Nummularine-C-Type Cyclopeptide Alkaloids

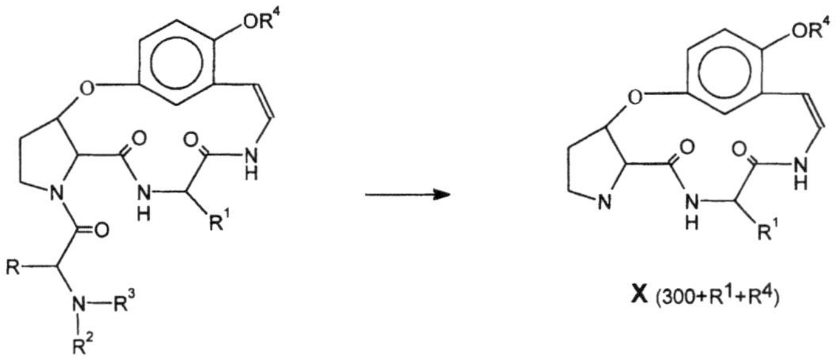

$\mathbf{M^+}$ $(355+R+R^1+R^2+R^3+R^4)$

$\mathbf{X}$ $(300+R^1+R^4)$

Fig. 17. Fragment **X** of 4(13)-nummularine-C-type cyclopeptide alkaloids

The fragmentation pattern of these alkaloids follows the lines of 5(13)-zizyphine-A type cyclopeptide alkaloids: fragments **a**, **b**, **r**, **s**, **t**, **v**, **w**, **x**, **y**, **z**, **n′**, **o′**, **p′**, **q′**, **r′**, and **s′** are the same. Fragments **c-e** are absent. Fragments **f-j**, **p**, **q** and **u** are the same but the alkyl residue of the ring-bound amino acid is designated as R^1 and the O-substituent as R^4.

f = $X\text{—}CHO^{+\bullet}$ $(\mathbf{X}+29) = (329+R^1+R^4)$

h = $\overset{+}{X}\text{—}H_2$ $(\mathbf{X}+2) = (302+R^1+R^4)$

g = $X\text{—}\overset{+}{C}{=}O$ $(\mathbf{X}+28) = (328+R^1+R^4)$

i = $X\text{—}H^{+\bullet}$ $(\mathbf{X}+1) = (301+R^1+R^4$

j = $\overset{+}{X}$ $(300+R^1+R^4)$

p $(152+R^1)$

q $(124+R^1)$

u $(232+R^1+R^4)$

Fragments **k-o** are absent.

4.7. Fragmentation of 4(15)-Mucronine-A-Type Cyclopeptide Alkaloids

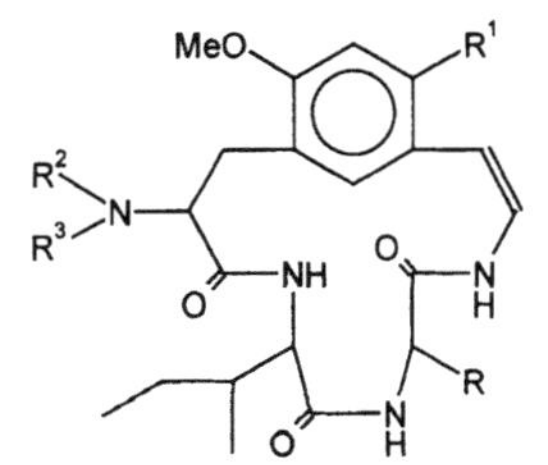

a = M^+-HNCO $(341+R+R^1+R^2+R^3)$

a′ = **a**-CH_3

a″ = **a**-C_2H_5

$\mathbf{M^+}$ $(384+R+R^1+R^2+R^3)$

R^1 = H or CH_3

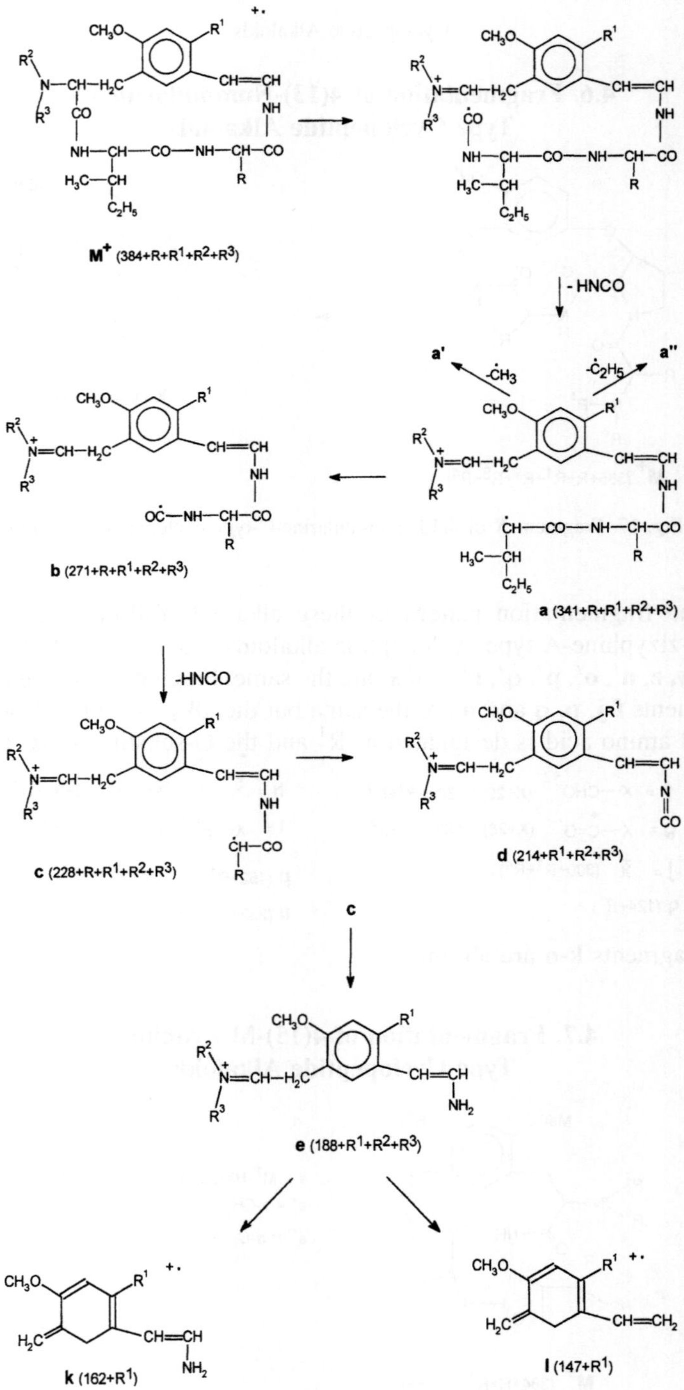

Fig. 18. Main fragments of 4(15)-mucronine-A-type cyclopeptide alkaloids

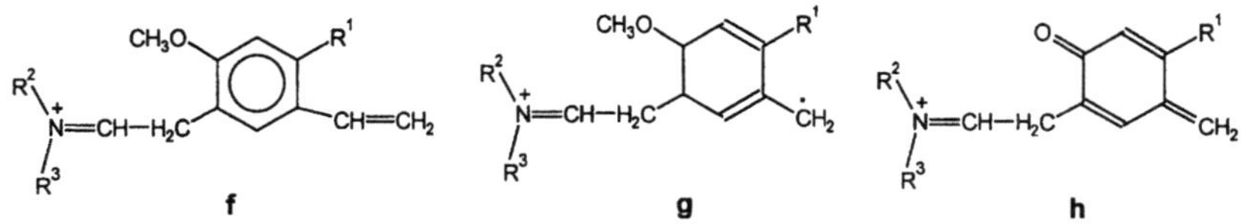

Fig. 19. Fragments of 4(15)-mucronine-A-type cyclopeptide dihydroalkaloids

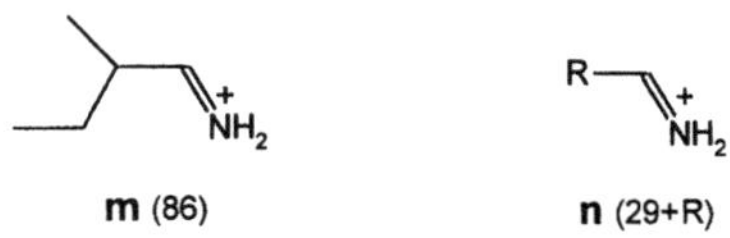

Fig. 20. Fragments **m**, **n** of 4(15)-mucronine-A-type cyclopeptide alkaloids

The different ring structure, compared with other cyclopeptide alkaloids, of this group leads to a different fragmentation pattern (Fig. 18). Fragments **a**, **a′**, **a″**, **b-e**, **k** and **l** are the most abundant.

Fragments **f**, **g** and **h** are encountered in the dihydro derivatives of this group (Fig. 19).

Dihydroalkaloids

In addition to the fragments shown in Fig. 19, fragments **n** and **m** (Fig. 20), which arise from the ring-bound amino acid and the intermediate amino acid respectively, are frequently observed. Because the latter is always an Ile, fragment **m** has a *m/z* value of 86. These fragments are similar to fragment **q** of 4(14)-type cyclopeptide alkaloids.

4.8. Fragmentation of Linear Peptide Alkaloids

These alkaloids, due to their open structure, fragment in a special way.

Celenamides

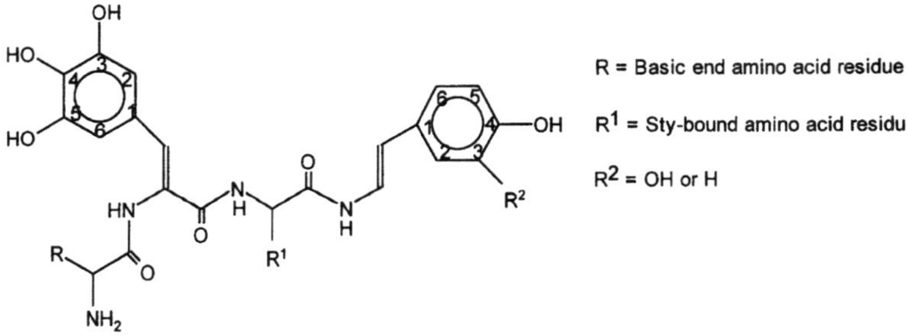

They exhibit fragments **a-l**, which are displayed in Fig. 21 (*60, 61*). We note the following: i) The celenamides are marine natural products of sponge origin and usually bear a Br atom in R^1 side chain; ii) some of the fragments, due to the presence of Br, exhibit two isotope peaks which

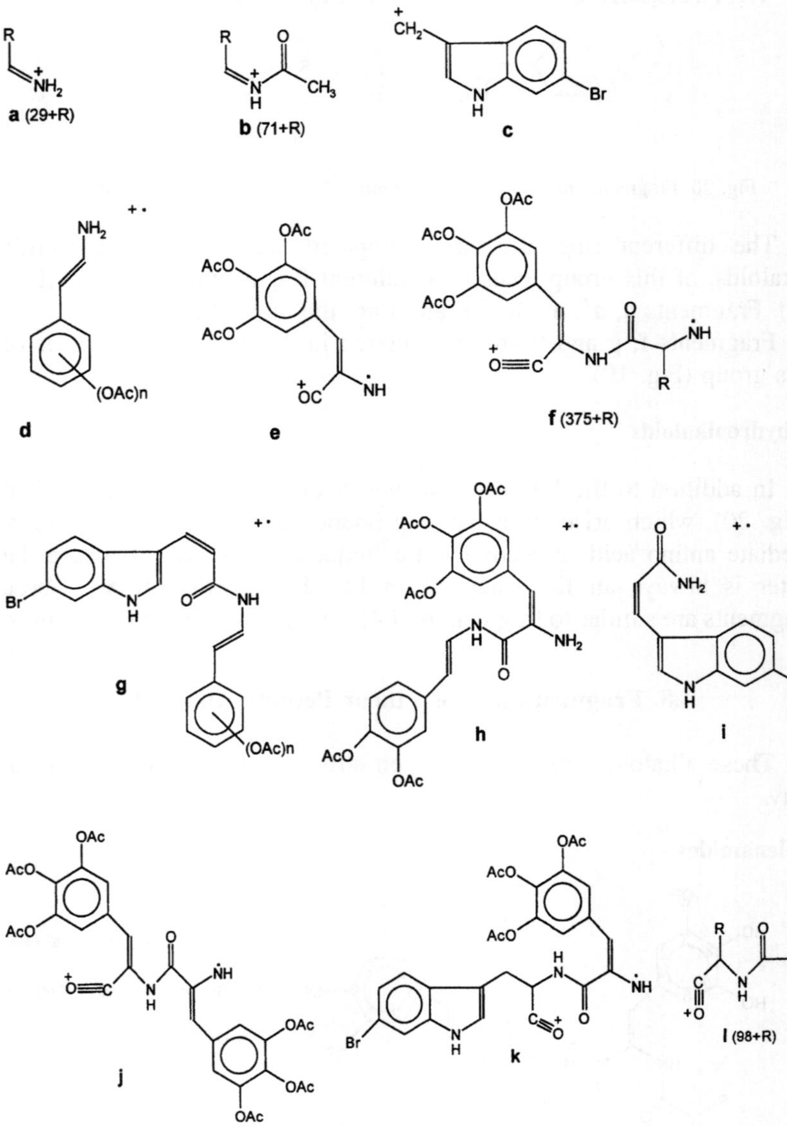

Fig. 21. Fragments **a-l** of celenamides

differ by 2 amu; and iii) many fragments arise after successive losses of Ac (in the form $COCH_2$), because these alkaloids have been isolated as acetylated derivatives.

Lasiodine-A

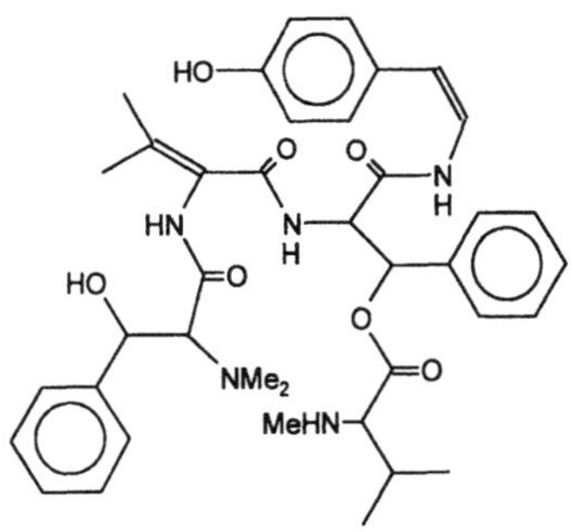

This alkaloid contains a unique structure. It is comprised of five units, four of them being attached to one side of the aminophenol, the fifth unit. The ring-bound amino acid bears an additional basic terminal amino acid. According to MARCHAND *et al.* (*62*) the fragments that are encountered are **i** ($m/z = 135$), **a** and **o** ($m/z = 86$ and 58 respectively). They originate from one basic terminal amino acid (N-Me-L-Val) in accord with the fragmentation behaviour of 4(14)-frangulanine-type cyclopeptide alkaloids.

Sanjoinine-G2

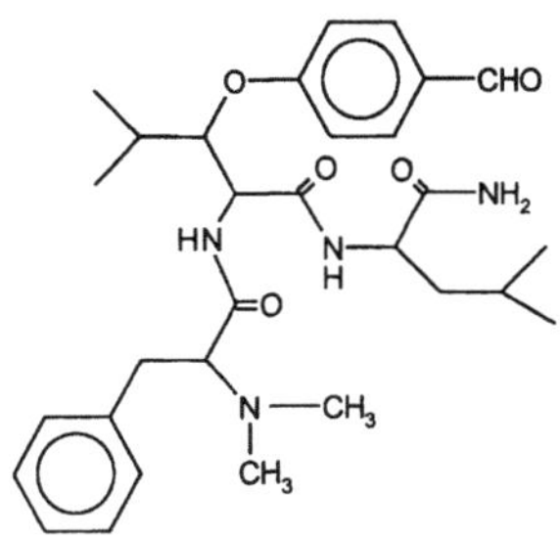

This alkaloid is in the form of an aminoaldehyde, where the normally ring-bound amino acid is no longer attached to the phenolic ring. Because it resembles 4(14)-frangulanine-type cyclopeptide alkaloids, it produces some fragments of this group: **a**-**d** and **l**-**n**. Fragment **k** of 4(14)-type cyclopeptide alkaloids is modified and includes the amino group: **k** + NH_2 ($m/z = 226$). Interestingly, it shows fragment **i**′ ($m/z =$ 121), like hymenocardine. Finally, three special fragments are noted: M^+-leucinamide ($m/z = 409$), **b**-**i**′-H^+ = **b**-122 and M^+-leucinamide-**i**′ = 287.

4.9. Fragmentation of Neutral Compounds Related to Cyclopeptide Alkaloids

The three neutral compounds that have been isolated so far display the fragmentation behaviour of the group which they resemble: scutianene-C and sanjoinenine fragment like 4(14)-frangulanine-, -integerrine- and -pandamine-type cyclopeptide alkaloids, while lotusanine-B fragments like 5(14)-scutianine-A-type cyclopeptide alkaloids.

The absence of an amino end residue leads to the absence of the corresponding fragments. Accordingly, instead of fragment **a** a common fragment is observed with $m/z = 131$ which arises from *trans*-cinnamic acid, the terminal fragment in all three alkaloids (*59*, *63*, *64*).

Scutianene-C and sanjoinenine afford the following 4(14)-frangulanine-, -integerrine- and -pandamine-type fragments: **e**, **f**, **h**, **i**, **l**, **m**, **q**, **x**, **y** and **z**.

Lotusanine-B displays the following 5(14)-scutianine-A-type fragments: **e**, **f**, **i**, **m**, **q**, **q″**. In addition, fragments **x**, **x**-CO, **y**, **y**-CO of 4(14)-frangulanine-, -integerrine- and -pandamine-type cyclopeptide alkaloids appear. Finally fragments **n** and **o**, analogues of **n** and **o** of 5(14)-amphibine-B-type cyclopeptide alkaloids occur (Fig. 22).

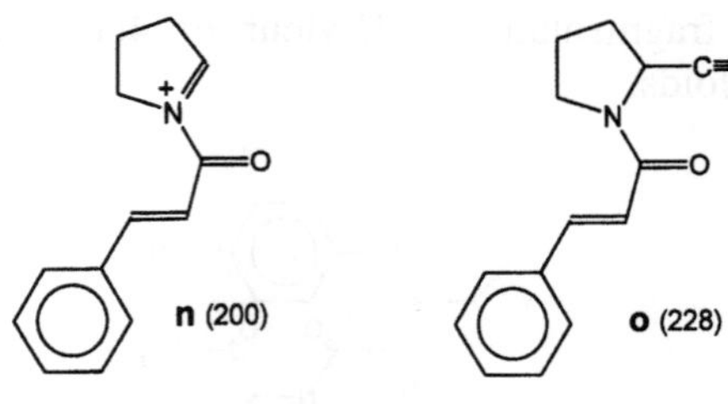

Fig. 22. Fragments **n**, **o** of lotusanine-B

4.10. Miscellaneous

There are three cyclopeptide alkaloids whose skeleton contains an additional amino acid. In nummularine-G and sativanine-B, which belong to the 4(14)-integerrine group of alkaloids, the additional ring is formed between the β-OH-amino acid and the basic terminal amino acid. In sativanine-D, a 5(13)-zizyphine-A-type cyclopeptide alkaloid, the additional ring is formed between the normal additional amino acid and the basic terminal amino acid. In these cases fragment **a** is converted to **a′**, containing the new additional ring and the basic end residue and is

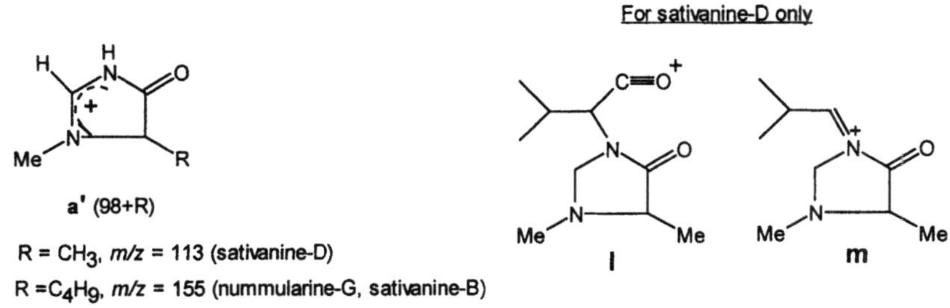

Fig. 23. Special fragments of nummularine-G, sativanine-B and sativanine-D

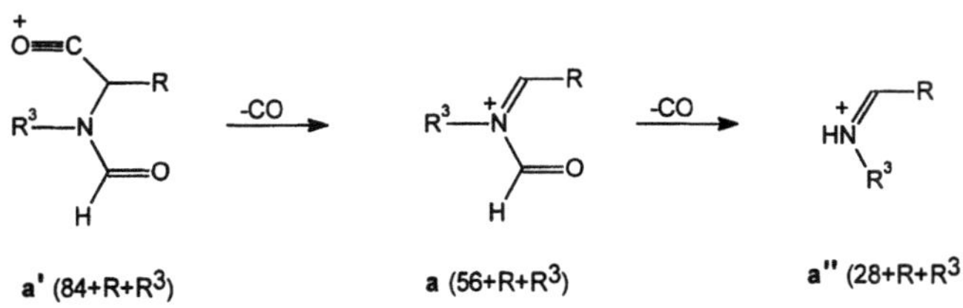

Fig. 24. Special fragments of *N*-formyl-containing cyclopeptide alkaloids

not the base peak anymore (*50*, *51*, *65-67*). The sativanine-D fragments **l** amd **m** are shown in Fig. 23 (*50*, *51*, *65*).

There are four cyclopeptide alkaloids containing an *N*-formyl basic terminal amino acid. Three of them, nummularine-T, rugosanine-A and sativanine-F, are 5(13)-zizyphine-A-types and sativanine-K is a 4(13)-nummularine-C-type. From these alkaloids fragment **a** has the usual structure, having the formyl residue as an N-substituent, and is not the base peak. The last corresponds to fragment **a**′, which gives birth to **a** and **a**″ (Fig. 24) (*52*, *54*, *56*, *68*).

In the case of the three 5(13)-zizyphine-A type cyclopeptide alkaloids a fragment **c**-CO appears, possibly due to stabilization of the leaving radical (*52*, *56*, *68*).

In some cases the fragment shown in Fig. 25 was observed. It originates from the β-OH-amino acid, is observed sporadically and only when the β-OH-amino acid is Leu ($m/z = 84 = 41 + 43$).

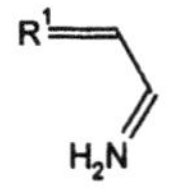

m/z = 41+R

Fig. 25

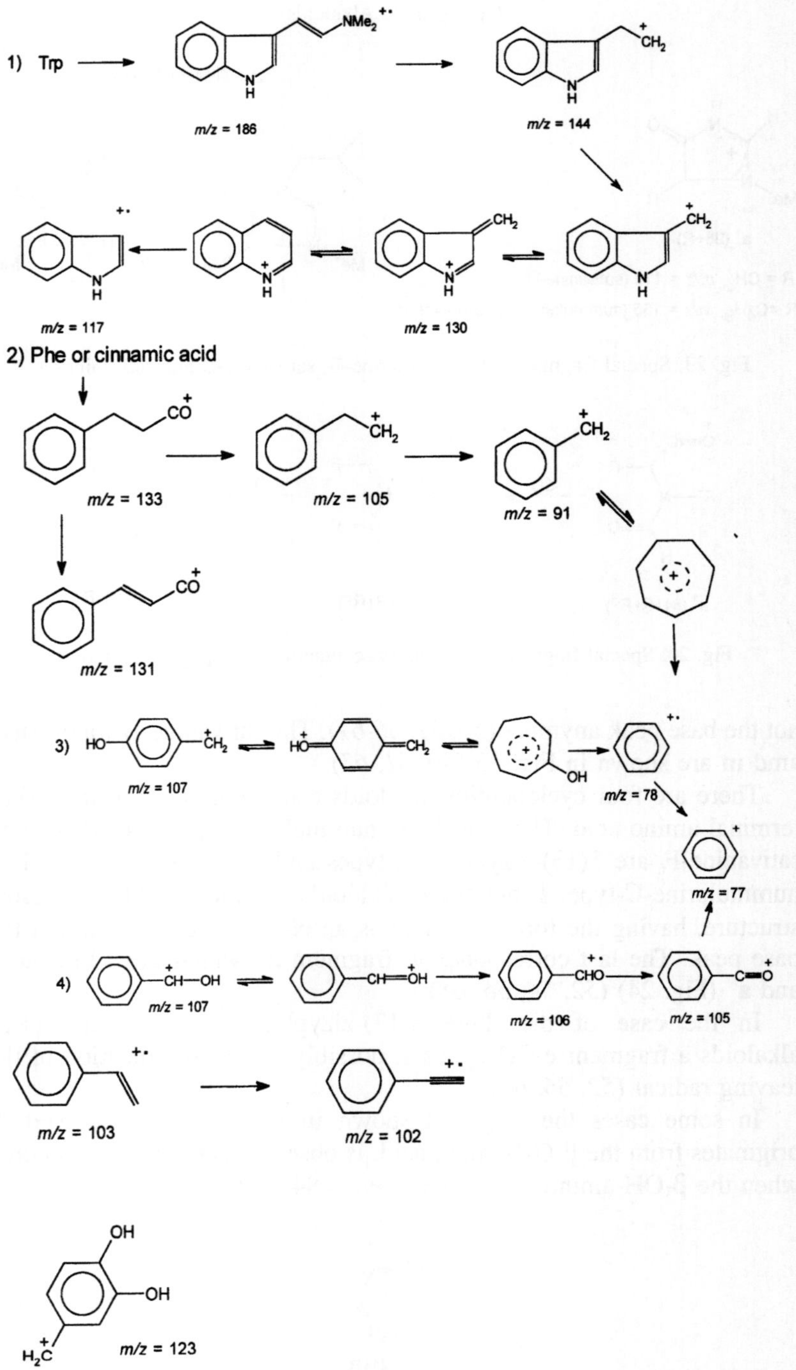

Fig. 26. Origin of common fragments of cyclopeptide alkaloids

4.11. Common Fragments

During the fragmentation of all cyclopeptide alkaloids there are some fragments that occur universally, no matter what the class is, and are due to the occurrence of some common structural features like tryptamine, phenylalanine and hydroxybenzyl units. These common fragments are depicted in Fig. 26 on previous page.

5. Identification Strategy

There are no standard protocols leading to structural elucidation of cyclopeptide alkaloids on the basis of spectroscopic and mass spectral data. The following strategy can be used as a guide for assigning an isolated compound to a specific cyclopeptide alkaloid group. It is presumed that the alkaloid studied is a pure compound.

The initial guide to classification of an alkaloid as a cyclopeptide is the IR spectrum. Bands at ca. 1685 and 1630–1655 cm^{-1} reveal the existence of secondary amido groups, while at ca. 1625 cm^{-1} the styryl double bond absorption is observed.

The UV spectrum will provide the first indication about the group to which the cyclopeptide alkaloid in question belongs. Absence of intense absorption points towards a 14-membered compound, while bands at 270 (log ε ~4.0) and 320 nm (log ε ~3.8) indicate a 13-membered compound. 15-Membered cyclopeptide alkaloids reveal a maximum around 275 nm (log ε ~4.2) and sometimes a shoulder at ca. 220 nm.

As regards the mass spectrum which, as discussed earlier contains much information for the identification process, the *m/z* 135 fragment is universal and thus of great diagnostic value. Also fragments of *m/z* 151 or 165, are important since they are similarly diagnostic of 5(13)-zizyphine-A-type cyclopeptide alkaloids. Moreover, generally, alkaloids having four or five rings have molecular weights between 430 and 600 or 560 and 730 respectively.

After the cyclopeptide nature of the alkaloid has been confirmed, our interest is to assign the compound to a certain subgroup: the coexistence of the pairs of fragments at *m/z* 190 (**f**) and 97 (**m**) as well as 247 (**t**) and 175 (**u**) reveals that the compound belongs to the frangulanine group, while the coexistence of pairs at 224 and 131 together with 281 and 209 is typical of the integerrine type. On the other hand, presence of the pairs 190–97 and 247–175 and absence of IR bands or NMR signals for the styrylamine double bond makes highly probable the occurrence of a 4(14)-pandamine-type cyclopeptide alkaloid.

Fragments at *m/z* 96 (**v**) and 68 (**w**), together with a strong molecular ion point towards the 5(13)-zizyphine-A- or 5(14)-amphibine-B-types.

These two types are distinguishable on the basis of either NMR spectroscopy, by a three-proton singlet near δ 3.8 indicative of the 5(13)-zizyphine-A-type, or mass spectrally, by the fragment of *m/z* 135 afforded only by the amphibine-B-type. Zizyphine-A-type cyclopeptide alkaloids afford fragments of 151 or 165 amu, depending on the presence of OH or OCH_3 in the aromatic ring of the styrylamine unit. A special sign for recognition of these two groups is the coexistence of fragments **b**-**j**. Hence, their mass spectrum is characteristic because it contains a cluster of 11 peaks differing by 26 amu (**b**-**c**, **g**-**h**), 2 amu (**c**-**d**), 41 amu (**d**-**e**) and 1 amu (**f**-**g**, **h**-**i** and **i**-**j**). Moreover the assignment of the amphibine-B-type becomes unambiguous if fragments having *m/z* 229, 203 and 186 occur. For the zizyphine-A-type the corresponding fragments are observed at *m/z* 245, 219 and 202 (for OH compounds) or 259, 233 and 216 (for OCH_3).

A molecular ion of high *m/z* (~650) with a fragmentation typical of 4(14)-cyclopeptide alkaloids and a fragment of m/z 70 (**q**′) reveal a 5(14)-scutianine-A-type alkaloid, further supported by the presence of two fragments (**c**-**d**) differing by 26 amu.

4(14)-Amphibine-F-type alkaloids are among the difficult cases because they closely follow the pattern of the 5(14)-amphibine-B-types. The lower molecular ion and the absence of fragments differing by 2 (**c**-**d**) and 41amu (**d**-**e**) are, however characteristic for this group. Another difficult case is the 4(13)-nummularine-C group whose fragmentation follows the 5(13)-zizyphine-A-type fragmentation pattern. The distinction follows the previous arguments.

4(15)-Mucronine-A-type alkaloids are easily distinguished due to the presence of three fragments of high *m/z* differing by 15 and 29 amu (**a**-**a**′ = 15, **a**′-**a**″ = 29). Moreover the fragment **m**, with *m/z* 86, is characteristic of this group. Note that fragments **a** and **b** differ by 70 amu.

Celenamides are the cyclopeptide alkaloids with the highest molecular weight (around 1000) and are isolated from sponges. The universal presence of fragment **e**, with *m/z* 319 is quite characteristic. The occurrence of fragments differing by 2 amu is due to the presence of Br in the R^1 side chain.

6. Physical and Spectral Data of Cyclopeptide Alkaloids and Related Compounds

The relative or absolute stereochemistry given below for some units of several alkaloids has been assigned either by the original authors of the references cited (denoted by superscript 1) or by us (denoted by superscript 2) from the NMR spectra described in the cited references, using ^{1}H ($J_{a,b}$ values of the β-OH-Leu or β-OH-Phe unit) and/or ^{13}C (chemical shift values of C-β of the β-OH-Leu unit) NMR data. The superscript 1,2 means that initially the configuration was insufficiently or erroneously attributed and is now confirmed or corrected by the chemical and spectral data described in the references cited.

Additionally, the nomenclature of cyclopeptide alkaloids has been kept unaltered by using the trivial names. The nomenclature proposed in (*45*) and (*46*), although chemically correct, is confusing especially when correlations or references to known compounds are made. Furthermore, since the present review article is centered on structure elucidation, some simplifications were adopted. Thus, in order to facilitate references or correlations with known compounds, the prefixes *N*- and *O*-desmethyl or *N*- and *O*-methyl have been used.

Abbreviations

Ala = Alanine, 6-Br-Trp = 6-bromotryptophan, Dide-Phe = α,β-didehydro-3,4,5-trihydroxyphenylalanine, Dide-Val = α,β-didehydro-valine, diOH-Sty = 3,4-dihydroxy-*trans*-styrylamine = [(*E*)-1-amino-2-(3,4-dihydroxyphenyl)ethene], Gly = Glycine, Ileu = Isoleucine, Leu = Leucine, N-Me = *N*-methyl, N,N-diMe = *N,N*-dimethyl-, β-OH = β-hydroxy, OH-Sty = 4-hydroxy-*trans*-styrylamine = [(*E*)-1-amino-2-(4-hydroxyphenyl)ethene], Phe = Phenylalanine, Phe-Et = Phenylethylamine, Pro = Proline, Sty = Styrylamine, Trp = Tryptophan, Tyr = Tyrosine, Val = Valine.

The physical and spectral data will be presented in the following order:

Mp in °C. If considerable variations exist in the literature, we report all the values. If the range of variation is limited we note the range and refer to the literature.

$[a]_D$ in degrees followed by the temperature in parentheses; the concentration and the solvent are given in parentheses after the numerical value.

UV spectral data. The solvent is in parentheses and the λ_{max} values in nm are then listed, each with loge in parentheses.

IR spectral data. Solvent used (or KBr) is in parentheses and v_{max} values in cm^{-1} follow.

Where CD (circular dichroism) data exist, the solvent used is in parentheses. The values are in Δε (positive or negative) and the corresponding λ_{max} (nm) is in parentheses.

MS data correspond to EI (70 eV), except otherwise stated (FD, CI/ NH_3 etc.). The values are in amu and the interpretation is in parentheses. Our preference was for interpretation rather than for relative intensities of the fragments, since the former is more convenient for identification purposes.

^1H and ^{13}C NMR spectral data. The field (MHz) and solvent are in parentheses. Chemical shifts (δ) are in ppm relative to TMS. The multiplicity and the *J* values (in Hz) for ^1H signals are given, with the assignment in parentheses. ^{13}C signals are displayed on the chemical formulae.

For NMR spectra we usually quote the data obtained with the most powerful instrument. For NMR data run on a different instrument or under other conditions we just quote the references. When we interpret more than one NMR spectrum for the same compound it means that they are complementary.

The reader will find references to NMR spectra obtained with special techniques (COSY, TOCSY, NOESY etc.) and literature citations about X-ray analysis.

In addition we include derivatives that are naturally occurring or have been synthesized. Finally, plant sources (genera, species, family, plant part) for the compound are given. The molecular formula and molecular weight of the alkaloid are given alongside the structure. When an accurate mass from high resolution MS exists it is noted in parentheses. The found, not the calculated, value is presented.

4(13)-Nummularine-C-Type Cyclopeptide Alkaloids

1. Subfraction II *(69)*

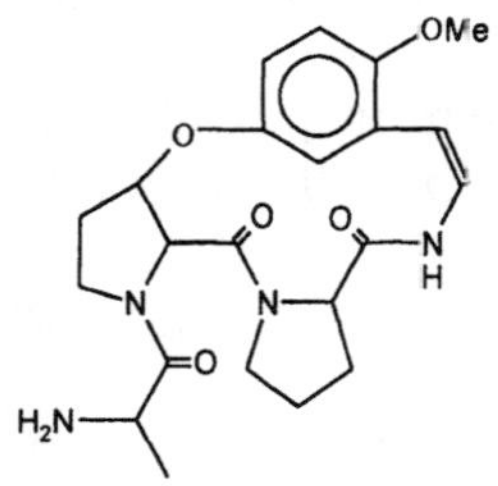

A = Ala
B = β-OH-Pro
C = Pro
$C_{22}H_{28}N_4O_5$, 428

Mp = 72 *(69)*
UV ($CHCl_3$) : 262, 310 *(69)*
IR ($CHCl_3$) : 3400, 2760, 1710 *(69)*
MS : 427($M-H^+$) *(69)*
Sources : *Sphaeranthus indicus* (Asteraceae)-flowers *(69)*

2. Daechuine-S7

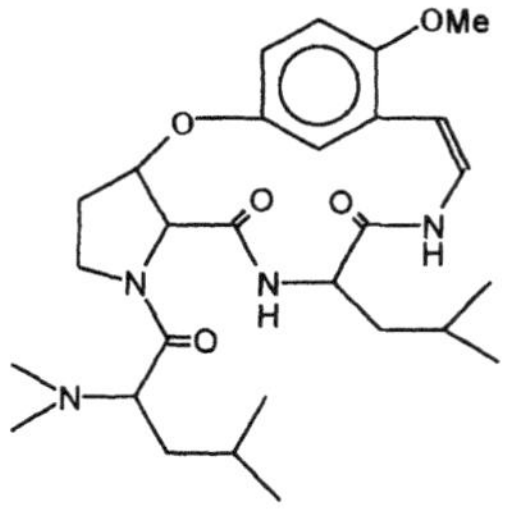

A = N,N-diMe-Leu
B = β-OH-Pro
C = Leu
$C_{28}H_{42}N_4O_5$, 514

Mp = 158 (*70*)
[a]$_D$ = -648.3 (*70*)
Sources : *Zizyphus jujuba* var. *inermis* (Rhamnaceae)-stem bark (*70*)

3. Compound 2 (*71*)

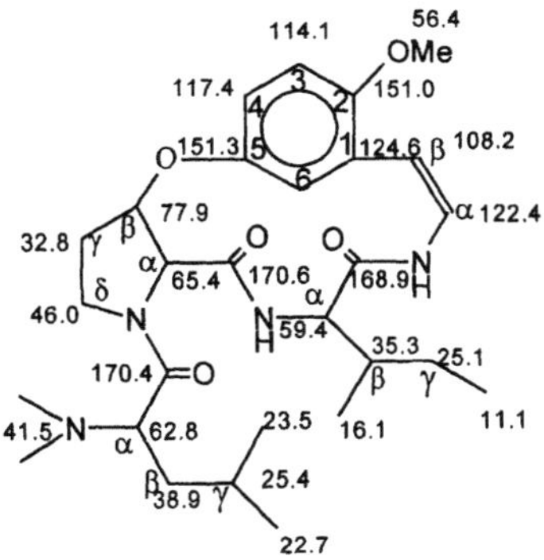

A = N,N-diMe-Leu
B = *trans*-β-OH-Pro[1] (*71*)
C = Ileu
$C_{28}H_{42}N_4O_5$, 514

Mp = 153-154 (*71*)
[a]$_D$ (20) = -418 (c=1.1, $CHCl_3$) (*71*)
UV (MeOH) : 268 (4.15), 320 (4.00) (*71*)
`MS : 514(M^+), 471(M^+-C_3H_7), 401(f), 372(j), 304(u), 233(s), 216(t), 209(p), 165(x), 148(z), 114(a=100%), 86(q') (*71*)
^{1}H-NMR (300MHz, DMSO-d_6) : 0.77t, 7.2 Hz, 3H (CH_2-Me-Ileu), 0.79d, 6.4 Hz, 3H (Me-N,N-diMe-Leu), 0.81d, 6.4 Hz, 3H (Me-N,N-diMe-Leu), 0.85d, 6.7 Hz, 3H (CH-Me-Ileu), 1.14m, 1H (γ-H-Ileu), 1.25m, 1H (γ-H-N,N-diMe-Leu), 1.30m, 2H (γ-H-Ileu + β-H-N,N-diMe-Leu), 1.60m, 1H (β-H-N,N-diMe-Leu), 1.82m, 1H (β-H-Ileu), 2.16m, 1H (γ-H-Pro), 2.20s, 6H (NMe_2), 2.50m, 1H (γ-H-Pro), 3.27dd, 7.6, 4.1 Hz, 1H (α-H-N,N-diMe-Leu), 3.58m, 1H (δ-H-Pro), 3.73s, 3H (OMe), 4.05dd, 8.2, 6.7 Hz, 1H (α-H-Ileu), 4.05m, 1H (δ-H-Pro), 4.49d, 1.8 Hz, 1H (α-H-Pro), 5.12bt, 6.1 Hz, 1H (β-H-Pro), 5.84d, 8.8 Hz, 1H (β-H-Sty), 6.71dd, 10.2, 8.8 Hz, 1H (α-H-Sty), 6.76d, 3.0 Hz, 1H (6-H-Sty), 6.82dd, 8.6, 3.0 Hz, 1H (4-H-Sty), 7.01d, 8.6 Hz, 1H (3-H-Sty), 8.23d, 8.8 Hz, 1H (NH-Ileu), 9.10d, 10.2 Hz, 1H (NH-Sty) (*71*)
^{13}C-NMR (75MHz, DMSO-d_6) : see figure (*71*)
^{1}D-TOCSY, COSY, FLOCK, ^{2}D-NOESY spectra (*71*)
Sources : *Zizyphus mucronata* (Rhamnaceae)-roots (*71*)

4. Sativanine-G

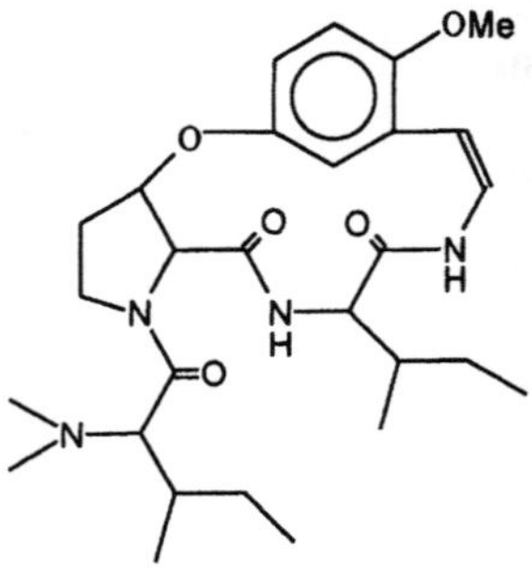

A = N,N-diMe-Ileu
B = *trans*-β-OH-Pro[1] (*72*)
C = Ileu
$C_{28}H_{42}N_4O_5$, 514.3168 (MS) (*72*)

Mp = 92 (*72*)
UV (MeOH) : 258, 320 (*72*)
IR (KBr) : 3380, 2835, 2780, 1670, 1635, 1610, 1230, 1040 (*72*)
MS : 514(M^+), 457(b), 401(f), 400(g), 374(h), 304(u), 259(r), 233(s), 216(t), 209(p), 165(x), 114(a=100%), 96(v), 86(p'/q') (*72*)
Sources : *Zizyphus sativa* (Rhamnaceae)-stem bark (72)

5. Sativanine-K

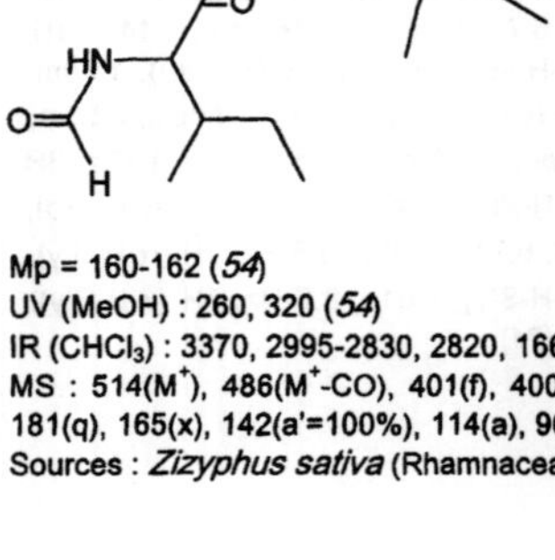

A = *N*-formyl-Ileu
B = β-OH-Pro
C = Ileu
$C_{27}H_{38}N_4O_6$, 514.2778 (MS) (*54*)

Mp = 160-162 (*54*)
UV (MeOH) : 260, 320 (*54*)
IR ($CHCl_3$) : 3370, 2995-2830, 2820, 1665, 1630, 1610, 1220, 1020 (*54*)
MS : 514(M^+), 486(M^+-CO), 401(f), 400(g), 374(h), 373(i), 372(j), 259(r), 233(s), 216(t), 209(p), 181(q), 165(x), 142(a'=100%), 114(a), 96(v), 86(q'/a''), 68(w) (*54*)
Sources : *Zizyphus sativa* (Rhamnaceae)-stem bark (*54*)

6. Nummularine-S

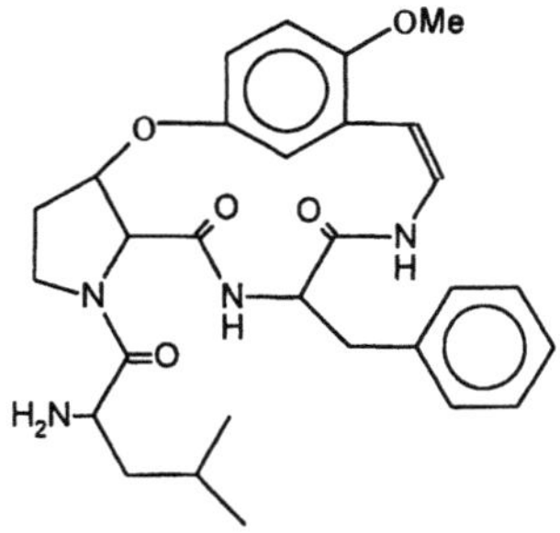

A = Leu
B = β-OH-Pro
C = Phe
$C_{29}H_{36}N_4O_5$, 520.2693 (MS) (*57*)

Mp=210-211 (*57, 73*)
UV (MeOH) : 268, 318 (*57*)
IR (KBr) : 3340, 2995-2830, 2820, 1670, 1640, 1610, 1220, 1030 (*57*)
MS : 520(M^+), 463(b), 435(f), 434(g), 408(h), 407(i), 406(j), 338(u), 259(r), 233(s), 216(t), 215(q), 165(x), 120(q'), 96(v), 86(a=100%), 68(w) (*57*)
Sources : *Zizyphus nummularia* (Rhamnaceae)-stem bark (*57*)

7. Tscheschamine

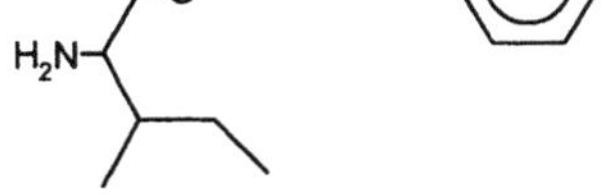

A = Ileu
B = β-OH-Pro
C = Phe
$C_{29}H_{36}N_4O_5$, 520.2692 (MS) (*74*)

Mp = 197-198 (*74*)
UV (MeOH) : 260, 320 (*74*)
IR (KBr) : 3300, 2820, 1680, 1640, 1615, 1230, 1040 (*74*)
MS : 520(M^+), 463(b), 435(f), 434(g), 408(h), 407(i), 406(j), 259(r), 243(p), 233(s), 216(t), 165(x), 120(q'), 96(v), 86(a=100%), 68(w) (*74*)
^{1}H-NMR (100MHz, $CDCl_3$) : 0.80-0.95 complex signal, 6H (2xMe), 3.80s, 3H (OMe), 4.30d, 5.8 Hz, 1H (α-H-Pro), , 5.50m, 1H (β-H-Pro), 5.90d, 9.0 Hz, 1H (β-H-Sty), 6.95dd, 12.0, 9.0 Hz, 1H (α-H-Sty), 6.70-8.60m, 11H (8xAr-H + 2xNH + 1 olefinic H), 7.30 and 8.25, 1H each-D_2O exchangeable (NH_2) (*74*)
Sources : *Zizyphus sativa* (Rhamnaceae)-stem bark (*74*)

8. Lotusine-F

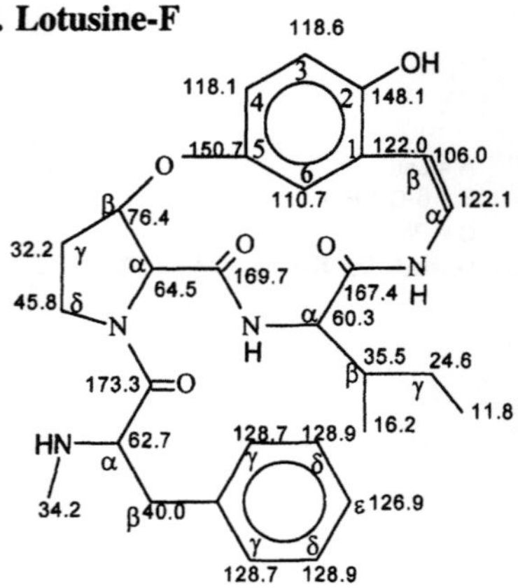

A = N-Me-Phe
B = β-OH-Pro
C = Ileu
$C_{29}H_{38}N_4O_5$, 520

$[a]_D$ = -244 (c=0.5, $CHCl_3$) *(75)*
UV (MeOH) : 210, 268, 323 *(75)*
UV (MeOH+NaOH) : 210, 263, 353 *(75)*
IR (KBr) : 3272, 2970, 1692, 1211 *(75)*
MS : 521(M^++H), 460(M^+-C_3H_7-H_2O), 429(b), 391(M^+-C_2H_5-R), 307(b'-R), 289(u-H^+), 273.(u-OH), 202(t), 181(q), 167(o), 154(p-R_1=100%), 151(x), 138(n-H^+), 134(a), 124(q-R_1), 107, 96(v), 91, 89, 86(q'), 77, 65 *(75)*
^{1}H-NMR (300MHz, $CDCl_3$) : 0.92t, 7.3 Hz, 3H (CH_2-Me), 1.06d, 6.9 Hz, 3H (CH-Me), 1.15-1.28m, 1H (γ-H-Ileu), 1.44ddd, 13.2, 7.3, 3.8 Hz, 1H (γ-H-Ileu), 2.06-2.16m, 2H (β-H-Ileu + H-Pro), 2.31-2.40m, 1H (γ-H-Pro), 2.38s, 3H (NH-Me), 2.51dt, 8.5, 6.5 Hz, 1H (δ-H-Pro), 2.74dd, 13.0, 9.3 Hz, 1H (β-H-N-Me-Phe), 3.05dd, 13.0, 4.9 Hz, 1H (β-H-N-Me-Phe), 3.60dd, 9.3, 4.9 Hz, 1H (α-H-N-Me-Phe), 3.67m, 1H (δ-H-Pro), 4.36dd, 4.7, 4.5 Hz, 1H (α-H-Ileu), 4.41d, 2.8 Hz, 1H (α-H-Pro), 5.33dt, 7.0, 2.8 Hz, 1H (β-H-Pro), 5.85d, 9.0 Hz, 1H (β-H-Sty), 6.57d, 2.9 Hz, 1H (6-H-Sty), 6.68dd, 8.8, 2.9 Hz, 1H (4-H-Sty), 6.80d, 8.8 Hz, 1H (3-H-Sty), 6.96dd, 11.3, 9.0 Hz, 1H (α-H-Sty), 7.02dd, 7.7, 2.0 Hz, 1H (γ-H-N-Me-Phe), 7.21m, 1H, (ε-H-N-Me-Phe), 7.25m, 2H (δ-H-N-Me-Phe), 7.39d, 5.1 Hz, 1H (NH-Ileu), 8.47d, 11.3 Hz, 1H (NH-Sty) *(75)*
^{13}C-NMR (75MHz, $CDCl_3$) : see figure *(75)*
COSY, HMBC, HMQC spectra *(75)*
Sources : *Zizyphus lotus* (Rhamnaceae)-root bark *(75)*

9. Subfraction I *(69)*

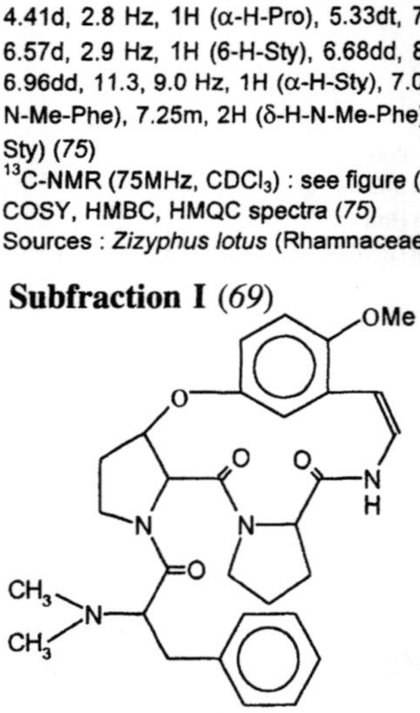

A = N,N-diMe-Phe
B = β-OH-Pro
C = Pro
$C_{30}H_{36}N_4O_5$, 532

Mp = 75 *(69)*
UV ($CHCl_3$) : 290, 355 *(69)*
IR ($CHCl_3$) : 3500, 2720, 1710 *(69)*
MS : 531(M-H^+) *(69)*
Sources : *Sphaeranthus indicus* (Asteraceae)-flowers *(69)*

10. Daechucyclopeptide-I (= Daechuine-S26)

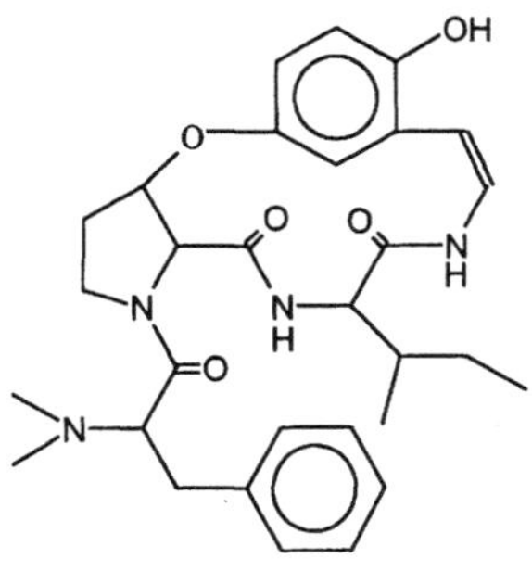

A = N,N-diMe-Phe
B = β-OH-Pro
C = Ileu
$C_{30}H_{38}N_4O_5$, 534

Mp = 114 (*70*)
Sources : *Zizyphus jujuba* var. *inermis* (Rhamnaceae)-fruits, stem bark (*70*)

11. Nummularine-C

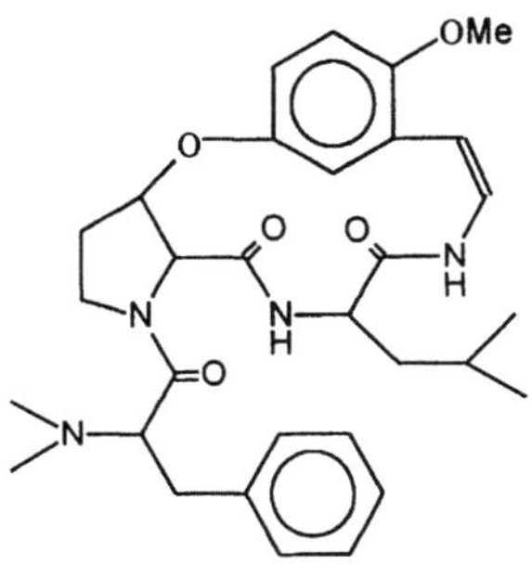

A = N,N-diMe-Phe
B = *trans*-β-OH-Pro[1] (*76*)
C = Leu
$C_{31}H_{40}N_4O_5$, 548.3002 (MS) (*76*)

Mp = 278-280 (*76*)
$[a]_D$ (20) = -371 (c=0.2, $CHCl_3$) (*76*)
UV (MeOH) : 270 (4.10), 320 (3.65) (*76*)
IR (KBr) : 3300, 2850, 2775, 1670, 1630, 1610, 1210, 1025 (*76*)
MS : 548(M^+), 533(M^+-CH_3), 505(M^+-C_3H_7), 457(b), 401(f), 400(g), 374(h), 373(i), 372(j), 304(u), 259(r), 233(s), 216(t), 209(p), 165(x), 148(a=100%), 96(v), 86(q'), 68(w) (*76*)
^{1}H-NMR (60 and 90MHz, $CDCl_3$) : 1.00d, 6H (2xC-Me), 2.40s, 6H (NMe_2), 3.75s, 3H (OMe), 4.38d, 5.0 Hz, 1H (β-H-Pro), 5.84d, 9.0 Hz, 1H (1 olefinic H), 6.66-8.32m, 11H (8xAr-H + 2xNH + 1 olefinic H) (*76*)
Derivatives : Dihydro-nummularine-C (*76*)
Sources : *Zizyphus nummularia* (Rhamnaceae)-root bark (*43*, *76*), stem bark (*66*)

12. Daechuine-S6

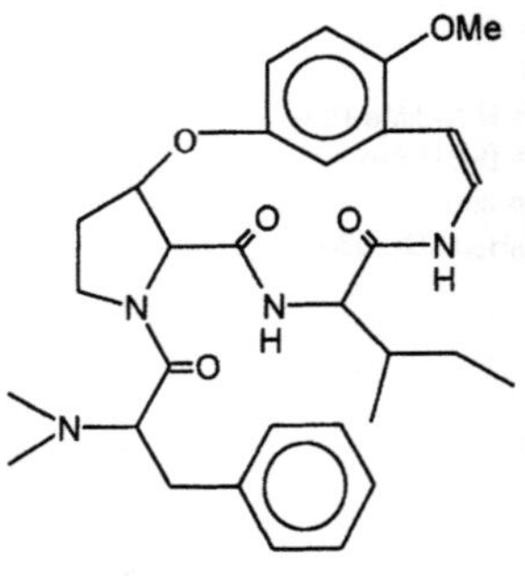

A = N,N-diMe-Phe
B = β-OH-Pro
C = Ileu
$C_{31}H_{40}N_4O_5$, 548

Mp = 192 (*70*)
[a]$_D$ = -393.5 (*70*)
Sources : *Zizyphus jujuba* var. *inermis* (Rhamnaceae)-stem bark (*70*)

13. Sativanine-E

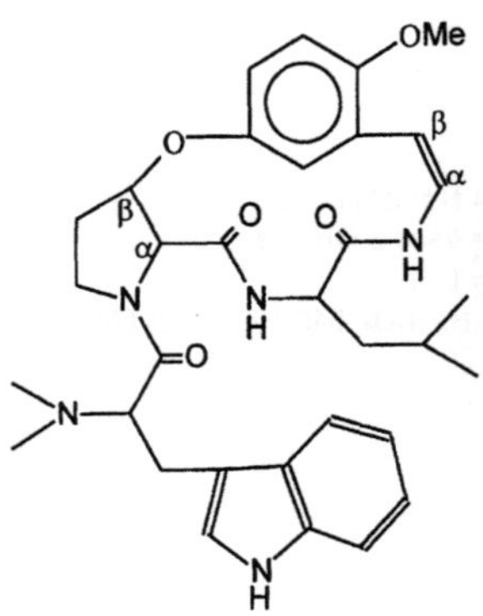

A = N,N-diMe-Trp
B = β-OH-Pro
C = Leu
$C_{33}H_{41}N_5O_5$, 587.3114 (MS) (*49*)

Mp = 127-128 (*49*)
[a]$_D$ (20) = -99 (c=0.2, $CHCl_3$) (*49*)
UV (MeOH) : 268, 272, 282, 289, 321 (*49*)
IR ($CHCl_3$) : 3385, 3250, 2775, 1670, 1630, 1590, 1490, 1195, 1040 (*49*)
MS : 587(M^+), 457(b=100%), 401(f), 400(g), 374(h), 373(i), 372(j), 304(u), 259(r), 233(s), 216(t), 187(a), 181(q), 165(x), 144, 130, 96(v), 86(q'), 68(w) (*49*)
^{1}H-NMR (90MHz, $CDCl_3$) : 0.91d, 7.0 Hz, 3H (Me-Leu), 0.97d, 7.0 Hz, 3H (Me-Leu), 2.48s, 6H (NMe_2), 3.79s, 3H (OMe), 4.22t, 4.4 Hz, 1H (CH-), 4.42d, 3.2 Hz, 1H (α-H-Pro), 5.32sextet, 6.4, 3.2 Hz, 1H (β-H-Pro), 5.92d, 9.0 Hz, 1H (β-H-Sty), 6.94q, 12.0, 9.0 Hz, 1H (α-H-Sty), 6.50-7.80m, 9H (3xAr-H + 5xIndole-H + 1 NH), 8.07 and 8.43, 2xs, 1H each-D_2O exchangeable (2xNH) (*49*)
Sources : *Zizyphus sativa* (Rhamnaceae)-stem bark (*49*)

14. Nummularine-R

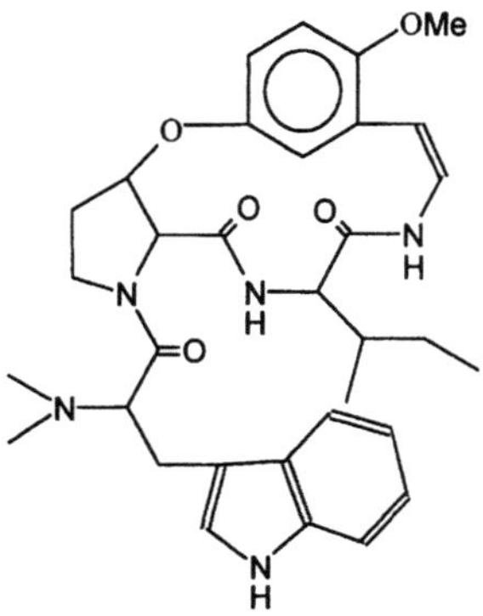

A = N,N-diMe-Trp
B = β-OH-Pro
C = Ileu
$C_{33}H_{41}N_5O_5$, 587.3116 (MS) (*77*)

Mp = 134-135 (*73, 77*)
UV (MeOH) : 268, 272 sh, 280 sh, 290 sh, 320 (*77*)
IR ($CHCl_3$) : 3385, 3251, 2775, 1670, 1630, 1592, 1490, 1200, 1040 (*77*)
MS : 587(M^+), 457(b=100%), 401(f), 400(g), 374(h), 373(i), 372(j), 304(u), 259(r), 233(s), 216(t), 187(a), 181(q), 165(x), 144, 130, 96(v), 86(q'), 68(w) (*77*)
Sources : *Zizyphus nummularia* (Rhamnaceae)-stem bark (*77*)

15. Daechuine-S10

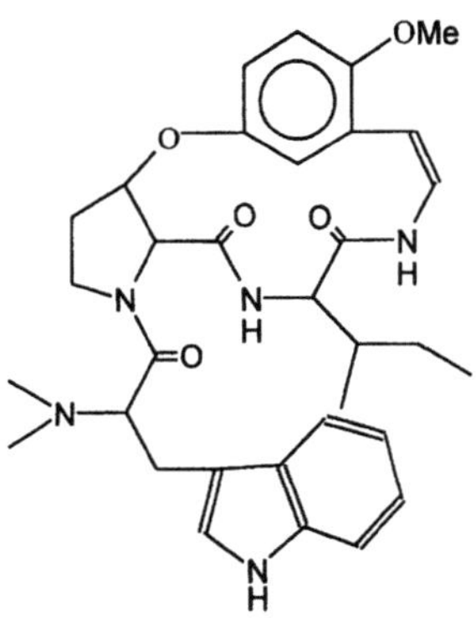

A = N,N-diMe-Trp
B = β-OH-Pro
C = Ileu
$C_{33}H_{41}N_5O_5$, 587

Mp = 126-128 (*70*)
$[a]_D$ = -381.5 (*70*)
Sources : *Zizyphus jujuba* var. *inermis* (Rhamnaceae)-stem bark (*70*)

16. Rugosanine-B

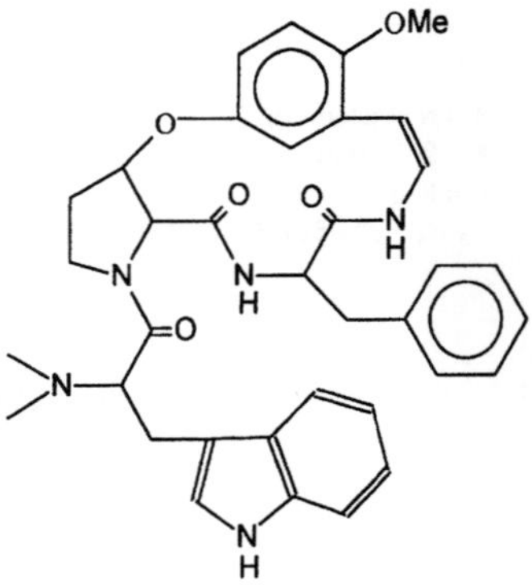

A = N,N-diMe-Trp
B = β-OH-Pro
C = Phe
$C_{33}H_{39}N_5O_5$, 621.3124 (MS) (*78*)

Mp = 216-218 (*73, 78*)
UV (MeOH) : 265, 270 sh, 280 sh, 292 sh, 322 (*78*)
IR ($CHCl_3$) : 3380, 3248, 2860, 2782, 1690, 1635, 1620, 1585, 1470, 1210, 1040 (*78*)
MS : 621(M^+), 491(b=100%), 435(f), 434(g), 408(h), 407(i), 406(j), 338(u), 259(r), 243(p), 233(s), 216(t), 187(a), 165(x), 144, 130, 96(v), 68(w) (*78*)
Sources : *Zizyphus rugosa* (Rhamnaceae)-stem bark (*78*)

5(13)-Zizyphine-A-Type Cyclopeptide Alkaloids

17. Sativanine-H

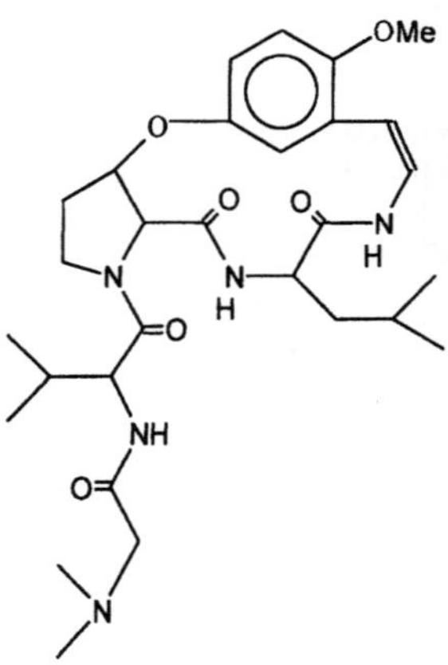

A = N,N-diMe-Gly
B = β-OH-Pro
C = Leu
E = Val
$C_{29}H_{43}N_5O_6$, 557.3200 (MS) (*79*)

Mp = 190-192 (*78, 79*)
UV (MeOH) : 260, 320 (*79*)
IR ($CHCl_3$) : 3380, 2835, 2785, 1670, 1640, 1610, 1230, 1040 (*79*)
MS : 557(M^+), 500(c), 498(d), 457(e), 401(f), 374(h), 373(i), 372(j), 233(s), 216(t), 209(p), 195(n), 185(l), 181(q), 165(x), 96(v), 86(q'), 58(a=100%) (*79*)
Sources : *Zizyphus sativa* (Rhamnaceae)-stem bark (*79*)

18. Nummularine-P

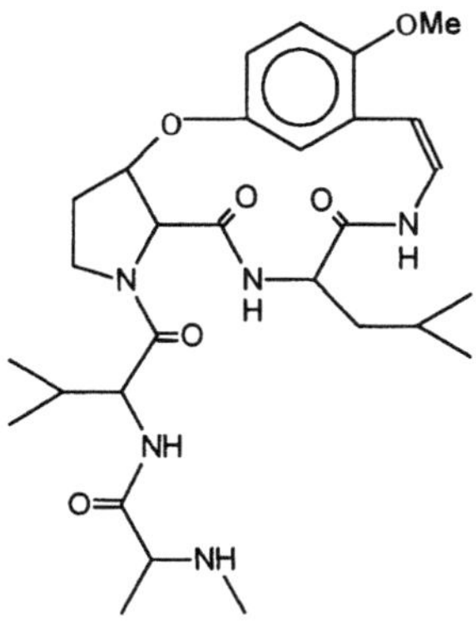

A = N-Me-Ala
B = β-OH-Pro
C = Leu
E = Val
$C_{29}H_{43}N_5O_6$, 557.3200 (MS) (*80*)

Mp = 143-144 (*80*), 179-180 (*56, 78*)
UV (MeOH) : 258 (3.99), 320 (3.90) (*80*)
IR ($CHCl_3$) : 3382, 2831, 2775, 1688, 1640, 1610, 1230, 1035 (*80*)
MS : 557(M^+), 542(b), 500(c), 498(d), 457(e), 401(f), 400(g), 374(h), 373(i), 372(j), 259(r), 233(s), 216(t), 209(p), 195(n), 185(l), 165(x), 157(m), 96(v), 86(q'), 72(q''), 58(a=100%) (*80*)
^{1}H-NMR (90MHz, $CDCl_3$) : 0.65d, 5.0 Hz, 12H (2xC-Me_2), 1.34d, 7.0 Hz, 3H (C-Me-N-Me-Ala), 1.70m, 2H (CH-Me_2), 2.47s, 3H (NH-Me), 2.60-3.65 complex pattern, 6H (3xCH_2-), 3.77s, 3H (OMe), 4.00-4.80 complex pattern, 5H (4xCO-CH-N- + 1NH), 5.60dt, 1H (β-H-Pro), 5.90d, 8.5 Hz, 1H (1 cis-olefinic-H), 6.79-8.61 complex pattern, 7H (3xAr-H + 3xNH + 1 cis-olefinic-H) (*80*)
Derivatives : *N*-Formyl-nummularine-P (*80*)
Sources : *Zizyphus nummularia* (Rhamnaceae)-stem bark (*80*)

19. Sativanine-C

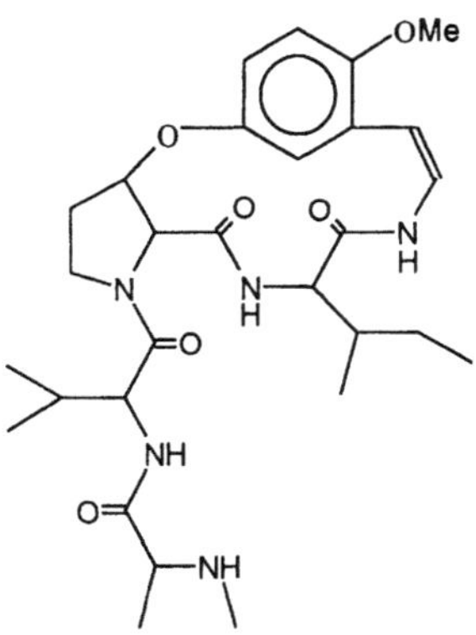

A = N-Me-Ala
B = β-OH-Pro
C = Ileu
E = Val
$C_{29}H_{43}N_5O_6$, 557.3200 (MS) (*81*)

Mp = 113-114 (*81*)
UV (MeOH) : 258, 320 (*81*)
IR ($CHCl_3$) : 3380, 2835, 2780, 1670, 1635, 1610, 1230, 1040 (*81*)
MS : 557(M^+), 542(b), 500(c), 498(d), 457(e), 401(f), 400(g), 374(h), 373(i), 372(j), 259(r), 233(s), 216(t), 209(p), 195(n), 185(l), 181(q), 165(x), 157(m), 96(v), 86(q'), 72(q''), 58(a=100%) (*81*)
Derivatives : *N*-Formyl-sativanine-C (*52, 81*)
Sources : *Zizyphus sativa* (Rhamnaceae)-stem bark (*81*)

20. Sativanine-D

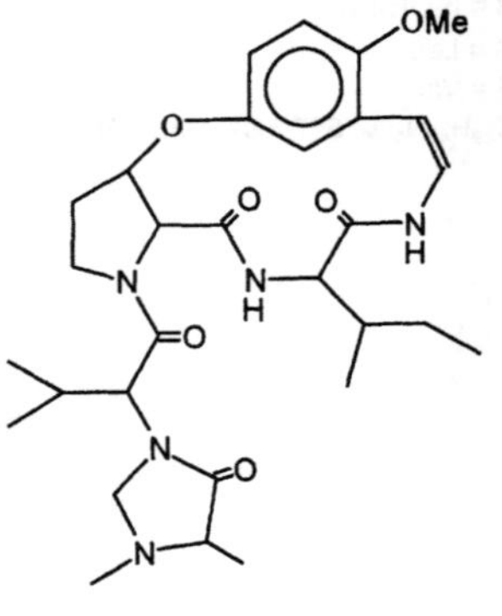

A = N-Me-Ala
B = β-OH-Pro
C = Ileu
E = Val
$C_{30}H_{43}N_5O_6$, 569.3220 (MS) (*51*)

Mp = 119-121 (*51*)
UV (MeOH) : 265, 320 (*51*)
IR ($CHCl_3$) : 2830, 1680, 1635, 1615, 1220, 1025 (*51*)
MS : 569(M^+), 457(e), 401(f), 400(g), 374(h), 373(i), 304(u), 233(s), 216(t), 209(p), 197(l), 195(n), 169(m), 165(x), 113(a'=100%), 96(v), 86(q'), 72(q''), 68(w) (*51*)
Sources : *Zizyphus sativa* (Rhamnaceae)-stem bark (*51*)

21. Rugosanine-A

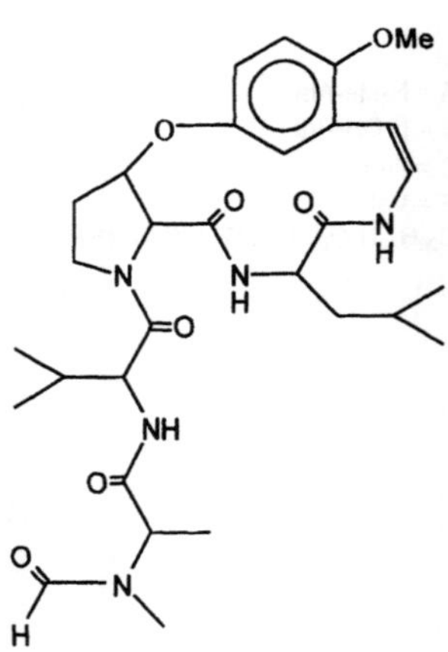

A = *N*-formyl-N-Me-Ala
B = β-OH-Pro
C = Leu
E = Val
$C_{30}H_{43}N_5O_7$, 585.3177 or 585.3187 (MS) (*56*)

Mp = 237-240 (*56*), 283-285 (*73*)
UV (MeOH) : 260 (3.68), 320 (3.64) (*56*)
IR ($CHCl_3$) : 3385, 2860, 2785, 1690, 1635, 1610, 1240-1205 (*56*)
MS : 585(M^+), 557(M^+-CO), 472(c-CO), 457(e), 401(f), 400(g), 374(h), 373(i), 372(j), 304(u), 259(r), 233(s), 221(o), 216(t), 213(l), 211(k-CO-R_1), 209(p), 195(n), 185(m), 181(q), 165(x), 114(a'=100%), 86(a/q'), 58(a'') (*56*)
Sources : *Zizyphus rugosa* (Rhamnaceae)-stem bark (*56*)

22. Nummularine-N

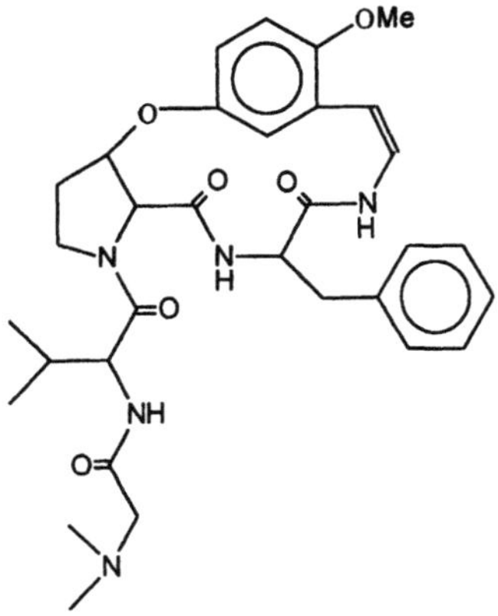

A = N,N-diMe-Gly
B = β-OH-Pro
C = Phe
E = Val
$C_{31}H_{41}N_5O_6$, 591

Mp = 243-245 (*82*)
UV (MeOH) : 267 (4.00), 320 (3.80) (*82*)
IR (KBr) : 3300, 2821, 1688, 1640, 1610, 1200, 1020 (*82*)
MS : 591(M^+), 548(M^+-C_3H_7), 534(c), 532(d), 491(e), 435(f), 434(g), 408(h), 406(j), 338(u), 259(r), 243(p), 233(s), 216(t), 165(x), 157(m), 96(v), 68(w), 58(a=100%) (*82*)
^{1}H-NMR ($CDCl_3$) : 0.68d, 5.0 Hz, 6H (2xC-Me), 2.48s, 6H (NMe_2), 3.80s, 3H (OMe), 6.00d, 7.5 Hz, 1H (1 olefinic-H), 6.70-8.60m, 12H (8xAr-H + 3xNH + 1 olefinic-H) (*82*)
Sources : *Zizyphus nummularia* (Rhamnaceae)-stem bark (*82*)

23. Nummularine-B (= Daechuine-S27 = *N*-Desmethyl-amphibine-H)

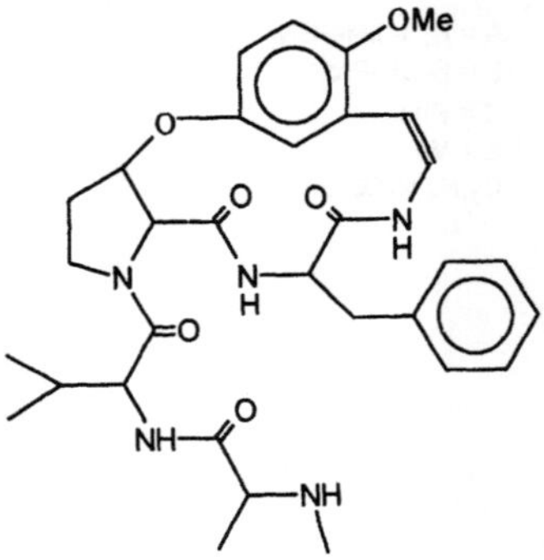

A = N-Me-Ala
B = *trans*-β-OH-Pro[1] (*76*)
C = Phe
E = Val
$C_{32}H_{41}N_5O_6$, 591.3059 (MS) (*76*)

Mp = 226-231 (*68, 70, 73, 76*)
[a]$_D$ (20) = -390 (c=0.2, $CHCl_3$) (*76,70*)
UV (MeOH) : 268 (4.05), 321 (3.90) (*76*)
IR (KBr) : 3300, 2820, 2770, 1690, 1640, 1610, 1220, 1020 (*76*)
MS : 591(M^+), 576(b), 171(m), 58(a) (*76*)
^{1}H-NMR (90MHz, $CDCl_3$) : 0.57d, 6.8 Hz, 3H (Me-Val), 0.68d, 6.8 Hz, 3H (Me-Val), 1.30d, 6.8 Hz, 3H (Me-N-Me-Ala), 1.75m, 6.8 Hz, 1H (β-H-Val), 2.40s, 3H (NH-Me-N-Me-Ala), 2.20-2.70m, 2H (2xγ-H-β-OH-Pro), 2.70-3.30q+m, 3H (α-H-N-Me-Ala + 2xβ-H-Phe), 3.30-3.80m, 1H (δ-H-β-OH-Pro), 3.80s, 3H (OMe), 4.14m, 1H (δ-H-β-OH-Pro), 4.40-4.80m, 3H (α-H-Val + α-H-Phe + α-H-β-OH-Pro), 5.52dt, 7.4, 4.4 Hz, 1H (β-H-β-OH-Pro), 5.97d, 9.0 Hz, 1H (β-H-Sty), 6.70-8.60m, 12H [8xAr-H + 3xNH + 1xcis-olefinic-H : 6.70-7.40 complex pattern, 8H (8xAr-H), 6.97dd, 12.0, 9.0 Hz, 1H (α-H-Sty), 7.34d, 1H (NH-Phe), 7.63d, 9.0 Hz, 1H (NH-Val), 8.48d, 12.0 Hz, 1H (NH-Sty)] (*83*).
^{1}H-NMR (60 and 90MHz, $CDCl_3$) : (*76*)
Derivatives : *N*-Methyl-dihydro-nummularine-B = Dihydro-amphibine-H (*76*)
N-Formyl-nummularine-B (*52*)
Nummularine-B-cycl. (*51, 65*)
Sources : Rhamnaceae
Zizyphus jujuba-stem bark (*84*)
Z. jujuba var. *inermis*-stem bark (*70*)
Z. nummularia-root bark (*43, 76*), stem bark (*66, 82*)
Z. sativa-stem bark (*67*)
Z. xylopyra-stem bark (*85*)

24. Zizyphine-B (= Zizyphinine = *N*-Desmethyl-zizyphine-A)

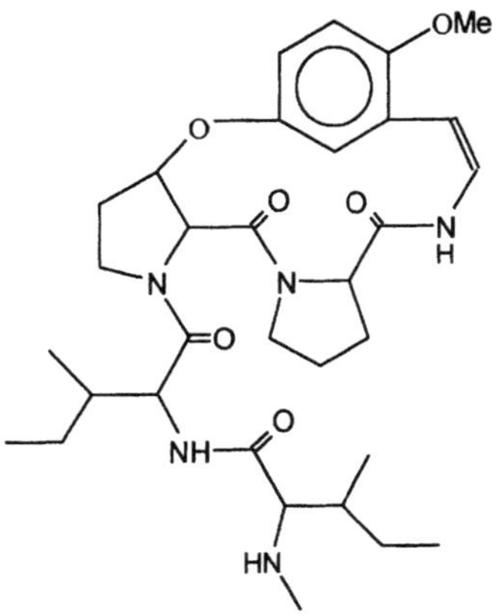

A = N-Me-Ileu
B = *trans*-β-OH-Pro[1] (*10, 38*)
C = *trans*-Pro[1] (*10*)
E = Ileu
$C_{32}H_{47}N_5O_6$, 597.74 (MS) (*10*)

Mp = amorphous (*10*)
[a]$_D$ (24) = -457 (c=1.0, $CHCl_3$) (*10*)
UV (EtOH) : 267 (4.00), 320 (3.88) (*10*)
IR (CH_2Cl_2) : 3745, 3484, 3012, 2924, 2882, 1704, 1653, 1600, 1513, 1418, 1325, 1225, 1186, 1053, 813 (*10,86*)
MS : 597(M^+), 554(M^+-C_3H_7), 498(c), 455(e), 426(e'), 385(f), 216(t), 202, 191(y), 114 (*10*)
MS : 100 (a=100%) (*87*)
Derivatives : *N*-Acetyl-zizyphine-B (*87*)
Sources : *Zizyphus oenoplia* (Rhamnaceae)-root bark (*10*), stem bark (*87*)

25. Zizyphine-F (= *O*-Desmethyl-zizyphine-A)

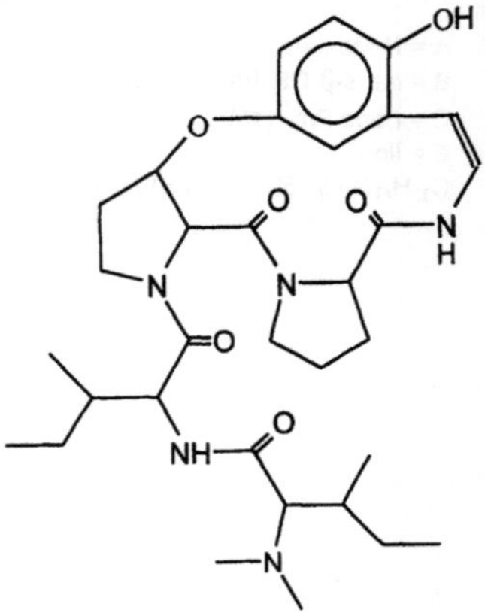

A = N,N-diMe-Ileu
B = *trans*-β-OH-Pro[1] (*40*)
C = Pro
E = Ileu
$C_{32}H_{47}N_5O_6$, 597

Mp = 235 (*40*)
$[a]_D$ (20) = -277 (c=0.15, MeOH) (*40*)
UV (MeOH) : 225 ((4.30), 265 (4.00), 322 (3.80) (*40*)
UV (MeOH+NaOH) : 225 (4.30), 268 (4.10), 351 (3.80) (*40*)
MS : 597(M^+), 554(M^+-C_3H_7), 540(b), 484(c), 482(d), 371(f), 370(g), 344(h), 343(i), 342(j), 324(k), 274(u'), 255(l), 245(r), 227(m), 219(s), 202(t), 193(p), 165(q), 114(a=100%) (*40*)
^{1}H-NMR ($CDCl_3$) : 0.95d, 6H (2xCH-Me), 2.23s, 6H (N-Me_2), 5.90-6.80m, 5H (3xAr-H + 2xolefinic-H) (*40*)
Derivatives : Dihydro-zizyphine-F (*40*)
Sources : Rhamnaceae
Zizyphus oenoplia-stem bark (*40*)
Z. spina-christi-stem bark (*88*)

26. Amphibine-H

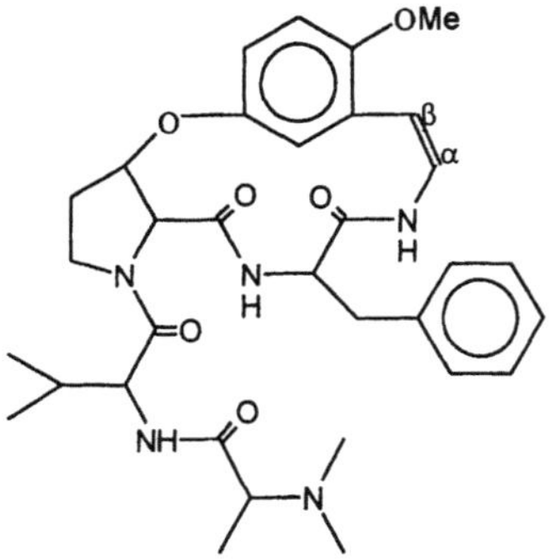

A = N,N-diMe-Ala
B = *trans*-β-OH-Pro[1] (*41*)
C = Phe
E = Val
$C_{33}H_{43}N_5O_6$, 605.3193 (MS) (*41*)

Mp = 201-205 [*41, 68, 73, 77*)
[a]$_D$ (20) = -570 (c=0.12, MeOH) (*41*)
UV (MeOH) : 268 (4.19), 321 (4.01) (*41*)
IR ($CHCl_3$) : 3320, 2820, 2770, 1680, 1640, 1610, 1210, 1025 (*41*)
MS : 605(M^+), 590(b), 534(c), 532(d), 491(e), 462(e'), 435(f), 434(g), 408(h), 407(i), 406(j), 358(f-77), 338(u), 259(r), 243(p), 233(s), 221(o), 216(t), 191(y), 171(m), 165(x), 164(x-H^+), 120(q'), 91, 72(a=100%/q''), 68(w) (*41*)
^{1}H-NMR (60 and 90MHz, $CDCl_3$) : 0.70d, 6.0 Hz, 6H (CH-Me_2-Val), 1.30d, 6.0 Hz, 3H (CH-Me-N,N-diMe-Ala), 2.30s, 6H (NMe_2), 3.80s, 3H (OMe), 5.95d, 9.0 Hz, 1H (β-H-Sty), 6.85q, 1H (α-H-Sty), 6.70-7.80m, 10H (8xAr-H + 2xNH), 8.55m, 1H (NH) (*41*)
Derivatives : Dihydro-amphibine-H (*41*)
Sources : Rhamnaceae
Zizyphus amphibia-stem bark (*41, 89*)
Z. jujuba-stem bark (*84*)
Z. nummularia-root bark (*43, 76*)
Z. spina-christi-stem bark (*88*)
Z. xylopyra-stem bark (*77*)

27. Zizyphine-A

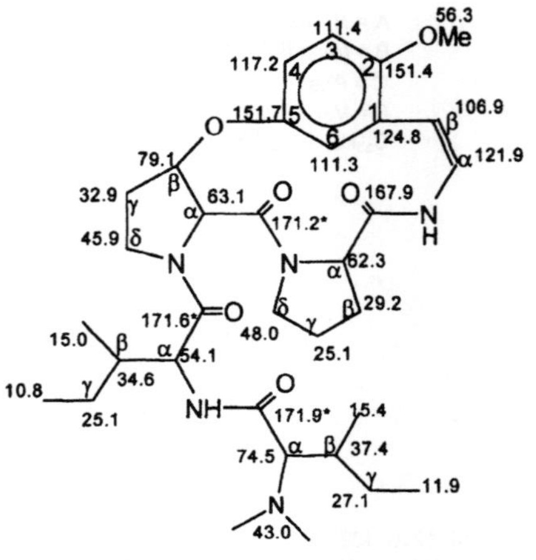

A = N,N-diMe-Ileu
B = *trans*-β-OH-Pro[1] (*10, 38*)
C = Pro
E = Ileu
$C_{33}H_{49}N_5O_6$, 611.76 (MS) (*10, 90*)

Mp = 121-126 (*10, 38*)
$[a]_D$ (20)= -411 (c=0.086, $CHCl_3$) (*38*)
$[a]_D$ (24) (*10*)
UV (EtOH) : 206 (4.58), 267 (4.03), 319 (3.91) (*10*)
UV (Dioxane) (*10*)
IR (CH_2Cl_2) : 3745, 3484, 3012, 2924, 2882, 2849, 1704, 1653, 1600, 1513, 1418, 1325, 1225, 1186, 1053, 813 (*10, 90*)
MS : 554(b), 498(c), 496(d), 385(f), 384(g), 358(h), 357(i), 356(j), 324(k), 288(u'), 259(r), 255(l), 235(o), 233(s), 227(m), 216(t), 209(n), 193(p), 165(q/x), 114(a=100%) (*38*)
MS : 611(M^+) (*10, 90*)
^{1}H-NMR (250MHz, $CDCl_3$) : 0.88t, 7.0 Hz, 3H (CH_2-Me-Ileu), 0.91d, 6.7 Hz, 3H (CH-Me-Ileu), 0.92d, 6.7 Hz, 3H (CH-Me-N,N-diMe-Ileu), 0.96d, 7.3 Hz, 3H (CH_2-Me-N,N-diMe-Ileu), 1.20-1.90m, 5H (β-H-N,N-diMe-Ileu + 2xγ-H-N,N-diMe-Ileu + 2xγ-H-Ileu), 1.95m, 5H (β-H-Ileu + 2xβ-H-Pro + 2xγ-H-Pro), 2.40, 1H (α-H-N,N-diMe-Ileu), 2.45s, 6H (N-Me_2), 2.50, 2H (γ-H-β-OH-Pro), 3.28, 1H (δ-H-Pro), 3.65, 1H (δ-H-β-OH-Pro), 3.80s, 3H (OMe), 4.25, 1H (δ-H-β-OH-Pro), 4.35, 1H (δ-H-Pro), 4.39d, 5.8 Hz, 1H (α-H-β-OH-Pro), 4.50dd, 9.0, 4.2 Hz, 1H (α-H-Pro), 4.50t, 8.5 Hz, 1H (α-H-Ileu), 5.26ddd, 9.5, 6.0, 5.8 Hz, 1H (β-H-β-OH-Pro), 5.95d, 8.7 Hz, 1H (β-H-Sty), 6.82d, 10.0 Hz, 1H (4-H-Sty), 6.82, 1H (6-H-Sty), 6.89d, 10.0 Hz, 1H (3-H-Sty), 6.93dd, 12.2, 8.7 Hz, 1H (α-H-Sty), 7.15d, 8.5 Hz, 1H (NH-Ileu), 8.36d, 12.2 Hz, 1H (NH-Sty) (*91*)
^{1}H-NMR (60MHz, $CDCl_3$) (*38, 90*)
Partially relaxed ^{1}H-NMR spectrum (250MHz, $CDCl_3$) (*91*)
^{13}C-NMR (62.9MHz, $CDCl_3$) (*91*) : see figure *values may be interchanged
^{13}C-NMR (62.9MHz, CD_3OD) (*91*)
^{13}C-NMR (20 and 25MHz, $CDCl_3$) (*8, 92*)
Derivatives : Dihydro-zizyphine-A (*10, 90*)
Zizyphine-A amido-aldehyde (*38*)
Zizyphine-A imido-aldehyde (*87*)
Sources : *Zizyphus oenoplia* (Rhamnaceae)-root bark (*10, 90*), stem bark (*87*)

28. Daechuine-S8-1

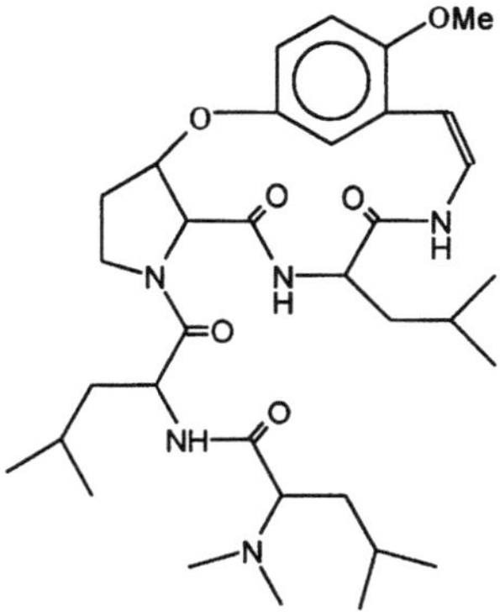

A = N,N-diMe-Leu
B = β-OH-Pro
C = Leu
E = Leu
$C_{33}H_{51}N_5O_6$, 613

Mp = 185-188 (*70*)
$[a]_D$ = -218.2 (*70*)
Sources : *Zizyphus jujuba* var. *inermis* (Rhamnaceae)-stem bark (*70*)

29. Nummularine-T

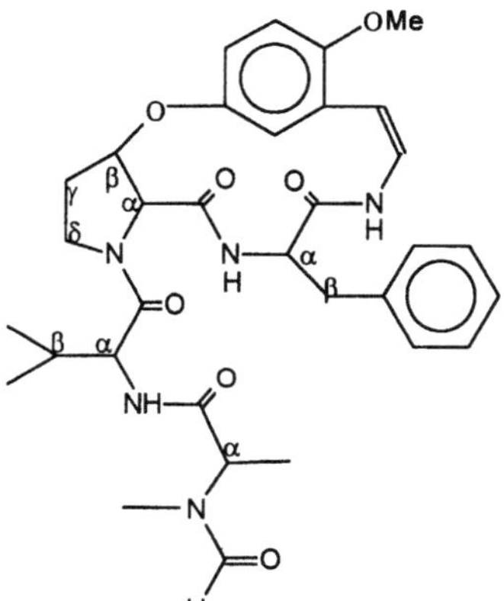

A = *N*-formyl-N-Me-Ala
B = β-OH-Pro
C = Phe
E = Val
$C_{33}H_{41}N_5O_7$, 619.2988 (MS) (*68*)

Mp = 188-190 (*68*)
UV (MeOH) : 267 (3.60), 318 (3.40) (*68*)
IR (KBr) : 3400, 2865, 2775, 1680, 1655, 1620, 1220, 1050 (*68*)
MS : 619(M^+), 591(M^+-CO), 506(c-CO), 491(e), 435(f), 434(g), 408(h), 407(i), 406(j), 259(r), 243(p), 233(s), 221(o), 216(t), 215(q), 213(l), 195(n), 185(m), 114(a'=100%), 96(v), 86(a=75%), 58(a'') (*68*)
^{1}H-NMR (90MHz, $CDCl_3$) : 0.56d, 7.0 Hz, 3H (Me-Val), 0.68d, 7.0 Hz, 3H (Me-Val), 1.23d, 7.0 Hz, 3H (CH-Me-N-Me-Ala), 1.38-1.70m, 1H (β-H-Val), 2.15-2.95m, 3H (2xγ-H-Pro + α-H-Ala), 2.88s, 3H (N-Me), 3.18-3.66m, 2H (2xβ-H-Phe), 3.78s, 3H (OMe), 4.22m, 2H (2xδ-H-Pro), 4.44-5.12m, 3H (α-H-Pro + α-H-Val + α-H-Phe), 5.37m, 1H (β-H-Pro), 5.98d, 9.0 Hz, 1H (β-H-Sty), 6.65-8.92m, 13H (8xAr-H + 3xNH + CHO + α-H-Sty) (*68*)
Sources : *Zizyphus nummularia* (Rhamnaceae)-stem bark (*68*)

30. Daechuine-S3

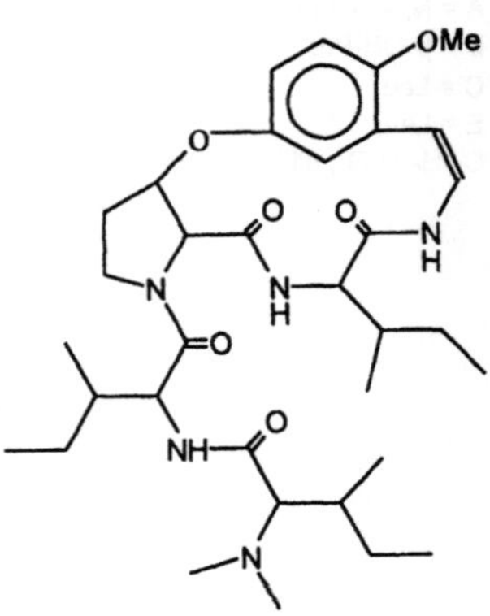

A = N,N-diMe-Ileu
B =β-OH-Pro
C = Ileu
E = Ileu
$C_{34}H_{53}N_5O_6$, 514

Mp = 192-194 (*70*)
$[a]_D$ = -440.0 (*70*)
Sources : *Zizyphus jujuba* var. *inermis* (Rhamnaceae)-stem bark (*70*)

31. Sativanine-F

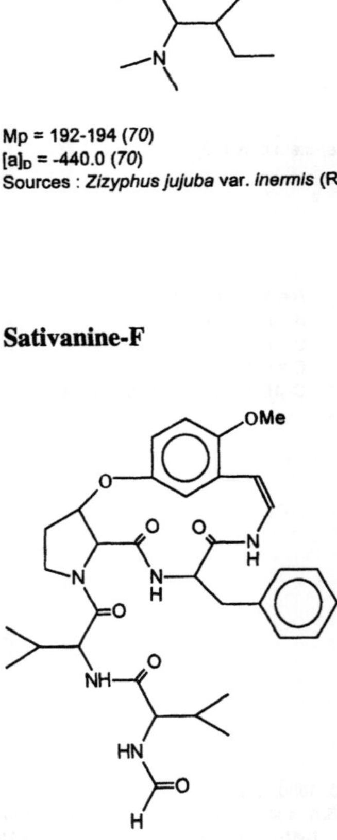

A = *N*-formyl-Val
B = β-OH-Pro
C = Phe
E = Val
$C_{34}H_{43}N_5O_7$, 633.3164 (MS) (*52*)

Mp = 139-141 (*52*)
UV : 258, 320 (*52*)
IR : 3400, 2820, 1670, 1630, 1615, 1190, 1030 (*52*)
MS : 633(M^+), 605(M^+-CO), 506(c-CO), 491(e), 435(f), 408(h), 407(i), 406(j), 259(r), 243(p), 233(s), 227(l), 221(o), 216(t), 215(q), 195(n), 165(x), 128(a'=100%), 120(q'), 100(a), 72(q''/a''), 68(w) (*52*)
Sources : *Zizyphus sativa* (Rhamnaceae)-stem bark (*52*)

32. Zizyphine-C

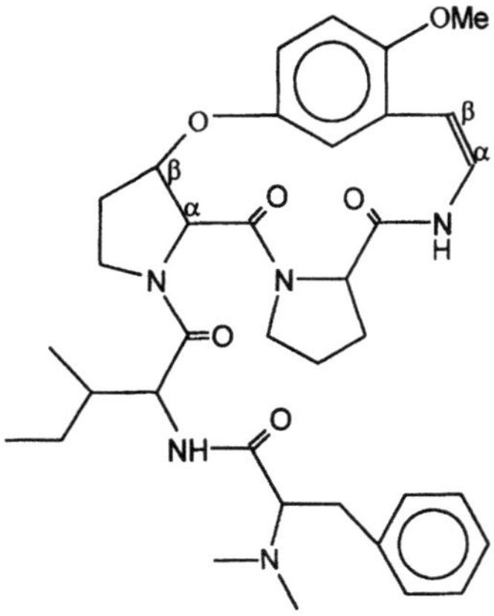

A = N,N-diMe-Phe
B = *trans*-β-OH-Pro[1] (*87*)
C = *trans*-Pro[1] (*87*)
E = Ileu
$C_{36}H_{47}N_5O_6$, 645.3524 (MS) (*87*)

Mp = amorphous (*87*)
$[a]_D$ (20) = -331 ± 5 (c=0.10, $CHCl_3$), -343 ± 5 (c=0.10, MeOH) (*87*)
UV (MeOH) : 266 (3.88), 318 (3.74) (*87*)
IR ($CHCl_3$) : 3400, 2820, 2785, 1690, 1645, 1597, 1257, 690 (*87*)
MS : 645(M^+), 148(a=100%) (*87*)
^{1}H-NMR (60 and 90MHz, $CDCl_3$) : 0.80-1.05m, 6H (2xC-Me), 2.36s, 6H (N-Me_2), 3.80s, 3H (OMe), 4.33d, 6.0 Hz, 1H (α-H-β-OH-Pro), 5.24m, 1H (β-H-β-OH-Pro), 5.95d, 8.9 Hz, 1H (β-H-Sty), 6.92dd, 11.4, 8.9 Hz, 1H (α-H-Sty), 7.25s, 5H (5xAr-H-N,N-diMe-Phe), 8.37d, 11.4 Hz, 1H (NH-Sty) (*87*)
Derivatives : Zizyphine-C amido-aldehyde (*87*)
Sources : *Zizyphus oenoplia* (Rhamnaceae)-stem bark (*87*)

33. Lotusine-E

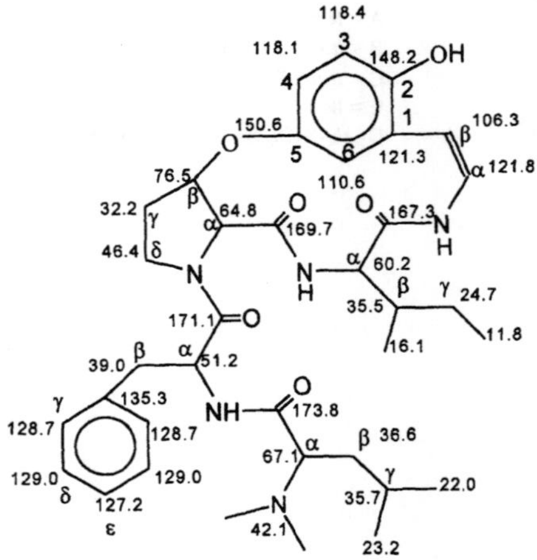

A = N,N-diMe-Leu

B = β-OH-Pro

C = Ileu

E = Phe

$C_{36}H_{49}N_5O_6$, 647

$[a]_D$ = -106 (c=1.0, $CHCl_3$) [75]
UV (MeOH) : 206, 268, 323 [75]
UV (MeOH+NaOH) : 212, 268, 353 [75]
IR (KBr) : 3257, 2954, 1684, 1645, 1219 [75]
MS : 647(M^+), 590(b), 442(M^+-a-R_1), 359(i), 289(l), 269(o), 261(m), 245(r), 243(n), 227(o-NCO), 209(p), 202(t), 190, 185(t-OH), 181(q), 177(y), 165(q-OH+H^+), 161(y-OH+H^+), 151(x), 148(s''+H^+), 134(z), 131(r''), 120(q''), 114(a=100%), 98(r'+H^+), 96(v), 91, 86(q'), 84(a-2Me), 72(o'), 69(w+H^+) [75]
^{1}H-NMR (300MHz, $CDCl_3$) : 0.88, d, 6.4 Hz, 3H (C-Me-N,N-diMe-Leu), 0.89, d, 6.4 Hz, 3H (C-Me-N,N-diMe-Leu), 0.92, t, 7.0 Hz, 3H (CH_2-Me-Ileu), 1.03, d, 6.9 Hz, 3H (CH-Me-Ileu), 1.20, m, 1H (γ-H-Ileu), 1.39, m, 1H (γ-H-Ileu), 1.48, m, 2H (β-H-N,N-diMe-Leu), 1.61, m, 1H (γ-H-N,N-diMe-Leu), 2.09, m, 1H (β-H-Ileu), 2.20, m, 1H (γ-H-Pro), 2.21, br. s, 6H (N-Me_2), 2.42, m, 1H (γ-H-Pro), 2.88, m, 1H (δ-H-Pro), 2.90, m, 1H (α-H-N,N-diMe-Leu), 2.96, br. d, 7.4 Hz, 2H (β-H-Phe), 4.03, ddd, 11.1, 8.4, 3.0 Hz, 1H (δ-H-Pro), 4.35, dd, 5.0, 4.5 Hz, 1H (α-H-Ileu), 4.40, d, 2.9 Hz, 1H (α-H-Pro), 4.96, q, 7.7 Hz, 1H (α-H-Phe), 5.34, dt, 7.0, 2.9 Hz, 1H (β-H-Pro), 5.87, d, 9.0 Hz, 1H (β-H-Sty), 6.58, d, 2.8 Hz, 1H (6-H-Sty), 6.67, dd, 8.8, 2.8 Hz, 1H (4-H-Sty), 6.80, d, 8.8 Hz, 1H (3-H-Sty), 6.94, dd, 11.2, 9.0 Hz, 1H (α-H-Sty), 7.08, dd, 7.6, 1.9 Hz, 2H (γ-H-Phe), 7.25, m, 1H, (ε-H-Phe), 7.26, m, 2H (δ-H-Phe), 7.27, m, 1H (NH-Ileu), 7.59, br. d, 7.7 Hz, 1H (NH-N,N-diMe-Leu), 8.46, d, 11.2 Hz, 1H (NH-Sty) [75]
^{13}C-NMR (75MHz, $CDCl_3$) : see figure [75]
COSY, HMBC, HMQC, HOHAHA spectra [75]
Sources : *Zizyphus lotus* (Rhamnaceae)-root bark [75]

34. Paliurine-B

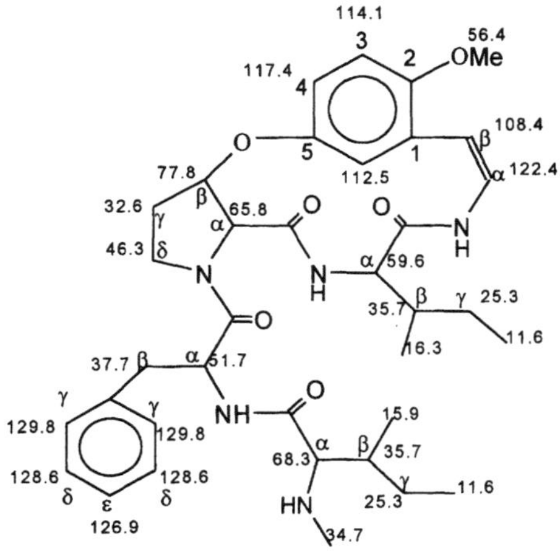

A = N-Me-Ileu

B = β-OH-Pro

C = Ileu

E = Phe

$C_{36}H_{49}N_5O_6$, 647

^{1}H-NMR (400MHz, DMSO-d_6) : 0.70d, 3H (CH-Me-N-Me-Ileu), 0.76t, 3H (CH_2-Me-N-Me-Ileu), 0.81t, 3H (CH_2-Me-Ileu), 0.89d, 3H (CH-Me-Ileu), 1.02m, 2H (γ-H-N-Me-Ileu), 1.16-1.40m, 2H (γ-H-Ileu), 1.45m, 1H (β-H-N-Me-Ileu), 1.81m, 1H (β-H-Ileu), 2.01s, 3H (NH-Me), 2.11m and 2.44m, 2H (γ-H-Pro), 2.74, 1H (α-H-N-Me-Ileu), 2.80 and 2.90, 2H (β-H-Phe), 3.49, 1H (δ-H-Pro), 3.73s, 3H (OMe), 3.99, 1H (δ-H-Pro), 4.11, 1H (α-H-Ileu), 4.46, 1H (α-H-Pro), 4.87, 1H (α-H-Phe), 5.11, 1H (β-H-Pro), 5.84d, 1H (β-H-Sty), 6.71dd, 1H (α-H-Sty), 6.76dd, 1H (4-H-Sty), 6.79d, 1H (6-H-Sty), 7.00d, 1H (3-H-Sty), 7.24, 2H (δ-H-Phe), 8.27d, 1H (NH-Ileu), 8.35d, 1H (NH-Phe), 9.15d, 1H (NH-Sty) (*93*)
^{13}C-NMR (100MHz, DMSO-d_6) : see figure (*93*)
COSY, ROESY, TOCSY and HETCOR NMR spectra (*93*)
Sources : *Paliurus ramosissimus* (Rhamnaceae)-roots, stems (*93*)

35. Nummularine-A (= *N*-Desmethyl-mucronine-D)

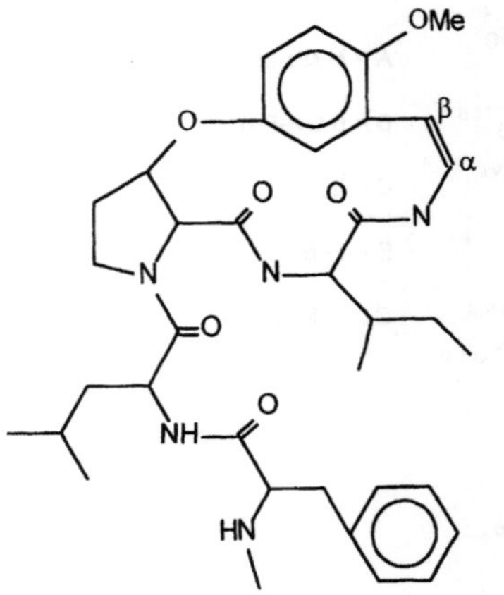

A = N-Me-Phe
B = *trans*-β-OH-Pro[1] (*76*)
C = Ileu
E = Leu
$C_{36}H_{49}N_5O_6$, 647.3669 (MS) (*76*)
Mp = 235-240 (*76*)
[a]$_D$ (20) = -397 (c=0.2, $CHCl_3$) (*76*)
UV (MeOH) : 268 (4.02), 322 (3.73) (*76*)
IR (KBr) : 3270, 2820, 2770, 1680, 1628, 1610, 1210, 1025 (*76*)
MS : 647(M^+), 556(b), 247(m), 134(a=100%) (*76*)
^{1}H-NMR (60 and 90MHz, $CDCl_3$) : 0.60-1.00m, 12H (4xC-Me), 2.31s, 3H (NH-Me), 3.72s, 3H (OMe), 5.83d, 9.0 Hz, 1H (β-H-Sty), 7.02-8.32m, (Ar-H.+ NH + 1 olefinic H) (*76*)
Derivatives : *N*-Methyl-dihydro-nummularine-A = Dihydro-mucronine-D (*76*)
Sources : Rhamnaceae
Zizyphus jujuba-stem bark (*84*)
Z. nummularia-root bark (*43, 76*), stem bark (*66*).

36. *O*-Desmethyl-mucronine-D

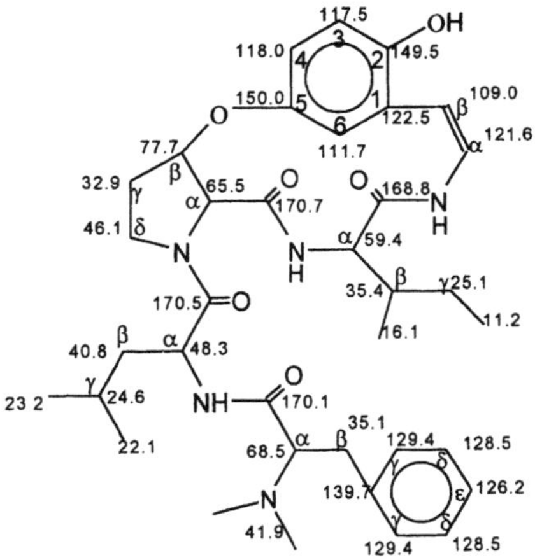

A = N,N-diMe-Phe
B = trans-β-OH-Pro[1] (*71*)
C = Ileu
E = Leu
$C_{36}H_{49}N_5O_6$, 647

$[a]_D$ (20) = -191 (c=0.3, $CHCl_3$) (*71*)
UV (MeOH) : 268 (3.68), 322 (3.48) (*71*)
MS : 647(M^+), 556(b), 499(c-H^+), 235(o), 202(t), 151(x), 148(a=100%), 86(q'/q'') (*71*)
^{1}H-NMR (300MHz, DMSO-d_6) : 0.75t, 7.2 Hz, 3H (CH_2-Me-Ileu), 0.81d, 6.2 Hz, 3H (Me-Leu), 0.82d, 6.9 Hz, 3H (Me-Leu), 0.83d, 6.9 Hz, 3H (CH-Me-Ileu), 1.11m, 1H (γ-H-Ileu), 1.32m, 1H (γ-H-Ileu), 1.37m, 2H (β-H-Leu), 1.51m, 1H (γ-H-Leu), 1.80m, 1H (β-H-Ileu), 2.08m, 1H (γ-H-Pro), 2.21s, 6H (N-Me_2), 2.50m, 1H (γ-H-Pro), 2.72dd, 13.6, 6.0 Hz, 1H (β-H-N,N-diMe-Phe), 2.94dd, 13.6, 8.9 Hz, 1H (β-H-N,N-diMe-Phe), 3.29dd, 8.9, 6.0 Hz, 1H (α-H-N,N-diMe-Phe), 3.50m, 1H (δ-H-Pro), 4.00m, 1H (δ-H-Pro), 4.00t, 8.0 Hz, 1H (α-H-Ileu), 4.40d, 1.9 Hz, 1H (α-H-Pro), 4.58dt, 8.7, 7.5 Hz, 1H (α-H-Leu), 5.05br. t, 6.2 Hz, 1H (β-H-Pro), 5.84d, 8.6 Hz, 1H (β-H-Sty), 6.65d, 3.0 Hz, 1H (6-H-Sty), 6.67dd, 8.6, 3.0 Hz, 1H (4-H-Sty), 6.67dd, 10.3, 8.6 Hz, 1H (α-H-Sty), 6.82d, 8.6 Hz, 1H (3-H-Sty), 7.09-7.26m, 5H (2xγ-H + 2xδ-H + ε-H-N,N-diMe-Phe), 8.14d, 8.5 Hz, 1H (NH-Ileu), 8.20d, 9.2 Hz, 1H (NH-Leu), 9.10d, 10.3 Hz, 1H (NH-Sty), 9.20br. s, 1H (OH) (*71*)
^{13}C-NMR (75MHz, DMSO-d_6) : see figure (*71*)
COSY, ROESY, TOCSY and FLOCK NMR spectra (*71*)
Sources : *Zizyphus mucronata* (Rhamnaceae)-roots (*71*)

37. Mucronine-D (= Daechuine-S9)

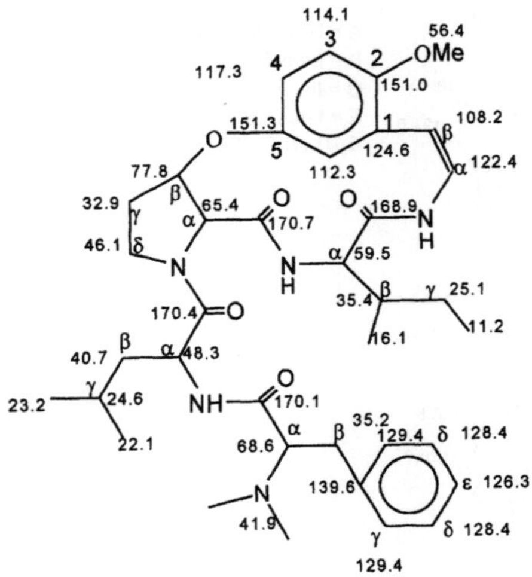

A = N,N-diMe-Phe

B = trans-β-OH-Pro[1] [37]

C = Ileu

E = Leu

$C_{37}H_{51}N_5O_6$, 661.3826 (MS) [37]

Mp = 115 [70], amorphous powder [37]
$[a]_D$ (20) = -487 (c=0.12, $CHCl_3$) [37], -457 (c=1.0, $CHCl_3$) [71]
UV (MeOH) : 268 (4.06), 320 (3.76) [37,71]
CD (Dioxane) : -14.02 (324), -18.28 (276), -29.53 (254), +3.13 (232), -14.50 (219) [37].
IR ($CHCl_3$) : 3400, 2820, 2770, 1680, 1640, 1610, 1210, 1025 [37]
MS : 661(M^+), 618(M^+-C_3H_7), 570(b), 514(c), 512(d), 471(e), 427(e-C_3H_7-H^+), 401(f), 400(g), 373(i), 372(j), 358(k), 304(u), 289(l), 285(i-$NHMe_2$-R_1), 261(m), 259(r), 235(o), 233(s), 216(t), 209(n/p), 165(x), 148(a=100%), 96(v), 86(q'/q'') [37]
MS [71]
^{1}H-NMR (300MHz, DMSO-d_6) : 0.76, t, 7.3 Hz, 3H (CH_2-Me-Ileu), 0.82, d, 6.4 Hz, 3H (Me-Leu), 0.83, d, 6.7 Hz, 3H (Me-Leu), 0.84, d, 6.4 Hz, 3H (CH-Me-Ileu), 1.11, m, 1H (γ-H-Ileu), 1.32, m, 1H (γ-H-Ileu), 1.38, m, 2H (β-H-Leu), 1.50, m, 1H (γ-H-Leu), 1.80, m, 1H (β-H-Ileu), 2.10, m, 1H (γ-H-Pro), 2.22, s, 6H (N-Me_2), 2.50, m, 1H (γ-H-Pro), 2.72, dd, 13.6, 6.0 Hz, 1H (β-H-N,N-diMe-Phe), 2.94, dd, 13.6, 8.9 Hz, 1H (β-H-N,N-diMe-Phe), 3.29, dd, 8.9, 6.0 Hz, 1H (α-H-N,N-diMe-Phe), 3.50, m, 1H (δ-H-Pro), 3.75, s, 3H (OMe), 4.01, m, 1H (δ-H-Pro), 4.05, t, 8.0 Hz, 1H (α-H-Ileu), 4.42, d, 1.9 Hz, 1H (α-H-Pro), 4.60, dt, 8.8, 7.5 Hz, 1H (α-H-Leu), 5.13, br. t, 6.2 Hz, 1H (β-H-Pro), 5.85, d, 8.9 Hz, 1H (β-H-Sty), 6.73, dd, 10.0, 8.9 Hz, 1H (α-H-Sty), 6.76, d, 3.0 Hz, 1H (6-H-Sty), 6.83, dd, 8.6, 3.0 Hz, 1H (4-H-Sty), 7.02, d, 8.6 Hz, 1H (3-H-Sty), 7.09-7.26, m, 5H (2xγ-H + 2xδ-H + ε-H- N,N-diMe-Phe), 8.15, d, 8.0 Hz, 1H (NH-Ileu), 8.17, d, 8.8 Hz, 1H (NH-Leu), 9.10, d, 10.0 Hz, 1H (NH-Sty) [71]
^{1}H-NMR (60 and 90MHz, $CDCl_3$) [37]
^{13}C-NMR (75MHz, DMSO-d_6) : see figure [71]
COSY, ROESY, TOCSY and FLOCK NMR spectra [71]
Derivatives : Dihydro-mucronine-D [37]
Mucronine-D amido-aldehyde [37]
Sources : <u>Rhamnaceae</u>
Zizyphus jujuba-stem bark [84]
Z. jujuba var. *inermis*-stem bark [70]
Z. mucronata-roots [71], stem bark [37]
Z. nummularia-root bark [43, 76]
Z. sativa-stem bark [67]

38. Nummularine-H (= *N*-Desmethyl-jubanine-A)

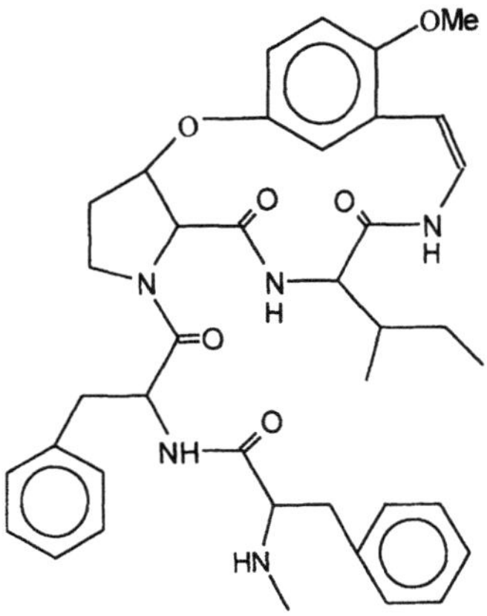

A = N-Me-Phe
B = *trans*-β-OH-Pro[1] (*66*)
C = Ileu
E = Phe
$C_{39}H_{47}N_5O_6$, 681.3539 (MS) (*66*)

Mp = 194-196 (*66*)
$[a]_D$ (20) = -343 (c=0.27, MeOH) (*66*)
UV (MeOH) : 268 (4.05), 320 (3.65) (*66*)
IR ($CHCl_3$) : 3345, 2835, 2795, 1670, 1630, 1610, 1230, 1020 (*66*)
MS : 681(M^+), 590(b), 548(c), 546(d), 505(e), 435(f), 434(g), 407(i), 406(j), 378(k), 309(l), 304(u), 281(m), 269(o), 259(r), 243(n), 233(s), 216(t), 209(p), 165(x), 134(a=100%), 120(q"), 96(v), 86(q') (*66*)
Derivatives : *N*-Acetyl-nummularine-H (*66*)
N-Methyl-dihydro-nummularine-H (*66*)
Sources : *Zizyphus nummularia* (Rhamnaceae)-stem bark (*66*)

39. Jubanine-A

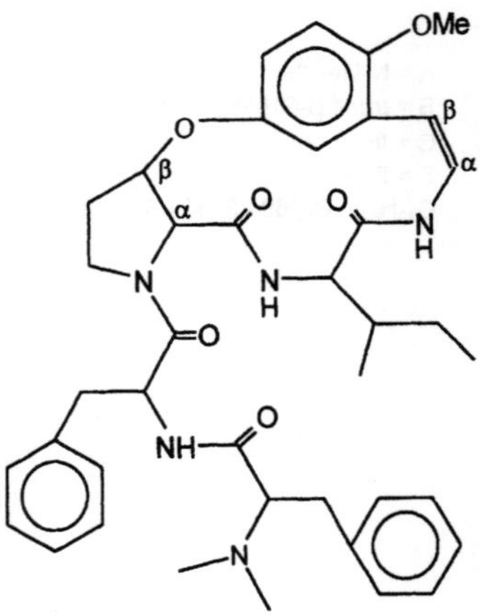

A = N,N-diMe-Phe
B = β-OH-Pro
C = Ileu
E = Phe
$C_{40}H_{49}N_5O_6$, 695.3579 (MS) (*84*)

Mp = Amorphous powder (*84*)
$[a]_D$ (20) = -326 (c=0.12, MeOH) (*84*)
UV (MeOH) : 265 (3.96), 320 (3.79) (*84*)
IR (KBr) : 3360, 2830, 2780, 1670, 1635, 1615, 1218, 1030 (*84*)
MS : 695(M^+), 604(b), 548(c), 546(d), 505(e), 435(f), 434(g), 407(i), 406(j), 392(k), 323(l), 304(u), 295(m), 269(o), 259(r), 243(n), 233(s), 216(t), 209(p), 165(x), 148(a=100%), 120(q''), 96(v), 86(q') (*84*)
^{1}H-NMR ($CDCl_3$) : 0.80-1.20 complex signal, 6H (2xC-Me), 2.25s, 6H (N-Me_2), 3.75s, 3H (OMe), 4.30d, 4.0 Hz, 1H (α-H-β-OH-Pro), 5.85d, 9.0 Hz, 1H (β-H-Sty) (*84*)
Sources : Rhamnaceae
Zizyphus jujuba-stem bark (*84*)
Z. nummularia-root bark (*94*)
Z. spina-christi-stem bark (*88*)

40. Nummularine-O (= *N*-Desmethyl-jubanine-B)

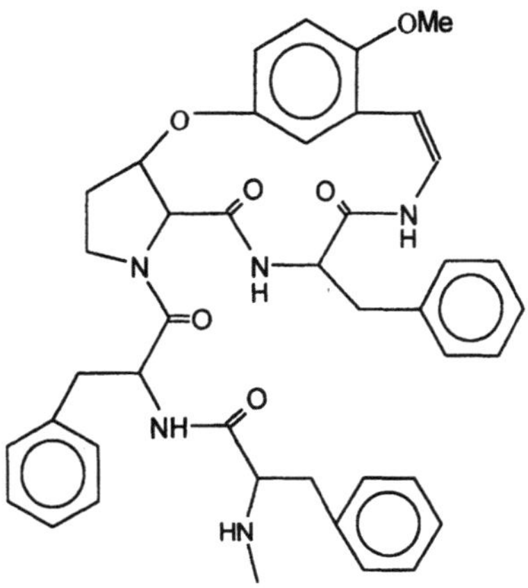

A = N-Me-Phe
B = β-OH-Pro
C = Phe
E = Phe
$C_{42}H_{45}N_5O_6$, 715.00 (MS) *(95)*

Mp = 159-161 *(95)*
$[a]_D$ (20) = -239 (c=0.2, MeOH) *(95)*
UV (MeOH) : 270 (3.54), 318 (3.36) *(95)*
IR (KBr) : 3360, 3000-2900, 2835, 2735, 1670, 1635, 1618, 1590, 1505, 1220, 1025 *(95)*
MS : 715(M^+), 624(b), 582(c), 580(d), 539(e), 435(f), 434(g), 408(h), 407(i), 406(j), 378(k), 338(u), 309(l), 281(m), 269(o), 259(r), 243(n/p), 233(s), 216(t), 165(x), 134(a=100%), 120(q'/q''), 96(v), 68(w) *(95)*
^{1}H-NMR ($CDCl_3$) : 2.40s, 3H (NH-Me), 3.67s, 3H (OMe) *(95)*
Derivatives : *N*-Formyl-nummularine-O *(95)*
Sources : *Zizyphus nummularia* (Rhamnaceae)-root bark *(94)*, stem bark *(95)*

41. Jubanine-B

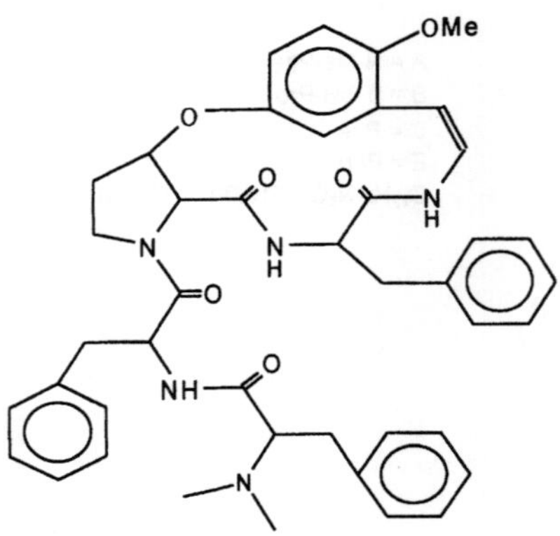

A = N,N-diMe-Phe
B = β-OH-Pro
C = Phe
E = Phe
$C_{43}H_{47}N_5O_6$, 729.3520 (MS) (*84, 95*)

Mp = Amorphous powder (*84*), 97-100 (*95*)
[a]$_D$ (20) = -215 (*95*) or -218 (*84*) (c=0.28, MeOH)
UV (MeOH) : 270 (3.54), 318 (3.36) (*84, 95*)
IR (KBr) : 3355, 2860, 2780, 1680, 1635, 1610, 1200, 1020 (*95*)
IR (KBr) (*84*)
MS : 729(M^+), 638(b), 582(c), 580(d), 539(e), 435(f), 434(g), 408(h), 407(i), 406(j), 392(k), 323(l), 295(m), 269(o), 259(r), 243(n/p), 216(t), 165(x), 148(a=100%), 120(q'/q''), 96(v) (*95*)
Sources : Rhamnaceae
Zizyphus jujuba-stem bark (*84*)
Z. nummularia-root bark (*94*), stem bark (*95*)

4(14)-Frangulanine-Type Cyclopeptide Alkaloids

42. Ceanothine-D

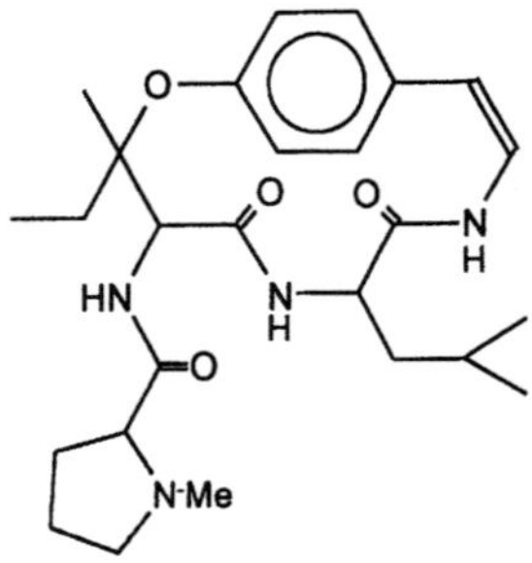

A = N-Me-Pro
B = β-OH-Ileu
C = Leu
$C_{26}H_{38}N_4O_4$, 470

Mp = 227-229 (*96*)
[a]$_D$ = -347 (*96*)
MS : 470(M^+), 441(M^+-C_2H_5), 387(g), 385(g-$2H^+$), 344(j), 303(e), 274(h), 190(f), 189(f-H^+), 182(l), 135(i), 97(m), 84(a'=100%) (*96*)
^{1}H-NMR (60MHz, $CDCl_3$) : 0.87t, 6.0 Hz, 3H (CH_2-Me-β-OH-Ileu), 0.98d, 6.5 Hz, 3H (Me-Leu), 1.27s, 3H (C-Me-β-OH-Ileu), 1.29d, 6.5 Hz, 3H (Me-Leu), 2.27s, 3H (N-Me-Pro) (*96*)
Sources : *Ceanothus americanus* (Rhamnaceae)-root bark (*96*)

43. Ceanothine-C

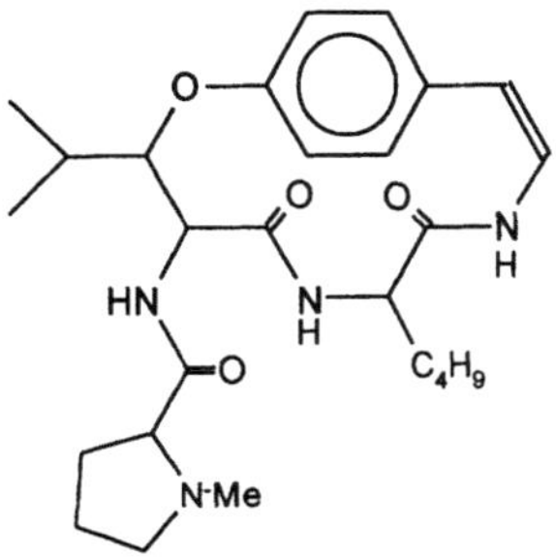

A = N-Me-Pro
B = β-OH-Leu
C = Leu or Ileu
$C_{26}H_{38}N_4O_4$, 470

Mp = 223-229 *(97)*
[a]D (25) = -368 (c=1.01, $CHCl_3$) *(97)*
MS : 470(M^+), 455(M^+-CH_3), 427(M^+-C_3H_7), 387(g), 385(g-2H^+), 344(j), 303(e), 274(h), 210(k), 190(f), 189(f-H^+), 182(l), 135(i), 134(i-H^+), 97(m), 86(q), 84(a'=100%) *(96)*
MS *(28, 97)*
^{1}H-NMR (60MHz, $CDCl_3$ and CD_3COOD) : little information *(97)*
Sources : *Ceanothus americanus* (Rhamnaceae)-root bark *(97)*

44. *N*-Desmethyl-myrianthine-C

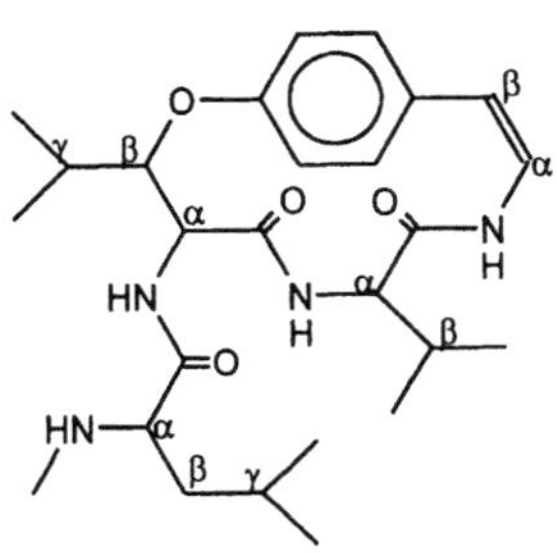

A = N-Me-Leu
B = *erythro*-β-OH-Leu[1] *(98)*
C = Val
$C_{26}H_{40}N_4O_4$, 472

Mp = amorphous powder *(99)*
[a]D (20) = -103 (c=1.0, $CHCl_3$) *(99)*
UV (MeOH) : 232 (3.95), 276sh. (3.58) *(99)*
IR (KBr) : 3275, 3030, 2960, 2920, 2880, 1680, 1630, 1510, 1385, 1375, 1285, 1245, 1185, 1175, 990, 880, 830, 795, 700 *(99)*
MS : 330(j), 135(i), 100(a=100%), 97(m), 72(p/q), 71(n), 58(o) *(99)*
MS (CI/NH_3) : 473(M+H^+), 330(j), 135(i), 100(a=100%) *(99)*
^{1}H-NMR (270MHz, DMSO-d_6) : 0.63d, 7.0 Hz, 3H (Me-Val), 0.68d, 7.0 Hz, 3H (Me-Val), 0.79d, 7.0 Hz, 3H (CH-Me-N-Me-Leu), 0.81d, 7.0 Hz, 3H (CH-Me-N-Me-Leu), 0.88d, 7.0 Hz, 3H (Me-β-OH-Leu), 1.13d, 7.0 Hz, 3H (Me-β-OH-Leu), 1.22t, 7.0 Hz, 2H (β-H-N-Me-Leu), 1.68m, 2H (β-H-Val + γ-H-N-Me-Leu), 2.12s, 3H (NH-Me), 2.14m, 1H (γ-H-β-OH-Leu), 2.93t, 7.0 Hz, 1H (α-H-N-Me-Leu), 3.75t, 9.0 Hz, 1H (α-H-Val), 4.46dd, 10.0, 8.0 Hz, 1H (α-H-β-OH-Leu), 4.82dd, 8.0, 2.0 Hz, 1H (β-H-β-OH-Leu), 6.19dd, 8.0, 4.0 Hz, 1H (α-H-Sty), 6.62d, 8.0 Hz, 1H (β-H-Sty), 6.95m, 4H (4xAr-H), 7.18d, 9.0 Hz, 1H-D_2O exchangeable (NH-Val), 7.68d, 4.0 Hz, 1H-D_2O exchangeable (NH-Sty), 8.21d, 10.0 Hz, 1H-D_2O exchangeable (NH-β-OH-Leu) *(99)*
Sources : *Plectronia odorata* (Rubiaceae)-aerial parts *(99)*

45. Pubescine-A

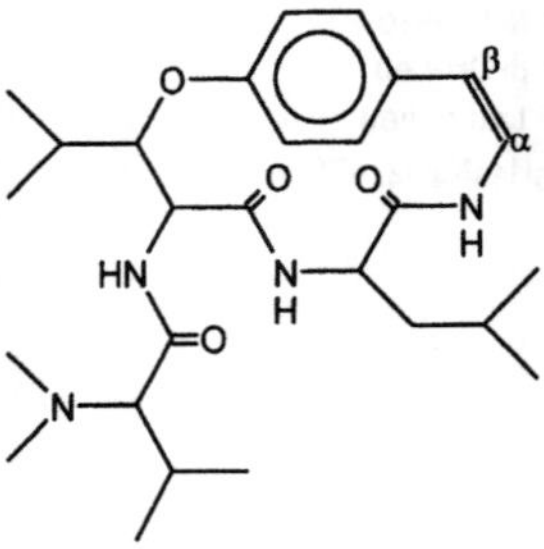

A = N,N-diMe-Val
B = L-*erythro*-β-OH-Leu[1] (*20*)
C = D-Leu[1] (*20*)
$C_{27}H_{42}N_4O_4$, 486.3188 (MS) (*20*)

Mp = 247-250 (*20*)
[a]$_D$ (20) = -230 (c=0.076, MeOH) (*20*)
UV (MeOH) : end absorption (*20*)
IR (KBr) : 3280, 2795, 1670, 1625, 1235, 985 (*20*)
MS : Identical with Melonovine-A (**43**) (*20*)
^{1}H-NMR ($CDCl_3$) : 0.64-1.11m and 1.15-2.17m, 18H (3xCH-Me_2 : Leu + β-OH-Leu + N,N-diMe-Val), 2.44s, 6H (N-Me_2), 6.39d, 8.0 Hz, 1H (α-H-Sty), 6.63d, 8.0 Hz, 1H (β-H-Sty), 6.33-6.83m and 6.96-7.33m, 6H (4xAr-H + 2xNH : D_2O exchangeable) (*20*)
Sources : *Discaria pubescens* (Rhamnaceae) (*20*)

46. Melonovine-A

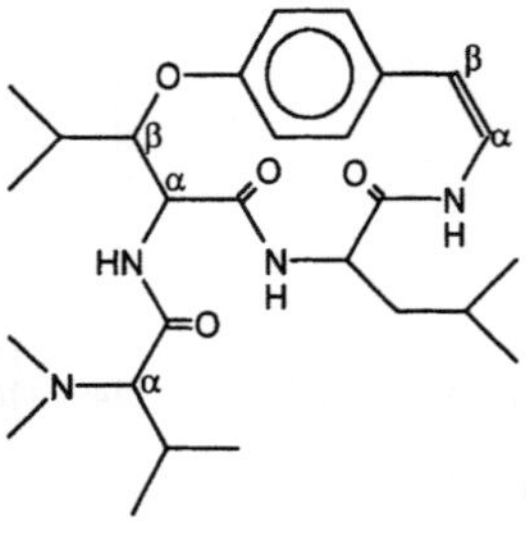

A = N,N-diMe-Val
B = *erythro*-β-OH-Leu[2]
C = Leu
$C_{27}H_{42}N_4O_4$, 486

Mp = 295 (*44*)
[a]$_D$ = -285 ($CHCl_3$) (*44*)
IR (Nujol) : 3380, 2780, 1680, 1605, 1240 (*44*)
MS : 486(M^+), 471(M^+-Me), 443(b), 344(j), 303(e), 210(k), 190(f), 182(l), 135(i), 100(a=100%), 97(m), 86(p/q), 85(n) (*44*)
^{1}H-NMR (100MHz, $CDCl_3$) : 0.74-1.32m, 18H (3xCH-Me_2), 2.20s, 6H (N-Me_2), 2.55d, 4.0 Hz, 1H (α-H-N,N-diMe-Val), 4.51dd, 10.0, 8.0 Hz, 1H (α-H-β-OH-Leu), 4.96dd, 8.0, 2.0 Hz, 1H (β-H-β-OH-Leu), 6.45d, 8.0 Hz, 1H (α-H-Sty), 6.60d, 8.0 Hz, 1H (β-H-Sty), 7.15m, 4H (4xAr-H) (*44*)
Sources : *Melochia tomentosa* (Sterculiaceae)-roots (*44*)

47. Daechuine-S5 (Melonovine-A or Pubescine-A or other diastereoisomer)

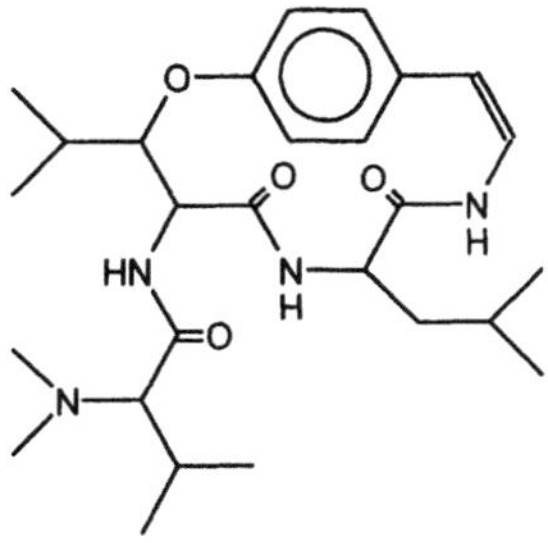

A = N,N-diMe-Val
B = β-OH-Leu
C = Leu
$C_{27}H_{42}N_4O_4$, 486

Mp = 233-235 (*70*)
[a]D = -421.3 (*70*)
Sources : *Zizyphus jujuba* var. *inermis* (Rhamnaceae)-stem bark (*70*)

48. Myrianthine-C

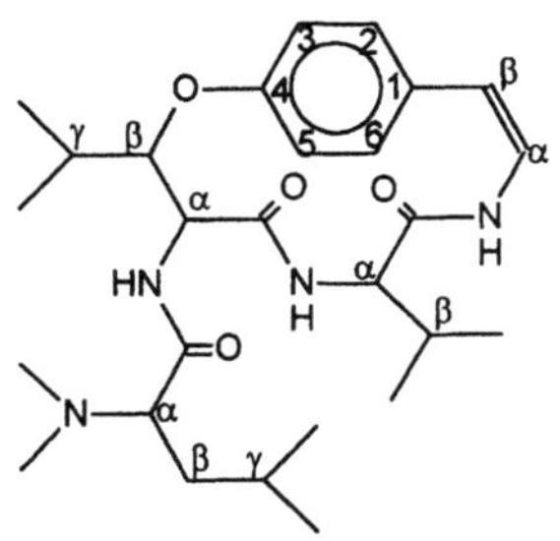

A = N,N-diMe-Leu
B = *erythro*-β-OH-Leu[1] (*98*)
C = Val
$C_{27}H_{42}N_4O_4$, 486

Mp = 287 (*98*), 294 (*100*)
[a]D (20) = -288 (c=1.0, $CHCl_3$) (*100*), -270 (c=1.6, $CHCl_3$) (*98*)
UV (MeOH) : 223 (3.85), 282sh. (3.22) (*98*)
UV (EtOH) (*100*)
IR (KBr) : 3260, 3020, 2955, 2860, 2820, 2780, 1680, 1650, 1510, 1385, 1370, 1280, 1240, 1210, 1190, 1170, 1050, 985, 950, 870, 820, 790, 765, 680 (*98*)
MS : 486(M^+), 429(b), 371(g-$2H^+$), 330(j), 289(e), 260(h), 196(k), 190(f), 168(l), 135(i), 114(a=100%), 97(m), 72(o/q) (*100*)
MS (*98*)
MS (CI/NH_3) : 487($M+H^+$), 330(j), 196(k), 168(l), 135(i), 114(a=100%), 97(m), 72(o/q) (*98*)
^{1}H-NMR (270MHz, DMSO-d_6) : 0.65d, 7.0 Hz, 3H (Me-Val), 0.70d, 7.0 Hz, 3H (Me-Val), 0.79d, 7.0 Hz, 3H (CH-Me-N,N-diMe-Leu), 0.82d, 7.0 Hz, 3H (CH-Me-N,N-diMe-Leu), 0.94d, 7.0 Hz, 3H (Me-β-OH-Leu), 1.12d, 7.0 Hz, 3H (Me-β-OH-Leu), 1.30t, 7.0 Hz, 2H (β-H-N,N-diMe-Leu), 1.44m, 1H (γ-H-N,N-diMe-Leu), 1.70m, 1H (β-H-Val), 2.21s, 6H (N-Me_2), 2.23m, 7.0, 2.0 Hz, 1H (γ-H-β-OH-Leu), 3.03t, 7.0 Hz, 1H (α-H-N,N-diMe-Leu), 3.73t, 9.0 Hz, 1H (α-H-Val), 4.38dd, 10.0, 8.0 Hz, 1H (α-H-β-OH-Leu), 4.81dd, 8.0, 2.0 Hz, 1H (β-H-β-OH-Leu), 6.14dd, 8.0, 4.0 Hz, 1H (α-H-Sty), 6.56d, 8.0 Hz, 1H (β-H-Sty), 6.90m, 4H (4xAr-H), 7.16d, 9.0 Hz, 1H-D_2O exchangeable (NH-Val), 7.78d, 4.0 Hz, 1H-D_2O exchangeable (NH-Sty), 8.20d, 10.0 Hz, 1H-D_2O exchangeable (NH-β-OH-Leu) (*99*)
^{1}H-NMR (60MHz, $CDCl_3$ + CD_3OD) (*100*)
Sources : *Myrianthus arboreus* (Urticaceae)-leaves (*100*)
Plectronia odorata (Rubiaceae)-aerial parts (*98, 99*)

49. Discarine-F (= *N*-Desmethyl-adouetine-X)

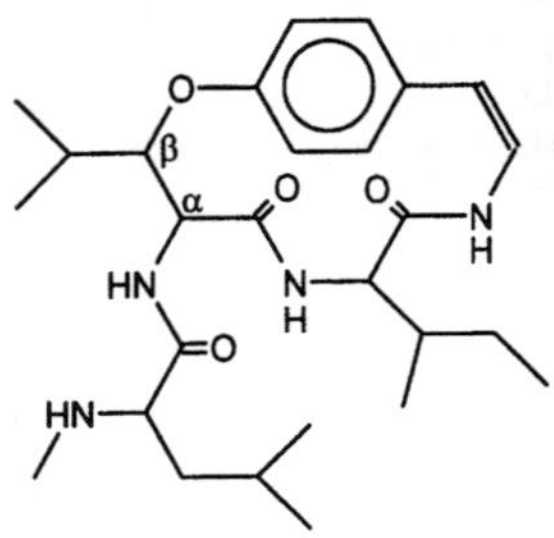

A = N-Me-Leu
B = *erythro*-β-OH-Leu[1] (*47*)
C = Ileu
$C_{27}H_{42}N_4O_4$, 486.67 (MS) (*47*)

Mp = 264 (*47*)
[a]D (20) = -191 ($CHCl_3$) (*47*)
UV (MeOH) : 255sh., 280sh. (*47*)
IR (KBr) : 3300, 2840, 1645, 1235 (*47*)
MS : 487($M+H^+$), 486(M^+), 485($M-H^+$), 471(M^+-CH_3), 443(M^+-C_3H_7), 429(b), 387(g), 345(j+H^+), 344(j), 303(e), 274(h), 210(k), 190(f), 189(f-H^+), 182(l), 181(c), 153(d), 136(i+H^+), 135(i), 134(i-H^+), 100(a=100%), 97(m), 86(q), 72(p), 71(n), 58(o) (*47*)
^{1}H-NMR (90MHz, $CDCl_3$) : 0.75-1.25m, 18H (6xC-Me), 2.21s, 3H (NH-Me), 4.40dd, 10.0, 8.0 Hz, 1H (α-H-β-OH-Leu), 4.95dd, 8.0, 2.0 Hz, 1H (β-H-β-OH-Leu) (*47*)
Sources : *Discaria febrifuga* (Rhamnaceae)-root bark (*47*)

50. Hovenine-A (= *N*-Desmethyl-frangulanine)

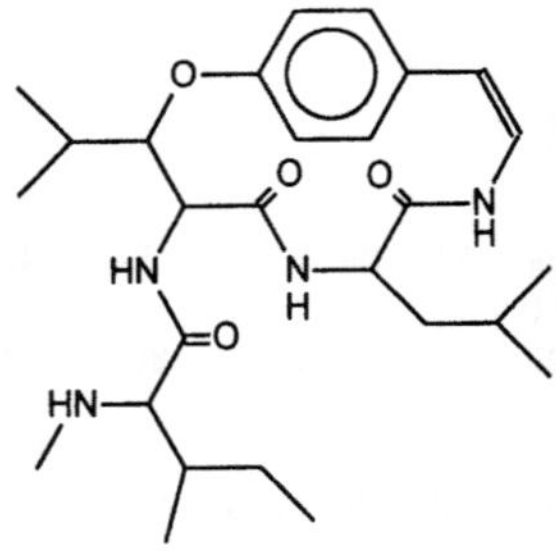

A = N-Me-Ileu
B = β-OH-Leu
C = Leu
$C_{27}H_{42}N_4O_4$, 486.3216 (MS) (*101*)

Mp = 215 (*101*)
IR (KBr) : 3270, 1628, 1237 (*101*)
MS : 486(M^+), 471(M^+-CH_3), 443(M^+-C_3H_7), 387(g), 344(j), 303(e), 210(k), 190(f), 182(l), 135(i), 101(a+H^+), 100(a=100%), 97(m) (*101*)
^{1}H-NMR (100MHz, C_5D_5N) : 0.75d, 7.0 Hz, 6H (Me-Leu), 0.85t, 7.0 Hz, 3H (CH_2-Me-N-Me-Ileu), 0.98d, 7.0 Hz, 3H (CH-Me-N,N-diMe-Ileu), 1.18, 2xd, 7.0 Hz, 6H (2xMe-β-OH-Leu), 2.38s, 3H (NH-Me), 2.99d, 6.0 Hz, 1H (α-H-N-Me-Ileu) (*101*)
Sources : Rhamnaceae
Hovenia dulcis-root bark (*101*)
H. tomentella-root bark (*101*)

51. Melofoline

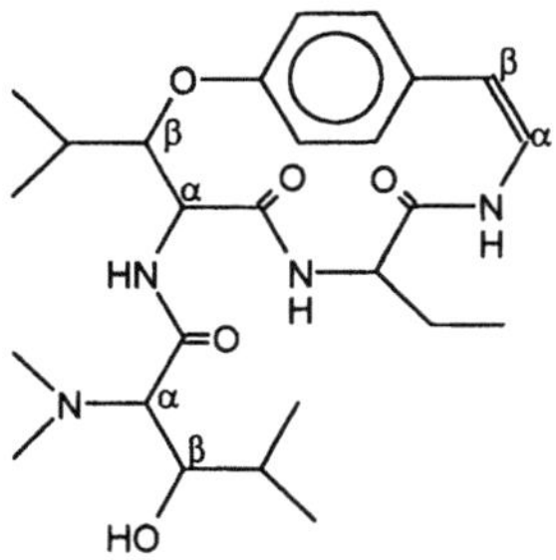

A = N,N-diMe-β-OH-Leu
B = β-OH-Leu
C = 2-aminobutyric acid
$C_{26}H_{40}N_4O_5$, 488

Mp = 305-307 (*53*)
[a]D (20) = -252 ($CHCl_3$) (*53*)
IR (KBr) : 3400, 3260, 2790, 1680, 1618, 1540, 1525, 1505, 1462, 1385, 1235 (*53*)
MS : 488(M^+), 275(e), 246 (h), 190 (f), 182(k), 154(l), 135(i), 130(a=100%), 115(a-Me), 112(a-H_2O), 100(a-2Me), 97(m), 82(a-H_2O-2Me), 58(q) (*53*)
^{1}H-NMR (80MHz, $CDCl_3$) : 0.75t, 6.0 Hz, 3H (Me-2-aminobutyric acid), 0.84d, 7.0 Hz, 3H (Me-β-OH-Leu), 1.04d, 7.0 Hz, 3H (Me-β-OH-Leu), 2.17s, 6H (N-Me_2), 3.42m, 1H (β-H-N,N-diMe-β-OH-Leu), 4.60dd, 10.0, 8.0 Hz, 1H (α-H-β-OH-Leu), 4.90dd, 8.0, 2.0 Hz, 1H (β-H-β-OH-Leu), 6.45d, 8.0 Hz, 1H (α-H-Sty), 6.62d, 8.0 Hz, 1H (β-H-Sty), 7.10m, 4H (4xAr-H), 8.25, 8.50 and 8.60, 3xbr. s, 3H (3xNH) (*53*)
Derivatives : *O*-Acetyl-melofoline (*53*)
Sources : *Melochia corchorifolia* (Sterculiaceae)-aerial parts (*53*)

52. Franganine (= Daechuine-S4)

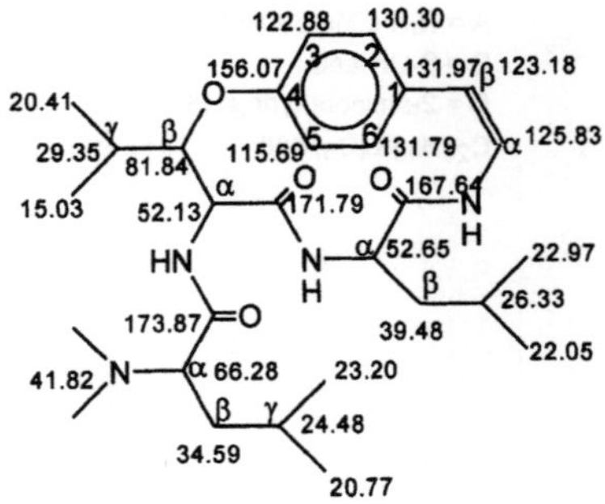

A = N,N-diMe-Leu
B = L-*erythro*-β-OH-Leu[1,2] (*55, 102*)
C = Leu
$C_{28}H_{44}N_4O_4$, 500

Mp = 239-248 (*70, 102-104*)
[a]D (22) = -302 (c=1.0, $CHCl_3$) (*102, 103*)
UV (EtOH) : end absorption (205-325) (*104*)
IR (KBr) : 3280, 1640, 1250 (*104*)
MS : 500(M^+), 485(M^+-CH_3), 471(M^+-C_2H_5), 457(M^+-C_3H_7), 443(b), 387(g), 344(j), 303(e), 274(h), 210(k), 195(c), 190(f), 182(l), 167(d), 135(i), 115(a+H^+), 114(a=100%), 97(m), 72(o) (*102*)
MS (*104*)
^{1}H-NMR (300MHz, $CDCl_3$) : 0.71d, 6.4 Hz, 3H (CH-Me-N,N-diMe-Leu), 0.81d, 6.4 Hz, 3H (CH-Me-N,N-diMe-Leu), 0.89d, 6.5 Hz, 3H (Me-Leu), 0.95d, 6.5 Hz, 3H (Me-Leu), 0.97d, 6.5 Hz, 3H (Me-β-OH-Leu), 1.28d, 6.5 Hz, 3H (Me-β-OH-Leu), 1.34m, 3H (β-H-N,N-diMe-Leu + β-H-Leu + γ-H-N,N-diMe-Leu), 1.60ddd, 13.5, 7.8, 5.7 Hz, 1H (β-H-N,N-diMe-Leu), 1.74m, 2H (β-H-Leu + γ-H-Leu), 1.93m, 1H (γ-H-β-OH-Leu), 2.22s, 6H (N-Me_2), 2.81dd, 7.8, 4.9 Hz, 1H (α-H-N,N-diMe-Leu), 4.07ddd, 11.1, 7.5, 3.6 Hz, 1H (α-H-Leu), 4.46dd, 9.9, 7.2 Hz, 1H (α-H-β-OH-Leu), 4.99dd, 7.2, 1.5 Hz, 1H (β-H-β-OH-Leu), 5.80d, 7.5 Hz, 1H (NH-Leu), 6.35d, 7.5 Hz, 1H (β-H-Sty), 6.48d, 9.9 Hz, 1H (NH-Sty), 6.66dd, 9.9, 7.5 Hz, 1H (α-H-Sty), 7.03m, 1H (5-H-Sty), 7.05m, 1H (3-H-Sty), 7.11m, 1H (6-H-Sty), 7.19m, 1H (2-H-Sty), 7.82d, 9.9 Hz, 1H (NH-β-OH-Leu) (*55*)
^{13}C-NMR (100MHz, $CDCl_3$) : see figure (*55*)
Spin echo and DEPT ^{13}C-NMR spectra (100MHz, $CDCl_3$+CD_3OD) (*55*)
Derivatives : Dihydro-franganine (*102*)
Sources : Celastraceae

Euonymus europaeus-root bark, roots (*104*)
Rhamnaceae
Discaria febrifuga-roots (*55*), stem bark (*105*)
Rhamnus frangula-stem bark (*102*)
Zizyphus jujuba var. *inermis*-stem bark (*70*)
Z. spina-christi-stem bark (*106*)
Sterculiaceae
Melochia corchorifolia-aerial parts (*103*), leaves (*107*)

53. Adouetine-X (= Ceanothamine-B)

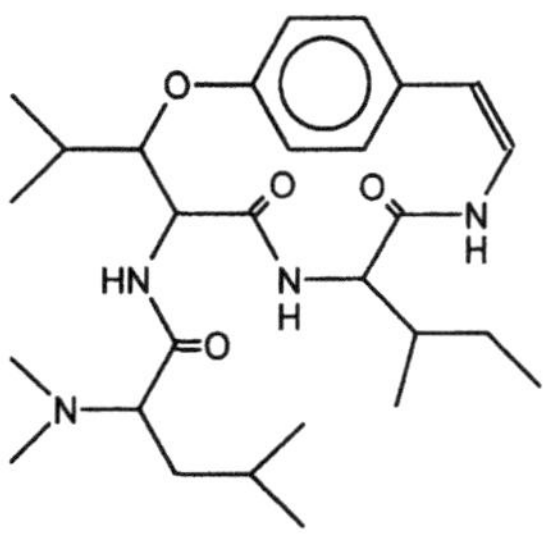

A = N,N-diMe-Leu
B = β-OH-Leu
C = Ileu
$C_{28}H_{44}N_4O_4$, 500

Mp = 277-280.5 (*9, 96, 97, 108*)
[a]D (20) = -316 (c=1.0, $CHCl_3$) (*9, 108*)
[a]D (*96, 97*)
UV (EtOH) : 250 (3.78) (*108*)
IR (KBr) : 3260, 1680, 1630, 1510, 1240 (*108*)
IR (KBr) (*9*)
MS : 443(b), 387(g), 344(j), 303(e), 274(h), 210(k), 195(c), 190(f), 182(l), 167(d), 135(i), 114(a=100%), 97(m) (*28*)
MS : 500 (M^+) (*96, 97*)
^{1}H-NMR (60MHz, TFA) : 0.70-1.10 complex, 18H (6xC-Me), 3.21d, 5.0 Hz, 6H (N-Me_2), 6.90-7.50m, 4H (4xAr-H) (*108*)
^{1}H-NMR (60MHz, TFA) : (*9*)
^{1}H-NMR (60MHz, CD_3COOD) : (*97*)
Sources : *Ceanothus americanus* (Rhamnaceae)-root bark (*96, 97*)
Waltheria americana (Sterculiaceae)-whole plant (*9, 108*)

54. Frangulanine (= Ceanothamine-A = Daechuine-S2)

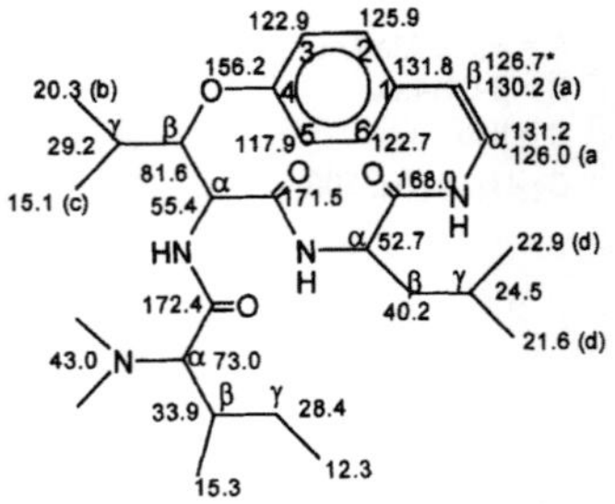

A = N,N-diMe-L-Ileu[1] (*27*)
B = L-*erythro*-β-OH-Leu[1,2] (*8, 12, 21, 27, 91, 104, 109, 110*)
C = L-Leu[1] (*27*)
$C_{28}H_{44}N_4O_4$, 500

According to Mp and $[a]_D$ values Daechuine-S2 may be a diastereoisomer of frangulanine.
Mp = 238-242 (*70*), 272-279 (*27, 33, 96, 97, 101, 104*)
$[a]_D$ (20) = -288 (*27*) or -296 (*33*) (c=0.1, $CHCl_3$)
$[a]_D$ (*70, 96*)
UV (MeOH) : 252 (3.74), 279 (3.13) (*27*)
UV (EtOH) : (*104*)
CD (Dioxane) : -20.2 (239), +2.0 (287) (*27*)
IR (KBr) : 3275, 2784, 1630, 1228 (*27*)
IR (KBr) (*33, 101, 104*)
MS : 500(M^+), 499($M-H^+$), 485(M^+-CH_3), 471($M^+-C_2H_5$), 457($M^+-C_3H_7$), 443(b), 387(g), 385($g-2H^+$), 344(j), 303(e), 274(h), 210(k), 195(c), 190(f), 189($f-H^+$), 182(l), 167(d), 135(i), 134($i-H^+$), 114(a=100%), 97(m), 86(p/q), 85(n) (*27*)
MS (*28, 97*)
^{1}H-NMR (100MHz, $CDCl_3$+10% CD_3OD) : 0.76d, 6.2 Hz, 3H (Me-Leu), 0.81d, 6.5** Hz, 3H (Me-Leu), 0.85d, 6.5** Hz, 3H (CH-Me-N,N-diMe-Ileu), 0.91t, 7.2** Hz, 3H (CH_2-Me-N,N-diMe-Ileu), 0.98d, 6.5 Hz, 3H (pro S-Me-β-OH-Leu), 1.21d, 6.5 Hz, 3H (pro R-Me-β-OH-Leu), 1.21 and 1.53, 2xm, 7.2** Hz, 2H (γ-H-N,N-diMe-Ileu), 1.21 and 1.53, 2xm, 8.0**, 6.5** Hz, 2H (β-H-Leu), 1.40m, 6.5**, 6.2** Hz, 1H (γ-H-Leu), 1.80m, 6.5**, 5.75** Hz, 1H (β-H-N,N-diMe-Ileu), 2.09m, 6.5, 2.0** Hz, 1H (γ-H-β-OH-Leu), 2.20s, 6H (N-Me_2), 2.58d, 6.0 Hz, 1H (α-H-N,N-diMe-Ileu), 3.98m, 8.0, 6.5** Hz, 1H (α-H-Leu), 4.42d, 8.0 Hz, 1H (α-H-β-OH-Leu), 4.83dd, 8.0, 2.0 Hz, 1H (β-H-β-OH-Leu), 6.28d, 7.2 Hz, 1H (α-H-Sty), 6.61d, 7.2 Hz, 1H (β-H-Sty), 7.01m, 4H (4xAr-H) (*110*)
** These assignments have been recorded on a 270MHz instrument.
^{1}H-NMR (250MHz, $CDCl_3$) : (*91*)
^{1}H-NMR (100MHz, $CDCl_3$ and DMSO-d_6, 35°C) : (*110*)
^{1}H-NMR (100MHz, DMSO-d_6, 80°C) (*109, 110*)
^{1}H-NMR (100MHz, $CDCl_3$) (*22, 104*)
^{1}H-NMR (100MHz, C_5D_5N) (*101*)
^{1}H-NMR (60MHz, $CDCl_3$) (*97*)
^{1}H-NMR (60MHz, CD_3COOD) (*97*)
^{1}H-NMR (C_5D_5N) (*27*)
Partially relaxed ^{1}H-NMR spectrum (100MHz, $CDCl_3$+10% CD_3OD) (*111, 112*)
^{13}C-NMR (25MHz, $CDCl_3$) : see figure (*110*)
*These assignments have been recorded on a 62.9MHz instrument. (a) Ambiguous assignments, (b) *Pro*-R, (c) *Pro*-S, (d) These were stereotopically indistinguishable.
^{13}C-NMR (25MHz, $CDCl_3$+10% or 30% CD_3OD) (*110*)
X-rays (tri-N-methyl-frangulanine methiodide) (*21, 22*)
Derivatives : Dihydro-frangulanine (*27, 97*)
Tri-*N*-methyl-frangulanine methiodide (*21, 22*)
Sources : Celastraceae
Euonymus europaeus-leaves, root bark, roots, stem (*104*)
Rhamnaceae
Ceanothus americanus-root bark (*96, 97*)
Discaria longispina-roots (*33*)
Hovenia dulcis-root bark (*101*)
H. tomentella-root bark (*101*)
Rhamnus frangula-stem bark (*27*)
Zizyphus jujuba var. *inermis*-stem bark (*70*)
Z. sativa-stem bark (*67*)

55. Discarine-E

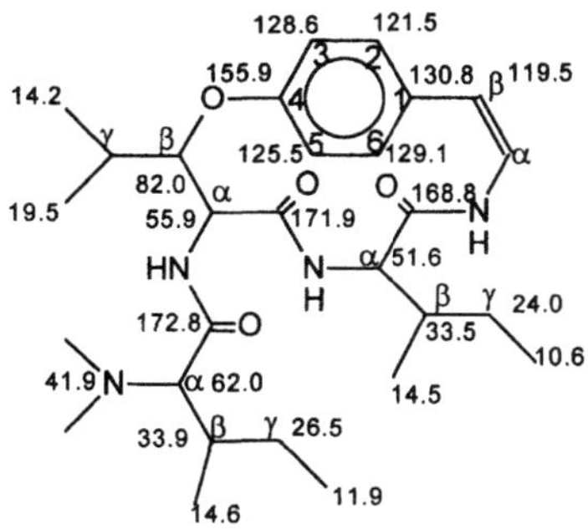

A = N,N-diMe-Ileu
B = β-OH-Leu
C = Ileu
$C_{28}H_{44}N_4O_4$, 500.68 (MS) (*48*)

Mp = 270-273 (*113*)
[a]D (25) = +236 (c=0.5, AcOH) (*113*)
UV (MeOH) : 250sh. (*48*)
IR (KBr) : 3280, 3030, 2960-2820, 2890, 1650, 1620, 1235 (*48*)
IR (KBr) (*113*)
MS : 500(M^+), 443(b), 387(g), 344(j), 303(e), 274(h), 210(k), 195(c), 190(f), 182(l), 167(d), 135(i), 114(a=100%), 97(m), 86(p/q) (*48*)
MS (*113*)
^{1}H-NMR (300MHz, $CDCl_3$) : 0.65d, 6.6 Hz, 3H (CH-Me-N,N-diMe-Ileu), 0.70d, 6.6 Hz, 3H (CH-Me-Ileu), 0.75t, 6.6 Hz, 3H (CH_2-Me-Ileu), 0.80t, 6.6 Hz, 3H (CH_2-Me-N,N-diMe-Ileu), 0.90d, 6.6 Hz, 3H (Me-β-OH-Leu), 1.10d, 6.6 Hz, 3H (Me-β-OH-Leu), 1.28m, 1H (β-H-Ileu), 1.40m, 2H (γ-H-Ileu), 1.50-1.60m, 2H (γ-H-N,N-diMe-Ileu), 1.70d, 1H (β-H-N,N-diMe-Ileu), 2.20s, 6H (N-Me_2), 2.25m, 1H (γ-H-β-OH-Leu), 2.75d, 1H (α-H-N,N-diMe-Ileu), 3.90dd, 1H (α-H-Ileu), 4.40dd, 1H (α-H-β-OH-Leu), 4.80dd, 1H (β-H-β-OH-Leu), 6.20dd, 7.5, 5.5 Hz, 1H (α-H-Sty), 6.65d, 7.5 Hz, 1H (β-H-Sty), 6.80-7.00dd, 4H (4xAr-H), 7.25d, 1H (NH-Ileu), 7.65d, 1H (NH-Sty), 8.10d, 1H (NH-β-OH-Leu) (*113*)
^{1}H-NMR (400MHz, CF_3COOD) : (*48*)
^{13}C-NMR (75MHz, $CDCl_3$) : see figure (*113*)
COSY, HETCOSY, DEPT and Spin-decoupling NMR spectra (*113*)
Sources : Rhamnaceae
Discaria febrifuga-stem bark (*48*)
D. longispina-root bark (*113*)

56. Ceanothine-B = 5-Benzyl-8-*N*-(*N*′-methyl-propyl)-9-isopropyl-phencyclopeptine

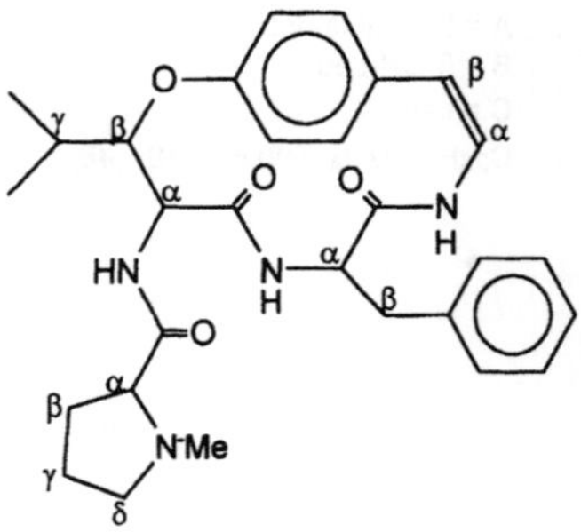

A = N-Me-Pro
B = β-OH-Leu
C = Phe
$C_{29}H_{36}N_4O_4$, 504.2795 (MS) (*46*)

Mp = 225 (*46*), 238.5-240.5 (*97, 114*)
[a]$_D$ (25) = -293 (c=0.68, $CHCl_3$) (*97, 114*)
UV : 250 (3.60) (*97, 114*)
IR : 839, 757, 707 (*114, 115*)
MS : 504(M^+), 489(M^+-CH_3), 475(M^+-C_2H_5), 461(M^+-C_3H_7), 421(g), 378(j), 337(e), 308(h), 244(k), 216(l), 195(c), 190(f), 167(d), 135(i), 120(q), 97(m), 91, 84(a'=100%) (*46*)
MS (*97, 114-116*)
^{1}H-NMR (270MHz, $CDCl_3$) : 0.91d, 6.7 Hz, 3H (Me-β-OH-Leu), 1.24d, 6.7 Hz, 3H (Me-β-OH-Leu), 1.63m, 1H (γ-H-β-OH-Leu), 1.70-1.90m, 2H (β-H-Pro + γ-H-Pro), 1.98s, 3H (N-Me), 2.10-2.20m, 1H (γ-H-Pro), 2.20-2.30m, 1H (β-H-Pro), 2.68dd, 10.6, 4.3 Hz, 1H (δ-H-Pro), 2.85dd, 14.7, 8.2 Hz, 1H (β-H-Phe), 3.01m, 1H (δ-H-Pro), 3.08dd, 14.7, 4.2 Hz, 1H (β-H-Phe), 4.30-4.40m, 1H (α-H-Phe), 4.34dd, 10.0, 7.0 Hz, 1H (α-H-β-OH-Leu), 4.93dd, 7.0, 2.0 Hz, 1H (β-H-β-OH-Leu), 5.99d, 6.9 Hz, 1H (β-H-Sty), 6.39d, 7.4 Hz, 1H (α-H-Pro), 6.40-6.50m, 1H (NH-Sty), 6.66m, 1H (α-H-Sty), 7.00-7.30m, 9H (9xAr-H), 7.75d, 10.0 Hz, 1H (NH-β-OH-Leu) (*46*)
^{1}H-NMR (60MHz, $CDCl_3$ and TFA) (*97, 114*)
Derivatives : Dihydro-ceanothine-B (*97, 114*)
Sources : Rhamnaceae
Ceanothus americanus-root bark (*97, 114*)
C. sanguineus-root bark (*46*)

57. Sanjoinine-B = *N*-Desmethyl-frangufoline

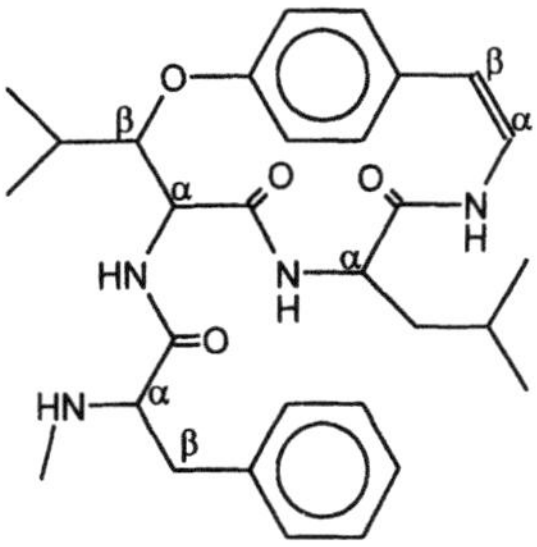

A = N-Me-Phe
B = β-OH-Leu
C = Leu
$C_{30}H_{40}N_4O_4$, 520

Mp = 212-214 (*63, 70*)
MS : 520(M^+), 505(M^+-CH_3), 429(b), 387(g), 344(j), 303(e), 274(h), 210(k), 182(l), 181(c), 153(d), 135(i), 134(a=100%), 97(m), 86(q) (*63*)
^{1}H-NMR (80MHz, $CDCl_3$) : 0.71d, 6.5 Hz, 3H (Me-Leu), 0.84d, 6.5 Hz, 3H (Me-Leu), 0.94d, 6.6 Hz, 3H (Me-β-OH-Leu), 1.26d, 6.6 Hz, 3H (Me-β-OH-Leu), 2.13s, 3H (NH-Me), 2.63, 2H (β-H-N-Me-Phe), 3.15, 1H (α-H-N-Me-Phe), 4.05m, 1H (α-H-Leu), 4.45dd, 10.0, 8.0 Hz, 1H (α-H-β-OH-Leu), 4.95dd, 8.0, 2.0 Hz, 1H (β-H-β-OH-Leu), 5.95d, 7.5 Hz, 1H (NH-Leu), 6.35d, 7.0 Hz, 1H (β-H-Sty), 6.65t, 1H (α-H-Sty), 7.10-7.42m, 10H (9xAr-H + NH-Sty), 7.90d, 10.0 Hz, 1H (NH-β-OH-Leu) (*63*)
Sources : *Zizyphus vulgaris* var. *spinosus* (Rhamnaceae)-seeds (*63, 70*)

58. *N*-Desmethyl-myrianthine-B = 5-*sec*-Butyl-8-*N*-(*N'*-methyl-phenyl-alanyl)-9-isopropyl-phencyclopeptine

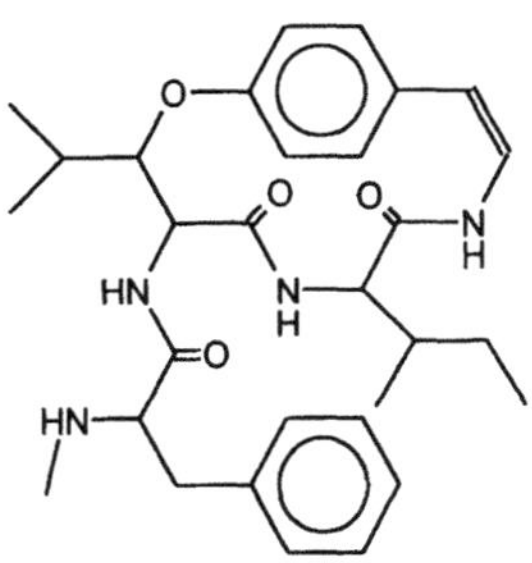

A = N-Me-Phe
B = β-OH-Leu
C = Ileu
$C_{30}H_{40}N_4O_4$, 520

Mp = 229 (*46*)
MS : 520(M^+), 463(M^+-C_4H_9), 429(b), 387(g), 344(j), 303(e), 274(h), 210(k), 190(f), 182(l), 181(c), 153(d), 135(i), 134(a=100%), 133(a-H^+), 97(m), 86(q) (*46*)
Sources : *Ceanothus sanguineus* (Rhamnaceae)-root bark (*46*)

59. Ceanothine-A (=*N*-Desmethyl-frangufoline or *N*-Desmethyl-myrianthine-B or diastereoisomer)

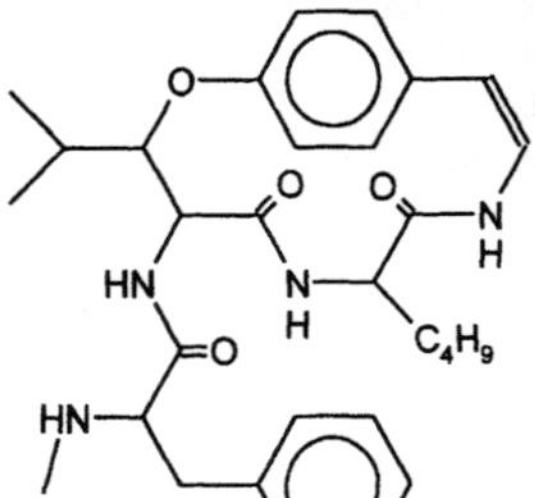

A = N-Me-Phe
B = β-OH-Leu
C = Leu or Ileu
$C_{30}H_{40}N_4O_4$, 520

Mp = 256-259 (*97*)
[a]D = -256 (c=0.5, $CHCl_3$) (*97*)
MS : 520(M^+), 134(a=100%), 91 (*97*)
^{1}H-NMR (60MHz, $CDCl_3$ and CD_3COOD) : little information (*97*)
Derivatives : *N*-Acetyl-ceanothine-A (*97*)
Sources : *Ceanothus americanus* (Rhamnaceae)-root bark (*97*)

60. Scutianine-C [reported as Scutianine-D in (*64*)]

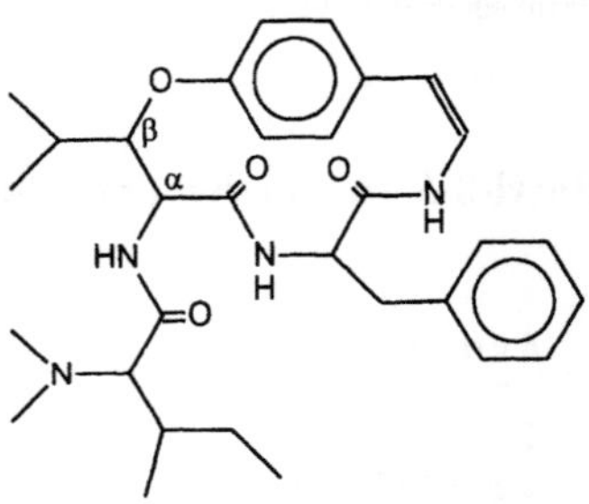

A = N,N-diMe-Ileu
B = L-*erythro*-β-OH-Leu[1] (*18, 64*)
C = L-Phe[1] (*18, 64*)
$C_{31}H_{42}N_4O_4$, 534.3191 (MS) (*18*)

Mp = 255-256 (*64*), 263-267 (*15, 18, 68*)
[a]D (20) = -234 (c=0.1, $CHCl_3$) (*18*)
[a]D (*15, 64*)
UV (MeOH) : end absorption (*18, 64*)
CD (Dioxane) : +2.10 (286.5), +1.48 (277), -17.45 (237.5) (*18*)
IR ($CHCl_3$) : 3370, 2770, 1680, 1620, 1240, 1040 (*18*)
MS : 534(M^+), 518($M-CH_4^+$), 477(b), 442($M-R_2-H^+$), 421(g), 419($g-2H^+$), 404(b'-CO), 392(g-HCO), 378(j), 337(e), 308(h), 257, 244(k), 216(l), 202, 195(c), 190(f), 175(u), 167(d), 160, 135(i), 120(q), 114(a=100%), 97(m), 91, 85(n) (*18*)
MS (*64*)
^{1}H-NMR (60 and 90MHz, $CDCl_3$) : 0.80-1.10m, 9H (3xC-Me), 1.26d, 7.0 Hz, 3H (Me-β-OH-Leu), 2.14s, 6H (N-Me_2), 4.92dd, 7.5, 2.0 Hz, 1H (β-H-β-OH-Leu), 6.00d, 8.0 Hz, 1H (NH), 6.20-6.60m, 3H (1xNH + 2xolefinic-H), 6.90-7.40m, 10H (9xAr-H + 1xNH) (*18*)
^{1}H-NMR (60 and 90MHz, C_5D_5N) : (*18*)
^{1}H-NMR (220MHz, DMSO-d_6 and C_5D_5N) : little information (*64*)
Derivatives : Dihydro-scutianine-C (*18*)
Sources : Rhamnaceae
Scutia buxifolia-roots (*64*), stem bark (*15, 18, 117, 118*)
Zizyphus nummularia-stem bark (*68*)

61. Frangufoline (= Sanjoinine-A = Daechuine-S1)

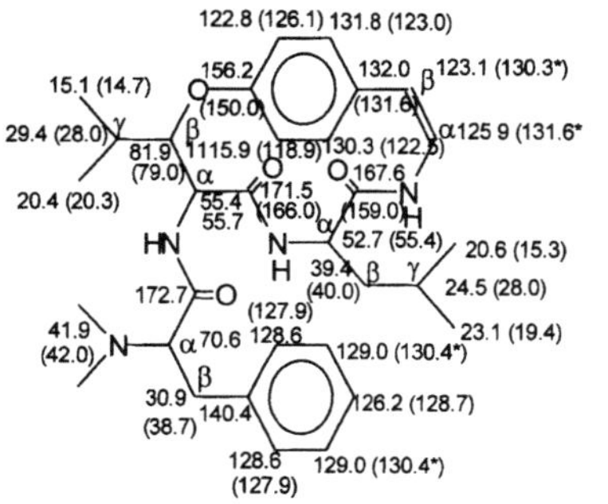

A = N,N-diMe-(S)-L-Phe[1] (*70*)
B = L-*erythro*-β-OH-Leu[1] (*59, 70*)
C = L-Leu[1] (*70*)
$C_{31}H_{42}N_4O_4$, 534.3170 (MS) (*104*) or 534.3200 (MS) (*17*)

Mp = 233-250 (*17, 63, 68, 70, 73, 77, 102, 104, 119*)
[a]D (27) = -316 (c=1.25, $CHCl_3$) (*63*)
[a]D (*17, 102*)
UV (MeOH) : 280 (4.49) (*59*)
UV (EtOH) : (*104*)
IR (KBr) : 3280, 2790, 1680-1630, 1610, 1240 (*63*)
IR (KBr) : (*17, 104, 119*)
IR ($CHCl_3$) : (*59*)
MS : 535($M+H^+$), 534(M^+), 489(w), 443(b), 420, 355(x), 327(x-CO), 300(y), 272(y-CO), 214(z), 190(f), 189($f-H^+$), 148(a=100%), 135(i), 131, 103, 91, 86(q), 84 (*59*)
MS (*63, 102, 104, 107, 119*)
^{1}H-NMR (500MHz, $CDCl_3$) : 0.57d, 6.6 Hz, 3H (Me-Leu), 0.71d, 6.6 Hz, 3H (Me-β-OH-Leu), 1.14d, 6.8 Hz, 3H (Me-Leu), 1.24d, 6.8 Hz, 3H (Me-β-OH-Leu), 2.02m, 1H (γ-H-Leu), 2.05m, 1H (γ-H-β-OH-Leu), 2.85s, 3H (N-Me), 2.97s, 3H (N-Me), 3.10m, 2H (β-H-Leu), 3.34br.d, 14.8 Hz, 1H (α-H-N,N-diMe-Phe), 4.05m, 1H (α-H-Leu), 4.65m, 1H (α-H-β-OH-Leu), 4.92dd, 7.5, 1.9 Hz, 1H (β-H-β-OH-Leu), 6.38d, 7.7 Hz, 1H (β-H-Sty), 7.00-7.50m, 9H (9xAr-H) (*59*)
^{1}H-NMR (80MHz, $CDCl_3$) : (*63*)
^{1}H-NMR ($CDCl_3$) (*17, 102*)
^{1}H-NMR (60MHz, CF_3COOD) (*104*)
^{13}C-NMR (20MHz, $CDCl_3$) : see figure (*63*)
* Values may be interchanged
^{13}C-NMR (125MHz, $CDCl_3$) : see figure values in parentheses (*59*)
Derivatives : Dihydro-frangulanine (*102*)
Frangufoline-dialdehyde (*63*)
Frangufoline-amido-aldehyde = Sanjoinine-G2 (*63*)
Sanjoinine-Ah1 = heat (220°C) induced epimer of sanjoinine-A (frangufoline) containing a N,N-diMe-(R)-D-Phe unit (*70*)
Sources : Celastraceae
Euonymus europaeus-leaves (*104*)
Rhamnaceae
Ceanothus sanguineus-root bark (*46*)
Discaria febrifuga-stem bark (*105*)
D. longispina-roots (*119*)
Rhamnus frangula-stem bark (*102*)
Zizyphus jujuba-stem bark (*77*)
Z. jujuba var. *inermis*-stem bark (*70*)
Z. lotus-aerial parts (*59*)
Z. mauritiana-stem bark (*35, 42*)
Z. nummularia-root bark (*43*), stem bark (*68, 73, 77*)
Z. vulgaris var. *spinosus*-seeds (*63, 70*)
Sterculiaceae
Melochia corchorifolia-leaves (*107*)
M. pyramidata-leaves (*17*)

62. Adouetine-Y′ = Myrianthine-B (= AM-1)

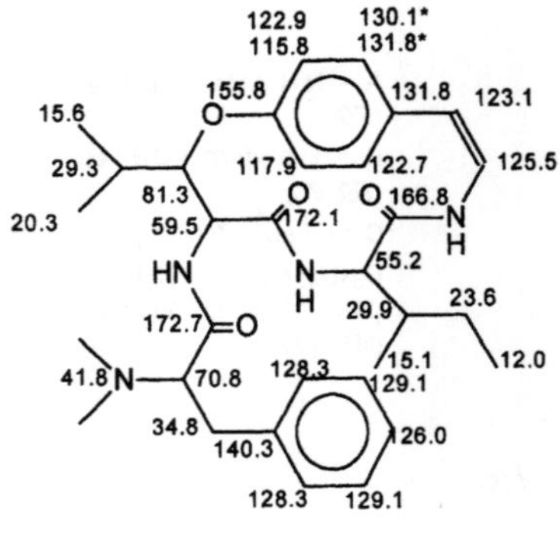

A = N,N-diMe-L-Phe[1] (*120*)
B = L-*erythro*-β-OH-Leu[1,2] (*113*, *120*)
C = L-Ileu[1] (*120*)
$C_{31}H_{42}N_4O_4$, 534

Mp = 289-302 (*53*, *100*, *107*, *113*, *119*, *120*)
[a]D (25) = -350 (c=0.1, $CHCl_3$) (*113*)
[a]D (20) (*53*, *100*, *107*, *119*, *120*)
UV (EtOH) : 250 (3.73) (*100*)
IR (KBr) : 3300, 1665, 1530, 1510, 1240, 1050 (*120*)
IR (KBr) : (*100*, *113*)
MS : 534(M^+), 519(M^+-CH_3), 443(b), 398(b'), 387(g), 370(b'-CO), 344(j), 303(e), 274(h), 210(k), 195(c), 190(f), 182(l), 167(d), 148(a=100%), 135(i), 133(a-CH_3^+), 97(m), 86(q) (*46*) MS (*100*, *107*, *108*, *113*, *119*, *120*)
MS (CI/CH_4) : 563(M+$C_2H_5^+$), 535(M+H^+=100%), 534(M^+), 447, 193 (*120*)
^{1}H-NMR (400MHz, $CDCl_3$) : 0.39d, 7.2 Hz, 3H (CH-Me-Ileu), 0.68m, 5H (CH_2-Me-Ileu), 1.02d, 7.2 Hz, 3H (Me-β-OH-Leu), 1.29d, 7.2 Hz, 3H (Me-β-OH-Leu), 1.95m, 1H (γ-H-β-OH-Leu), 2.28s, 6H (N-Me_2), 2.84dd, 13.5, 4.5 Hz, 1H (β-H-N,N-diMe-Phe), 3.15dd, 13.5, 7.2 Hz, 1H (β-H-N,N-diMe-Phe), 3.90dd, 8.1, 3.3 Hz, 1H (α-H-Ileu), 4.48dd, 9.9, 7.2 Hz, 1H (α-H-β-OH-Leu), 4.50dd, 7.2, 4.5 Hz, 1H (α-H-N,N-diMe-Phe), 5.05dd, 7.2, 2.0 Hz, 1H (β-H-β-OH-Leu), 5.77d, 8.1 Hz, 1H-D_2O exchangeable (NH-Ileu), 6.35d, 7.7 Hz, 1H-D_2O exchangeable (NH-Sty), 6.42d, 9.9 Hz, 1H (β-H-Sty), 6.66dd, 9.9, 7.7 Hz, 1H (α-H-Sty), 7.04-7.25m, 9H (9xAr-H), 7.79d, 9.9 Hz, 1H-D_2O exchangeable (NH-β-OH-Leu) (*120*)
^{1}H-NMR (300MHz, $CDCl_3$) : (*113*)
^{1}H-NMR (60MHz, $CDCl_3$ + CD_3OD) (*100*)
^{13}C-NMR (100.6MHz, $CDCl_3$) : see figure (*120*)
* Values may be interchanged
^{13}C-NMR (75MHz, $CDCl_3$) (*113*)
Derivatives : Dihydro-myrianthine-B (*120*)
Sources : Euphorbiaceae
Antidesma montana-leaves, terminal branches (*120*)
Rhamnaceae
Ceanothus sanguineus-root bark (*46*)
Discaria febrifuga-stem bark (*105*)
D. longispina-root bark (*113*), roots (*119*)
Sterculiaceae
Melochia corchorifolia-aerial parts (*53*), leaves (*107*)
Waltheria americana-whole plant (*108*)
Urticaceae
Myrianthus arboreus-leaves (*100*)

63. Lotusanine-A

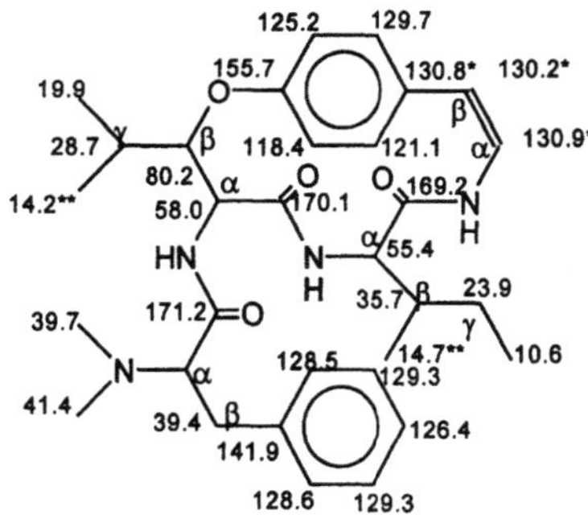

A = N,N-diMe-Phe
B = L-*erythro*-β-OH-Leu[1] (*59*)
C = Ileu
$C_{31}H_{42}N_4O_4$, 534

Reported as separate compound (Lotusanine-A) without mention of or reference to Adouetine-Y' or Myrianthine-B. Lotusanine-A is probably a diastereoisomer of myrianthine-B, although they have the same absolute configuration (L-*erythro*-) of the β-OH-Leu moiety.
Mp = colorless amorphous solid (*59*)
UV (MeOH) : 208 (4.20) (*59*)
IR ($CHCl_3$) : 3260, 1622, 1219 (*59*)
MS : 489(w), 443(b), 355(x), 327(x-CO), 300(y), 272(y-CO), 214(z), 190(f), 189(f-H^+), 148(a=100%), 135(i), 131, 103, 91, 86(q), 84, 83, 56 (*59*)
MS (FD) : 534(M^+) (*59*)
FAB/negative-ion and linked-scan mass spectra (*59*)
^{1}H-NMR (500MHz, $CDCl_3$) : 0.40d, 6.6 Hz, 3H (CH-Me-Ileu), 0.43t, 7.2 Hz, 3H (CH_2-Me-Ileu), 0.73d, 6.6 Hz, 3H (Me-β-OH-Leu), 0.98d, 6.6 Hz, 3H (Me-β-OH-Leu), 1.40m, 1H (β-H-Ileu), 1.79m, 1H (γ-H-Ileu), ~1.80m, 1H (γ-H-β-OH-Leu), 2.42s, 6H (N-Me_2), 3.54t, 1.8 Hz, 1H (α-H-N,N-diMe-Phe), 4.18d, 8.1 Hz, 1H (α-H-β-OH-Leu), 4.20m, 1H (α-H-Ileu), 4.64d, 7.9 Hz, 1H (β-H-β-OH-Leu), 6.00m, 1H (NH-Sty), 6.01br.s, 1H (β-H-Sty), 6.43d, 7.2 Hz 1H (α-H-Sty), 6.77-7.01m, 11H (9xAr-H + NH-Ileu + NH-β-OH-Leu) (*59*)
^{13}C-NMR (125MHz, $CDCl_3$) : see figure (*59*)
* and ** Values may be interchanged
COSY, HMBC and HMQC NMR spectra (*59*)
Sources : *Zizyphus lotus* (Rhamnaceae)-aerial parts (*59*)

64. Melonovine-B

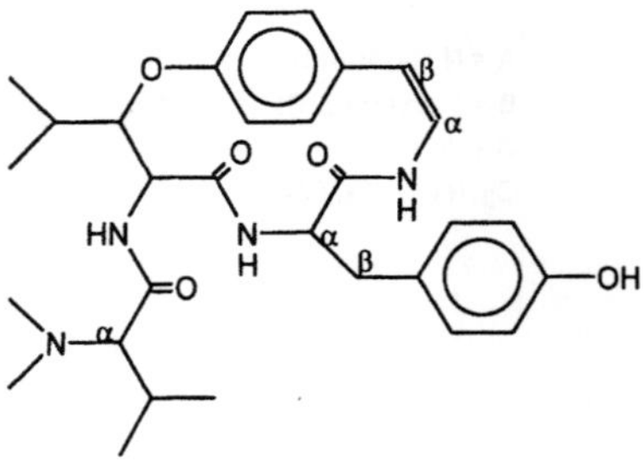

A = N,N-diMe-Val
B = β-OH-Leu
C = Tyr
$C_{30}H_{40}N_4O_5$, 536

Mp = 200-206 (*44*)
IR (Nujol) : 3400, 3280, 2780, 1680, 1610, 1250 (*44*)
MS : 536(M^+), 521(M^+-Me), 493(b), 394(j), 353(e), 260(k), 232(l), 190(f), 136(q), 135(i), 107, 100(a=100%), 97(m), 85(n), 78, 77 (*44*)
^{1}H-NMR (100MHz, $CDCl_3$) : 0.93d, 8.0 Hz, 6H (CH-Me_2-N,N-diMe-Val), 1.06d, 8.0 Hz, 3H (Me-β-OH-Leu), 1.23d, 8.0 Hz, 3H (Me-β-OH-Leu), 2.18s, 6H (N-Me_2), 2.51d, 4.0 Hz, 1H (α-H-N,N-diMe-Val), 2.60dd, 14.0, 6.0 Hz, 1H (β-H-Tyr), 3.10dd, 14.0, 5.0 Hz, 1H (β-H-Tyr), 4.30m, 1H (α-H-Tyr), 4.47dd, 10.0, 8.0 Hz, 1H (α-H-β-OH-Leu), 4.94dd, 8.0, 2.0 Hz, 1H (β-H-β-OH-Leu) (*44*)
Sources : *Melochia tomentosa* (Sterculiaceae)-roots (*44*)

65. Americine

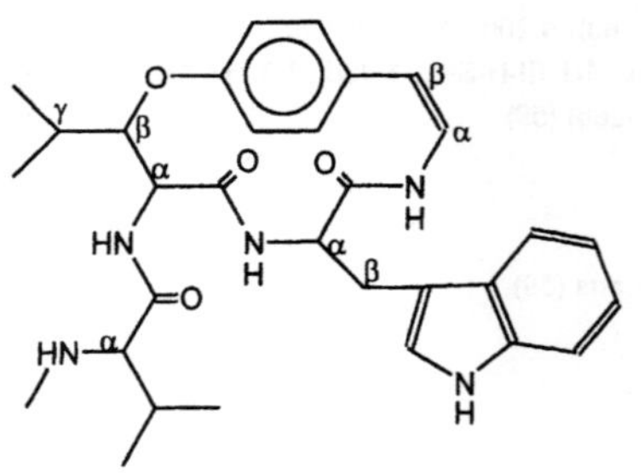

A = N-Me-Val
B = β-OH-Leu
C = Trp
$C_{31}H_{39}N_5O_4$, 545

Mp = 135.5-137 or 142-182 (*121*)
$[a]_D$ (20) = -198 (c=0.51, MeOH) (*121*)
UV (MeOH) : 221 (4.60), 273 (3.80), 280 (3.80), 290 (3.70) (*121*)
IR (KBr) : 3260, 1625, 1575, 1480, 1440, 1325, 1225, 1130, 1060, 960, 850, 733 (*121*)
MS : 545(M^+), 502(b), 460(g), 438(M^+-107), 417(j), 366(x-CO), 304(v), 135(i), 130, 97(m), 86(a=100%) (*121*)
^{1}H-NMR (100MHz, DMSO-d_6) : 0.70d, 6.2 Hz, 6H (CH-Me_2-N-Me-Val), 0.88d, 6.2 Hz, 3H (Me-β-OH-Leu), 1.08d, 6.2 Hz, 3H (Me-β-OH-Leu), 1.48m, 2H (β-H-N-Me-Val + γ-H-β-OH-Leu), 1.80s, 3H (NH-Me), 2.68m, 2H (β-H-Trp), 3.50m, 2H (α-H-N-Me-Val + α-H-Trp), 6.01m, 1H (α-H-Sty), 6.42d, 9.0 Hz, 1H (β-H-Sty), 6.86-8.11m, 12H (9xAr-H + 3xNH) (*121*)
Derivatives : Dihydro-americine (*121*)
Dihydro-methyl-americine (*121*)
Sources : *Ceanothus americanus* (Rhamnaceae)-root bark (*121*)

66. Scutianine-H

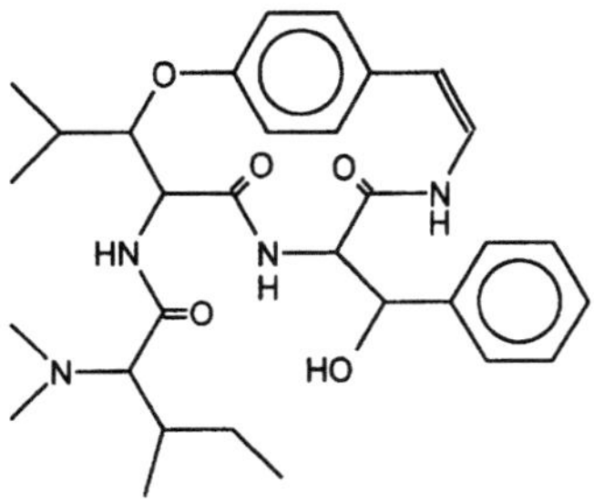

A = N,N-diMe-Ileu
B = β-OH-Leu
C = β-OH-Phe
$C_{31}H_{42}N_4O_5$, 550.3184 (MS) (*15*)

Mp = 242-243 (*15*)
[a]$_D$ (20) = -233 (c=0.1, $CHCl_3$) (*15*)
UV (MeOH) : No absorption (*15*)
IR : 3580, 3285, 2780, 1640, 1620, 1500, 1250 (*15*)
MS : 550(M^+), 535(M^+-CH_3), 507(M^+-C_3H_7), 493(b), 444(M^+-106), 401(M^+-C_3H_7-106), 394(j), 353(e), 342(x-CO+H^+), 331(g-106), 329(M^+-R_2-a), 190(f), 177, 135(i), 120, 114(a=100%), 107, 106, 105, 97(m), 85(n), 77 (*15*)
^{1}H-NMR (100MHz, C_5D_5N) : 0.84t, 6.0 Hz, 3H (CH_2-Me-N,N-diMe-Ileu), 0.94d, 6.0 Hz, 3H (CH-Me-N,N-diMe-Ileu), 1.20d, 8.0 Hz, 3H (Me-β-OH-Leu), 1.28d, 6.0 Hz, 3H (Me-β-OH-Leu), 2.32s, 6H (N-Me_2) (*15*)
^{1}H-NMR (100MHz, $CDCl_3$) : (*15*)
Derivatives : Dihydro-scutianine-H (*15*)
Oxo-dihydro-scutianine-H (*15*)
Sources : *Scutia buxifolia* (Rhamnaceae)-stem bark (*15, 118*)

67. Sanjoinine-F

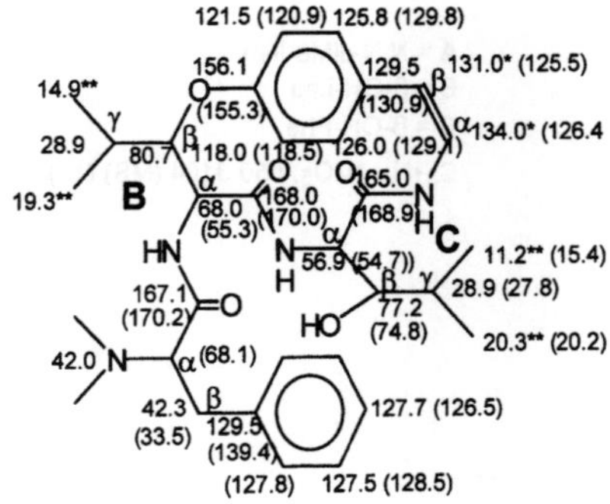

A = N,N-diMe-Phe
B = L-*erythro*-β-OH-Leu[2] (*59, 63*)
C = β-OH-Leu
$C_{31}H_{42}N_4O_5$, 550.3177 (MS) (*63*)

Mp = 228-229 (*63, 70*)
[a]$_D$ (25) = -107 (c=0.05, $CHCl_3$) (*59*)
[a]$_D$ (26) = -215 (c=0.28, $CHCl_3$) (*63, 70*)
UV (MeOH) : 280sh. (4.40) (*59*)
IR (KBr) : 3390, 3300, 2780, 1680-1620, 1604, 1235 (*63*)
IR ($CHCl_3$) : (*59*)
MS : 550(M^+), 535(M^+-CH_3), 507(M^+-C_3H_7), 505(w), 459(b), 371(x), 360(j), 343(x-CO), 316(y), 288(y-CO), 214(z), 190(f), 189(f-H^+), 148(a=100%), 135(i), 131, 103, 102(q), 97(m), 91, 84 (*59*)
MS (*63*)
MS (FD) : 550(M^+) (*59*)
^{1}H-NMR (500MHz, $CDCl_3$) : 0.76d, 6.6 Hz, 3H (Me-[B]-β-OH-Leu), 0.89d, 6.6 Hz, 3H (Me-[B]-β-OH-Leu), 1.17d, 6.5 Hz, 3H (Me-[C]-β-OH-Leu), 1.22d, 7.4 Hz, 3H (Me-[C]-β-OH-Leu), 2.05m, 2H (2xγ-H : [B]-β-OH-Leu + [C]-β-OH-Leu), 2.88s, 3H (N-Me), 2.91s, 3H (N-Me), 3.04d, 2.0 Hz, 2H (β-H-N,N-diMe-Phe), 3.44t, 1.8 Hz, 1H (α-H-N,N-diMe-Phe), 4.05m, 1H (β-H-[C]-β-OH-Leu), 4.40m, 1H (α-H-[C]-β-OH-Leu), 4.66m, 1H (α-H-[B]-β-OH-Leu), 4.87d, 7.9 Hz, 1H (β-H-[B]-β-OH-Leu), 6.20m, 1H (α-H-Sty), 6.70d, 7.4 Hz, 1H (β-H-Sty), 7.00-7.50m, 9H (9xAr-H) (*59*)
^{1}H-NMR (80MHz, $CDCl_3$) (*63*)
^{13}C-NMR (125MHz, $CDCl_3$) : see figure (*59*)
* and ** Values may be interchanged.
^{13}C-NMR (20MHz, $CDCl_3$) : see figure values in parentheses (*63*)
Sources : Rhamnaceae
Zizyphus lotus-aerial parts (*59*)
Z. vulgaris var. *spinosus*-seeds (*63, 70*)

68. *N*-Methyl-americine (= 5-β-Indolylmethyl-8-*N*,*N*-dimethylvalyl-9-isopropyl-phencyclopeptine)

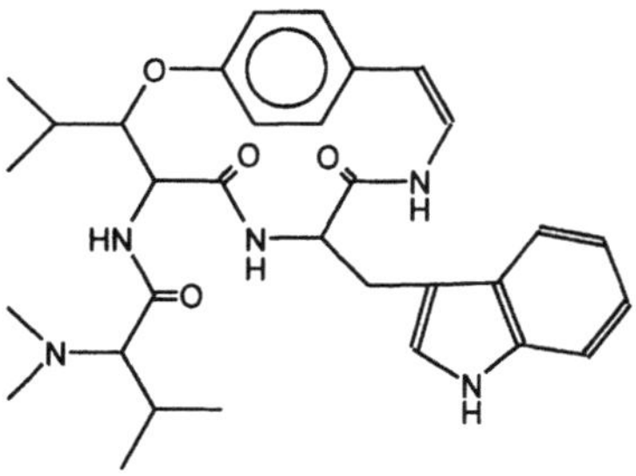

A = N,N-diMe-Val
B = β-OH-Leu
C = Trp
$C_{32}H_{41}N_5O_4$, 559.3146 (MS) (*45*)

Mp = 229 (*46*), 233 (*45*)
MS : 559(M^+), 516(b), 417(j), 376(e), 347(h), 283(k), 255(l), 195(c), 190(f), 170(r), 167(d), 159(q), 135(i), 130, 117, 100(a=100%), 97(m), 85(n), 72(o) (*45*)
MS (*46*)
^{1}H-NMR (270MHz, $CDCl_3$) : 0.84d, 6.8 Hz, 3H (Me-β-OH-Leu), 0.93d, 6.8 Hz, 3H (Me-β-OH-Leu), 0.96d, 6.9 Hz, 3H (C-Me-N,N-diMe-Val), 1.18d, 6.9 Hz, 3H (C-Me-N,N-diMe-Val) (*45*)
Sources : Rhamnaceae
Ceanothus integerrimus var. *integerrimus*-root bark (*45*)
C. sanguineus-root bark (*46*)

69. Homoamericine (= *N*-Desmethyl-texensine or *N*-Desmethyl-discarine-B or diastereoisomer)

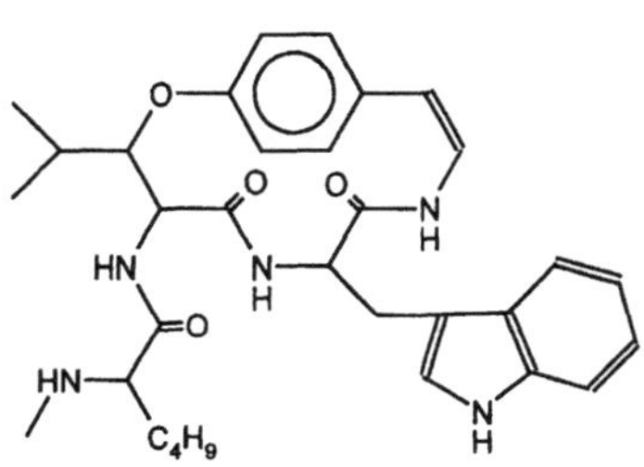

A = N-Me-Leu or -Ileu
B = β-OH-Leu
C = Trp
$C_{32}H_{41}N_5O_4$, 559

Has only been reported as a contaminant (4%) of americine and has never been isolated in pure state (*39*, *121*)
Mp = 135.5-137 or 142-182 (*121*)
MS : 559(M^+), 502(b), 460(g), 438(M^+-121), 417(j), 366(x-CO), 304(v), 135(i), 130, 100(a=100%), 97(m) (*121*)
Sources : *Ceanothus americanus* (Rhamnaceae)-root bark (*121*)

70. Scutianine-B

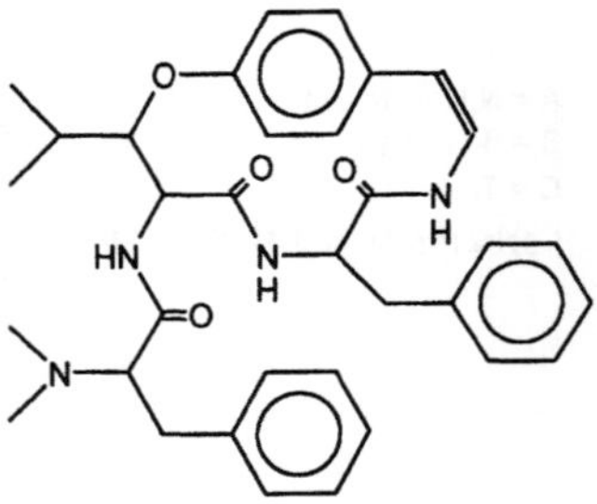

A = N,N-diMe-Phe
B = β-OH-Leu
C = Phe
$C_{34}H_{40}N_4O_4$, 568

Mp = 235-236 (*119*), 248-250 (*15, 44, 122*)
[a]D (20) = -290 (*15*) or -296 (*122*) or -308 (*119*) (c=0.1, $CHCl_3$)
UV (MeOH) : 275 (3.30) (*122*)
CD (Dioxane) : +1.40 (289), -24.00 (241), -10.10 (220) (*122*)
IR ($CHCl_3$) : 3390, 2785, 1685, 1620, 1230, 1030 (*122*)
MS : 568(M^+), 553(M^+-CH_3), 525(M^+-C_3H_7), 477(b), 421(g), 378(j), 337(e), 308(h), 244(k), 216(l), 195(c), 190(f), 167(d), 148(a=100%), 135(i), 120(q), 97(m), 91 (*122*)
MS (*44, 119*)
^{1}H-NMR ($CDCl_3$) : 0.93d, 3H (Me-β-OH-Leu), 1.25d, 3H (Me-β-OH-Leu), 2.17s, 6H (N-Me_2), 5.95d, 1H (1 olefinic-H), 6.47d, 1H (1 olefinic-H), 6.75-7.25m, 14H (14xAr-H) (*122*)
Sources : Rhamnaceae
Discaria febrifuga-stem bark (*105*)
Scutia buxifolia-roots (*119*), stem bark (*15, 18, 19, 117, 118, 122*)
Sterculiaceae
Melochia tomentosa-roots (*44*)

71. Texensine

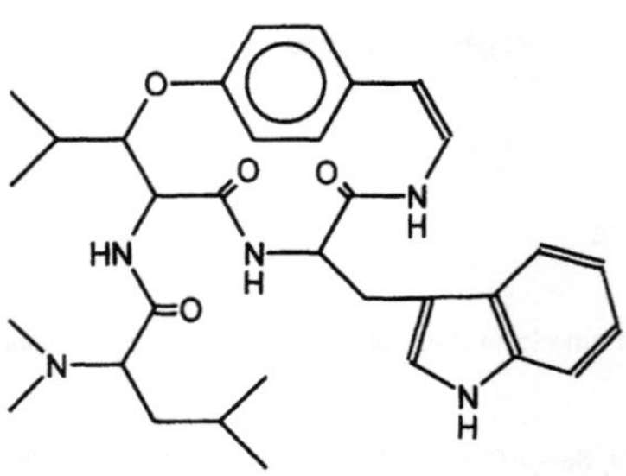

A = N,N-diMe-Leu
B = β-OH-Leu
C = Trp
$C_{33}H_{43}N_5O_4$, 573.3318 (MS) (*39*)

Mp = 249-252 (*39*)
[a]D (25) = -114 (c=0.50, $CHCl_3$) (*39*)
UV : 281, 290 (*39*)
IR : 3475, 3385, 1685, 1495 (*39*)
MS : 573(M^+), 558(M^+-CH_3), 516(b), 417(j), 376(e), 283(k), 255(l), 195(c), 190(f), 170(r), 167(d), 159(q), 135(i), 130, 114(a=100%), 97(m) (*39*)
^{1}H-NMR : 0.89d, 6.0 Hz, 3H (Me-β-OH-Leu), 0.99d, 6.5 Hz, 6H (2xMe-N,N-diMe-Leu), 1.24d, 6.0 Hz, 3H (Me-β-OH-Leu), 2.27s, 6H (N-Me_2) (*39*)
Derivatives : Dihydro-texensine (*39*)
Sources : *Colubrina texensis* (Rhamnaceae)-aerial parts (39)

72. Discarine-B

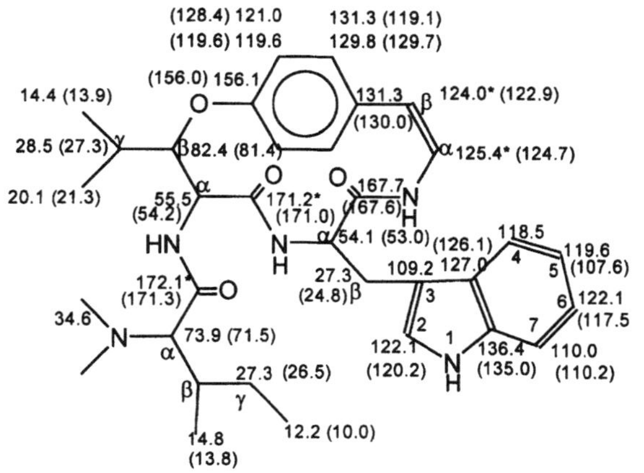

A = N,N-diMe-Ileu
B = L-*erythro*-β-OH-Leu[1,2] (*12, 15, 16, 109, 113*)
C = Trp
$C_{33}H_{43}N_5O_4$,
573.3264 (MS) (*46*),
573.3269 (MS) (*33*),
573.3297 (MS) (*45*)

Mp = 233-236 (*33, 45, 46*), 246-248 (*113*)
[a]D (25) = -154 (c=0.5, $CHCl_3$) (*113*)
[a]D (*33*)
UV (EtOH) : 226 (4.68), 273 (3.95), 284 (3.95), 294 (3.82) (*33*)
UV (EtOH) : (*113*)
IR (KBr) : 3400, 2800, 1650, 1620, 1230 (*113*)
IR (KBr) (*33*)
MS : 573(M^+), 516(b), 460(g), 417(j), 376(e), 347(h), 304(y), 283(k), 255(l), 195(c), 190(f), 170(r), 135(i), 130, 114(a=100%), 97(m) (*113*)
MS (*33, 45, 46*)
^{1}H-NMR (300MHz, DMSO-d_6) : 0.75d, 6.5 Hz, 3H (CH-Me-N,N-diMe-Ileu), 0.90t, 6.5 Hz, 3H (CH_2-Me-N,N-diMe-Ileu), 1.05d, 3H (Me-β-OH-Leu), 1.20m, 1H (γ-H-N,N-diMe-Ileu), 1.25d, 3H (Me-β-OH-Leu), 1.55m, 1H (γ-H-N,N-diMe-Ileu), 1.80d, 1H (β-H-N,N-diMe-Ileu), 2.10m, 1H (γ-H-β-OH-Leu), 2.30s, 6H (N-Me_2), 2.63d, 1H (α-H-N,N-diMe-Ileu), 2.90m, 1H (β-H-Trp), 3.10m, 1H (β-H-Trp), 4.20m, 1H (α-H-Trp), 4.52dd, 1H (α-H-β-OH-Leu), 4.90dd, 1H (β-H-β-OH-Leu), 6.20dd, 1H (α-H-Sty), 6.40d, 1H (β-H-Sty), 6.50d, 1H (NH-Trp), 6.85d, 1H (NH-Sty), 6.80-7.45m, 4H (4xAr-H-Sty), 7.05t, 1H (6-H-Trp), 7.14d, 1H (2-H-Trp), 7.28t, 1H (5-H-Trp), 7.40d, 1H (7-H-Trp), 7.50d, 1H (4-H-Trp), 7.80d, 1H (NH-β-OH-Leu), 10.25s, 1H (1-H-Trp) (*113*)
^{1}H-NMR (220MHz, $CDCl_3$, 18°C or 48°C) : (*109*)
^{1}H-NMR (220MHz, DMSO-d_6, 18°C) (*33, 109*)
^{1}H-NMR (220MHz, DMSO-d_6, 80°C) (*109*)
^{1}H-NMR (220MHz, $CDCl_3$ + 4% or 8% or 11% or 14% DMSO-d_6) (*109*)
^{13}C-NMR (15MHz, $CDCl_3$ + CD_3OD 2:1 v/v) : see figure (*16*)
^{13}C-NMR (75MHz, $CDCl_3$) : see figure values in parentheses (*113*)
N.B. There is a large difference in ^{13}C-NMR chemical shifts between the two reports.
Derivatives : Dihydro-discarine-B (*33*)
Sources : Rhamnaceae
Ceanothus integerrimus var. *californicus*-root bark (*45*)
C. integerrimus var. *integerrimus*-root bark (*45*)
C. sanguineus-root bark (*46*)
Discaria longispina-root bark (*113*), roots (*33*)

73. Discarine-X

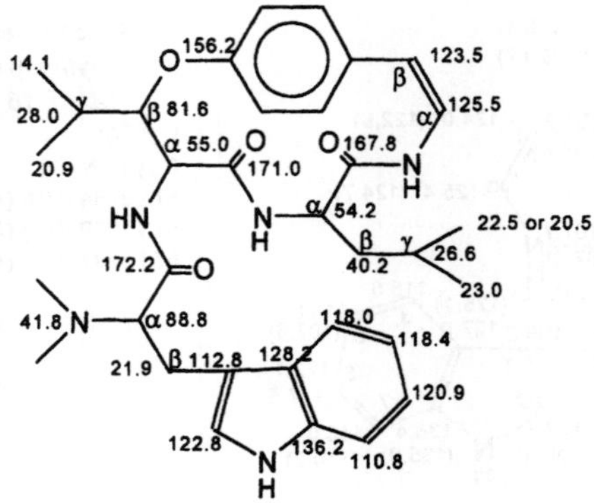

A = N,N-diMe-Trp
B = β-OH-Leu
C = Leu
$C_{33}H_{43}N_5O_4$, 573

Mp = 295-298 (*113*)
[a]D (25) = -184 (c=0.5, MeOH) (*113*)
UV (EtOH) : 218, 270, 280, 288 (*113*)
IR (KBr) : 3270, 2800, 1670, 1620, 1230 (*113*)
MS : 573(M^+), 530(M^+-C_3H_7), 443(b), 387(g), 344(j), 303(e), 274(h), 210(k), 195(c), 190(f), 187(a=100%), 182(l), 135(i), 130, 97(m), 86(q) (*113*)
^{1}H-NMR (300MHz, DMSO-d_6) : 0.60d, 6.7 Hz, 3H (Me-Leu), 0.62d, 6.7 Hz, 3H (Me-Leu), 1.10d, 6.7 Hz, 3H (Me-β-OH-Leu), 1.25d, 6.7 Hz, 3H (Me-β-OH-Leu), 1.30m, 1H (β-H-Leu), 1.74m, 1H (β-H-Leu), 1.80m, 1H (γ-H-Leu), 2.15s, 3H (N-Me), 2.20s, 3H (N-Me), 2.20m, 1H (γ-H-β-OH-Leu), 2.95d, 2H (β-H-N,N-diMe-Trp), 3.40dd, 1H (α-H-N,N-diMe-Trp), 3.70dd, 1H (α-H-Leu), 4.44dd, 1H (α-H-β-OH-Leu), 4.80dd, 1H (β-H-β-OH-Leu), 6.15dd, 7.4, 4.3 Hz, 1H (α-H-Sty), 6.35d, 7.0 Hz, 1H (β-H-Sty), 6.80-7.30m, 5H (5xAr-H-Trp), 7.05-7.50m, 4H (4xAr-H-Sty), 7.38d, 1H (NH-Leu), 7.70d, 1H (NH-Sty), 8.24d, 1H (NH-β-OH-Leu), 10.78br.s, 1H (NH-Trp) (*113*)
^{13}C-NMR (75MHz, $CDCl_3$) : see figure (*113*)
COSY NMR spectrum (*113*)
Sources : *Discaria longispina* (Rhamnaceae)-root bark (*113*)

74. Nummularine-K

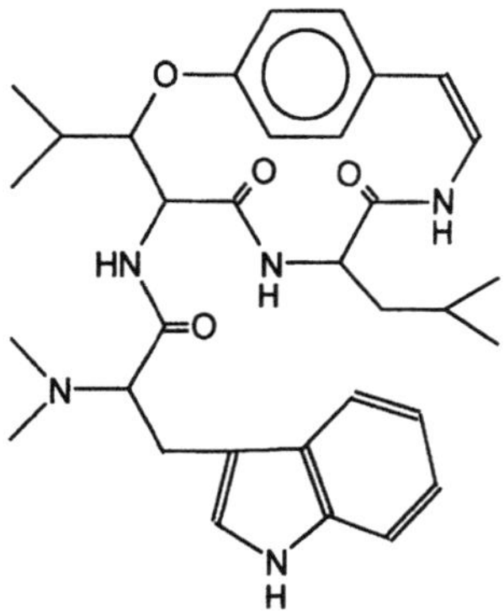

A = N,N-diMe-Trp
B = β-OH-Leu
C = Leu
$C_{33}H_{43}N_5O_4$, 573.3320 (MS) *(66)*

Mp = 235-239 *(66, 73, 77)*
[a]D (20) = -45 (c=0.04, MeOH) *(66)*
UV (MeOH) : 273 (3.81), 281 (3.79), 290 (3.71) *(66)*
IR (KBr) : 3235, 2790, 1675, 1625, 1610, 1235, 1045 *(66)*
MS : 573(M^+), 530(M^+-C_3H_7), 443(b), 387(g), 303(e), 274(h), 195(c), 190(f), 187(a=100%), 167(d), 144, 135(i), 130, 97(m), 86(q) *(66)*
^{1}H-NMR (90MHz, $CDCl_3$) : 0.41d, 7.0 Hz, 3H (Me-Leu), 0.86d, 7.0 Hz, 3H (Me-Leu), 0.97d, 7.0 Hz, 3H (Me-β-OH-Leu), 1.25d, 7.0 Hz, 3H (Me-β-OH-Leu), 2.31s, 6H (N-Me_2), 6.33-7.26m (Ar-H + NH + 1 olefinic-H) *(66)*
Derivatives : Dihydro-nummularine-K *(66)*
Sources : Rhamnaceae
Zizyphus nummularia-stem bark *(66)*
Z. xylopyra-stem bark *(77)*

75. Discarine-A

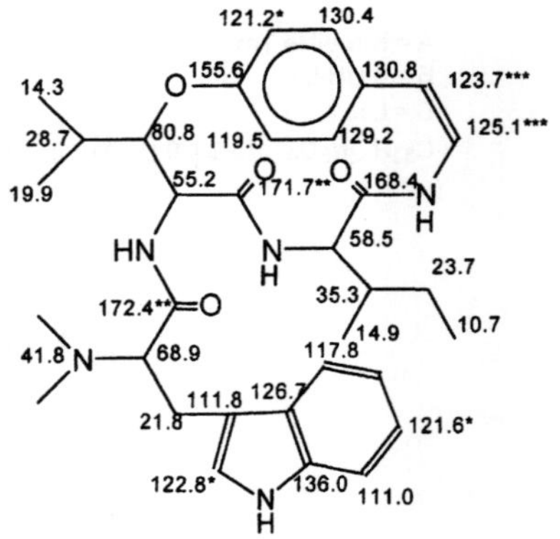

A = N,N-diMe-Trp
B = L-*erythro*-β-OH-Leu[1,2] (*8, 12, 15, 16*)
C = Ileu
$C_{33}H_{43}N_5O_4$, 573.3427 (MS) (*33*)

Mp = 229-231 (*33*)
$[a]_D$ = -282 (c=0.05, $CHCl_3$) (*33*)
UV (EtOH) : 223 (4.56), 273 (3.82), 284 (3.83), 294 (3.73) (*33*)
IR (KBr) : 3350, 1645 (*33*)
MS : 573(M^+), 443(b), 303(e), 190(f), 187(a), 182(l), 135(i), 130, 97(m), 86(q) (*33*)
^{1}H-NMR (220MHz, DMSO-d_6) : 0.61 asymmetric d, 6H (2xMe-Ileu), 0.86d, 7.0 Hz, 3H (Me-β-OH-Leu), 1.08d, 7.0 Hz, 3H (Me-β-OH-Leu), 2.26s, 6H (N-Me_2) (*33*)
^{13}C-NMR (15MHz, $CDCl_3$ + CD_3OD 2:1 v/v) : see figure (*16*)
*, ** and *** Values may be interchanged
Derivatives : Dihydro-discarine-A (*33*)
Sources : *Discaria longispina* (Rhamnaceae)-roots (*33*)

76. Amphibine-A

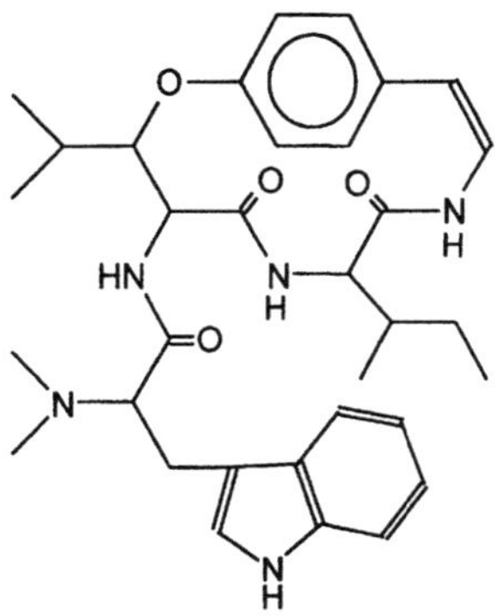

A = N,N-diMe-Trp
B = β-OH-Leu
C = Ileu
$C_{33}H_{43}N_5O_4$, 573

Mp = 236-239 (*34*, *123*)
[a]$_D$ (20) = -310 (c=0.021, MeOH) (*34*)
UV (MeOH) : 273 (3.82), 281 (3.83), 290 (3.73) (*34*)
MS : 573(M^+), 530(M^+-C_3H_7), 443(b), 387(g), 359(g-CO), 303(e), 274(h), 195(c), 190(f), 187(a=100%), 167(d), 144, 135(i), 130, 97(m), 86(q) (*34*)
^{1}H-NMR (TFA) : 3.20, 6H (N-Me_2) (*34*)
Sources : Rhamnaceae
Zizyphus amphibia-stem bark (*34*, *89*)
Z. nummularia-root bark (*43*)
Z. spina-christi-stem bark (*123*)

77. Scutianine-D [reported as Scutianine-C in (*119*)]

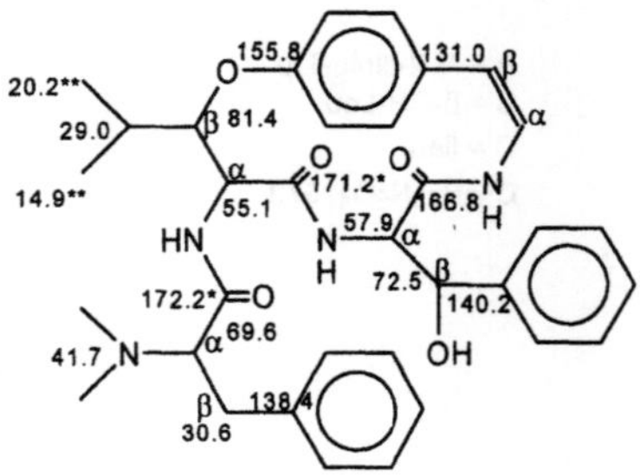

A = N,N-diMe-Phe
B = L-*erythro*-β-OH-Leu[1] (*18, 119*)
C = L-*threo*-β-OH-Phe[1] (*18, 119*)
$C_{34}H_{40}N_4O_5$, 584

Mp = 202-204 (*119*), 217-220 (*15, 18*)
[a]D (20) = -196 (*18*) or -202 (*15*) (c=0.1, $CHCl_3$)
[a]D (*119*)
UV (MeOH) : end absorption (*15, 18*)
CD (Dioxane) : +3.50 (284), +1.92 (263.5), -15.15 (236.5) (*18*)
IR ($CHCl_3$) : 3590, 3370, 2780, 1680, 1660, 1615, 1240, 1045 (*18*)
IR (*15, 119*)
MS : 584(M^+), 493(b), 478(M^+-106), 387(M^+-R-R_2), 331(g-106), 270, 254, 232(l), 223, 212(z-$2H^+$), 204, 195(c), 190(f), 189(f-H^+), 167(d), 148(a=100%), 135(i), 106, 105, 97(m), 91, 77 (*18*)
MS (*15, 119*)
^{1}H-NMR (220MHz, DMSO-d_6) : 0.88d, 7.0 Hz, 3H (Me-β-OH-Leu), 1.03d, 7.0 Hz, 3H (Me-β-OH-Leu), 2.13s, 6H (N-Me_2), 2.68m, 2H (β-H-N,N-diMe-Phe), 3.12dd, 7.0, 6.0 Hz, 1H (α-H-N,N-diMe-Phe), 4.10t, 9.0 Hz, 1H (α-H-β-OH-Phe), 4.23dd, 10.0, 8.0 Hz, 1H (α-H-β-OH-Leu), 4.34dd, 9.0, 5.0 Hz, 1H (β-H-β-OH-Phe), 4.65dd, 8.0, 2.0 Hz, 1H (β-H-β-OH-Leu), 5.48d, 5.0 Hz, 1H-disappearing on addition of D_2O (OH), 6.01dd, 7.0, 2.0 Hz, 1H (α-H-Sty), 6.66d, 7.0 Hz, 1H (β-H-Sty) (*119*)
^{1}H-NMR (60 and 90MHz, $CDCl_3$) (*18*)
^{13}C-NMR (25.2MHz, $CDCl_3$) : see figure (*15*)
* and ** Values may be interchanged.
Derivatives : *O*-Acetyl-scutianine-D (*15, 18*)
Dihydro-scutianine-D[#] (*18, 119*)
Oxo-dihydro-scutianine-D[#] (*119*)
[#]Reported as derivatives of Scutianine-C
Sources : *Scutia buxifolia* (Rhamnaceae)-roots (*119*), stem bark (*18, 19, 117, 118*)

78. Scutianine-E

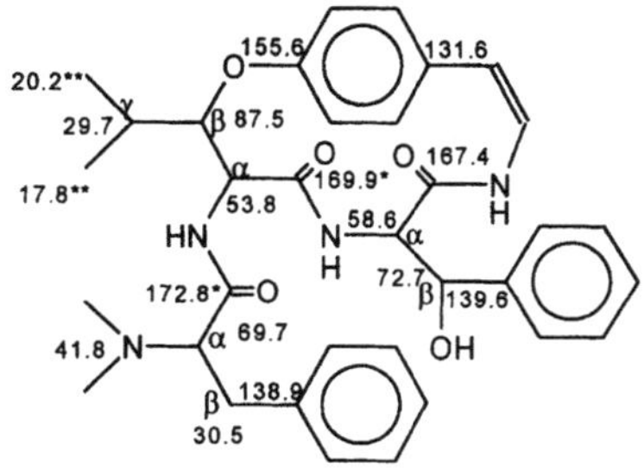

A = N,N-diMe-Phe
B = D-*erythro*-β-OH-Leu[1] (*15*, *18*)
C = D-*threo*-β-OH-Phe[1] (*15*, *18*)
$C_{34}H_{40}N_4O_5$, 584

Mp = 110-110.2 (*15*), 121 (*18*)
[a]D (20) = -21 (*15*) or -22.2 (*18*) (c=0.1, $CHCl_3$)
UV (MeOH) : end absorption (*15*, *18*)
CD (Dioxane) : -1.21 (276.5), +16.70 (235) (*18*)
IR ($CHCl_3$) : 3670, 3385, 2790, 1670, 1620, 1240, 1040 (*18*)
IR (*15*)
MS : Identical with that of Scutianine-D (*15*, *18*)
^{1}H-NMR (60 and 90MHz, $CDCl_3$) : 1.10d, 6.5 Hz, 3H (Me-β-OH-Leu), 1.12d, 6.5 Hz, 3H β-OH-Leu), 2.33s, 6H (N-Me_2), 2.52m, 7.0, 6.5 Hz, 1H (γ-H-β-OH-Leu), 2.90-3.50m, 2H N,N-diMe-Phe), 4.04dd, 7.0, 6.4 Hz, 1H (β-H-β-OH-Leu), 4.29dd, 5.6 Hz, 1H (α-H-β-OH-4.46dd, 6.4 Hz, 1H (α-H-β-OH-Leu), 4.99d, 5.6 Hz, 1H (β-H-β-OH-Phe), 6.34d, 7.5 Hz, 1 H-Sty), 6.54dd, 7.5, 1H (α-H-Sty), 6.64d, 11.0 Hz, 1H (NH), 6.80-7.50m, 15H (14xA 1xNH), 7.75d, 9.0 Hz, 1H (NH) (*18*)
^{13}C-NMR (25.2MHz, $CDCl_3$) : see figure (*15*)
* and ** Values may be interchanged.
Derivatives : O-acetyl-scutianine-E (*15*, *18*)
Dihydro-scutianine-E (*18*)
Sources : *Scutia buxifolia* (Rhamnaceae)-stem bark (*18*, *19*, *117*, *118*)

79. Scutianine-G

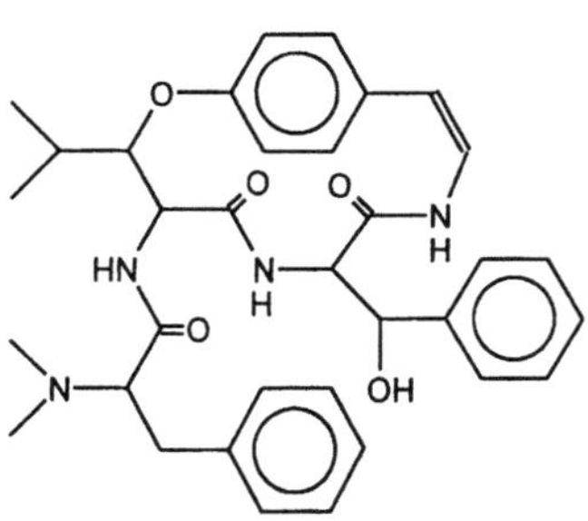

A = N,N-diMe-Phe
B = L-*threo*-β-OH-Leu[1] (*19*)
C = L-*erythro*-β-OH-Phe[1] (*19*)
$C_{34}H_{40}N_4O_5$, 584.2991 (MS) (*19*)

Mp = 162 (*19*)
[a]D (20) = -112 (c=0.02, MeOH) (*19*)
UV (MeOH) : end absorption (*19*)
IR (KBr) : 3600-3300, 3200, 1640, 1605, 1500, 1235, 1045 (*19*)
MS : 584(M^+), 493(b), 353(e), 195(c), 190(f), 148(a=100%), 135(i), 97(m) (*19*)
^{1}H-NMR ($CDCl_3$) : 0.95d, 6.2 Hz, 3H (Me-β-OH-Leu), 1.22d, 6.2 Hz, 3H (Me-β-OH-Leu), 2.15s, 6H (N-Me_2), 6.37-6.81m, 6H (3xNH + 2xolefinic-H + 1xOH), 7.03-7.58m, 14H (14xAr-H) (*19*)
Sources : *Scutia buxifolia* (Rhamnaceae)-stem bark (*19*)

80. Scutianine-J

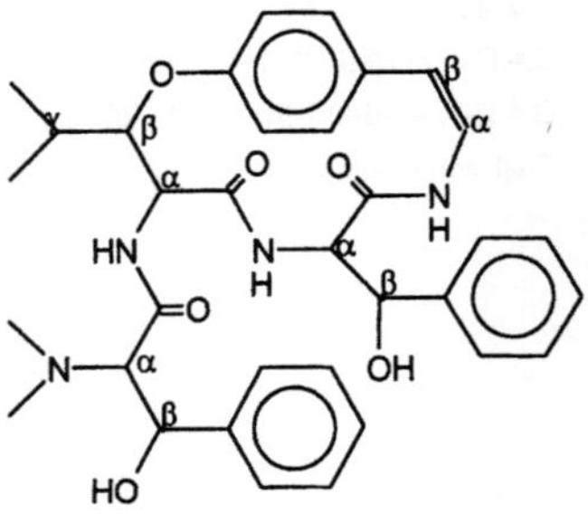

A = N,N-diMe-β-OH-Phe
B = β-OH-Leu
C = β-OH-Phe
$C_{34}H_{40}N_4O_6$, 600

Mp = amorphous (*118*)
UV (MeOH) : No absorption (*118*)
IR : 3600, 3280, 1650, 1625, 1250 (*118*)
MS : 493(b), 387(b-106), 385(b-R/R$_2$-H$^+$), 331(g-106), 195(c), 164(a=100%), 135(i), 107, 106, 105, 77 (*118*)
MS (FAB+) : 601(M+H$^+$) (*118*)
^{1}H-NMR (250MHz, CDCl$_3$) : 0.95d, 6.4 Hz, 3H (Me-β-OH-Leu), 1.20d, 6.4 Hz, 3H (Me-β-OH-Leu), 2.10m, 1H (γ-H-β-OH-Leu), 2.20s, 6H (N-Me$_2$), 4.35m, 3H (α-H-β-OH-Leu + α-H-β-OH-Phe + OH), 4.80m, 2H (β-H-β-OH-Leu + α-H-N,N-diMe-β-OH-Phe), 6.20d, 9.5 Hz, 1H (NH-β-OH-Phe), 6.28m, 1H (α-H-Sty), 6.55d, 9.5 Hz, 1H (NH-Sty), 7.60d, 7.6 Hz, 1H (NH-β-OH-Leu), 7.75d, 1H (β-H-Sty), 6.75-6.90m, 14H (14xAr-H) (*118*)
Sources : *Scutia buxifolia* (Rhamnaceae)-stem bark (*118*)

4(14)-Integerrine-Type Cyclopeptide Alkaloids

81. Desbenzoyl-aralionine-A

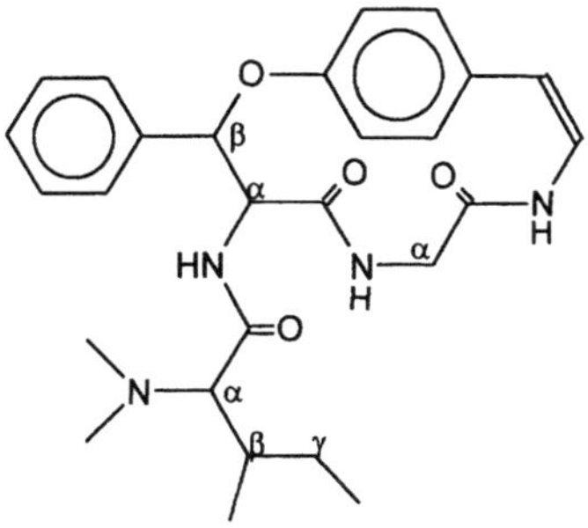

A = N,N-diMe-Ileu
B = *threo*-β-OH-Phe[2]
C = Gly
$C_{27}H_{34}N_4O_4$, 478.2564 (MS) *(30)*

Mp = 101-104 *(30)*
$[a]_D$ (20) = +100 (c=0.16, MeOH) *(30)*
UV (MeOH) : 217sh. (4.34), 271sh. (3.54) *(30)*
CD (MeOH) : -0.26 (290), +0.09 (276), -0.24 (260), +3.91 (235) *(30)*
IR ($CHCl_3$) : 3430, 3390, 1685, 1225 *(30)*
MS : 478(M^+), 463(M^+-CH_3), 421(b), 376(b'), 365(g), 363(g-2H^+), 348(b-CO), 281(e), 229(c), 224(f), 218(h), 201(d), 188(k), 160(l), 135(i), 131(m), 114(a=100%), 91, 85(n) *(30)*
^{1}H-NMR (60MHz, $CDCl_3$) : 0.48d, 7.0 Hz, 3H (CH-Me-N,N-diMe-Ileu), 0.80-1.85m, 6H (CH_2-Me- + β-H- + γ-H-N,N-diMe-Ileu), 2.30s, 6H (N-Me_2), 2.72d, 4.5 Hz, 1H (α-H-N,N-diMe-Ileu), 3.20dd, 17.0, 6.0 Hz, 1H (α-H-Gly), 4.12dd, 17.0, 7.0 Hz, 1H (α-H-Gly), 4.61d, 7.5 Hz, 1H (α-H-β-OH-Phe), ~6.20m, 3H (3xNH), 6.36d, 7.5 Hz, 1H (α-H-Sty), 6.49s, 1H (β-H-β-OH-Phe), 6.69d, 7.5 Hz, 1H (β-H-Sty), 6.90-7.80m, 9H (9xAr-H) *(30)*
Sources : *Araliothamnus vaginatus* (Rhamnaceae)-stem bark *(117)*

82. Sativanine-B

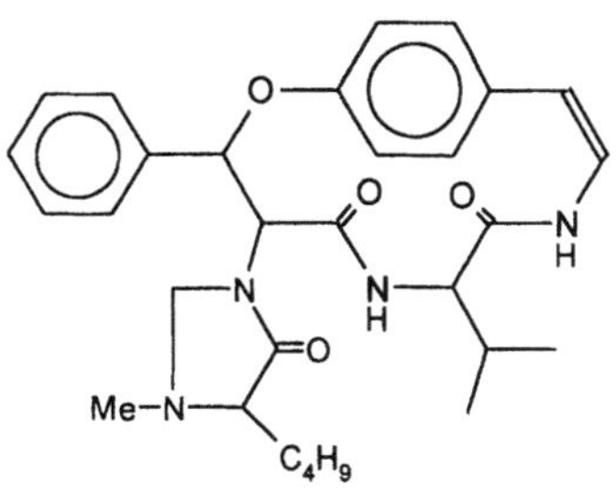

A = N-Me-Leu or -Ileu
B = β-OH-Phe
C = Val
$C_{30}H_{38}N_4O_4$, 518.2888 (MS) *(50, 67)*

Mp = amorphous *(50, 67)*
UV (MeOH) : 280sh. *(50, 67)*
IR ($CHCl_3$) : 3330, 2960-2870, 2800, 1675, 1610, 1585-1485, 1220, 1025 *(50)*
MS : 518(M^+), 491, 461(b), 364(j), 323(e), 260(h), 230(k), 227(c), 224(f), 202(l), 201(d), 155(a'), 135(i=100%), 131(m), 72(q) *(67)*
MS *(50)*
Sources : *Zizyphus sativa* (Rhamnaceae)-stem bark *(50, 67)*

83. 5-Isobutyl-8-*N*-methyl-isoleucyl-9-phenyl-phencyclopeptine

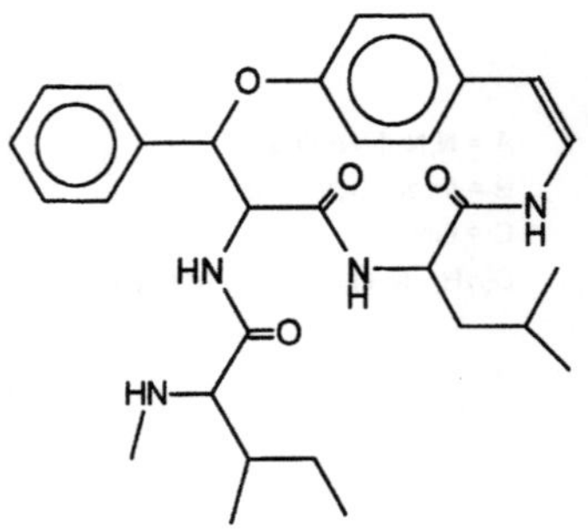

A = N-Me-Ileu
B = β-OH-Phe
C = Leu
$C_{30}H_{40}N_4O_4$, 520.3053 (MS) (*45*)

Mp = 213 (*45*)
MS : 520(M^+), 505(M^+-CH_3), 491(M^+-C_2H_5), 477(M^+-C_3H_7), 463(b), 421(g), 378(j), 337(e), 274(h), 244(k), 224(f), 216(l), 215(c), 187(d), 135(i), 131(m), 100(a=100%), 97(r), 86(q), 72(p), 58(o), 57($C_4H_9^+$) (*45*)
^{1}H-NMR (270MHz, $CDCl_3$) : 0.57d, 6.9 Hz, 3H (CH-Me-Ileu), 0.66m, 6H (CH_2-Me-Ileu + Me-Leu), 0.76d, 6.5 Hz, 3H (Me-Leu) (*45*)
Sources : Rhamnaceae
Ceanothus integerrimus var. *californicus*-root bark (*45*)
C. integerrimus var. *integerrimus*-root bark (*45*)

84. Nummularine-D (= *N*-Desmethyl-integerrenine)

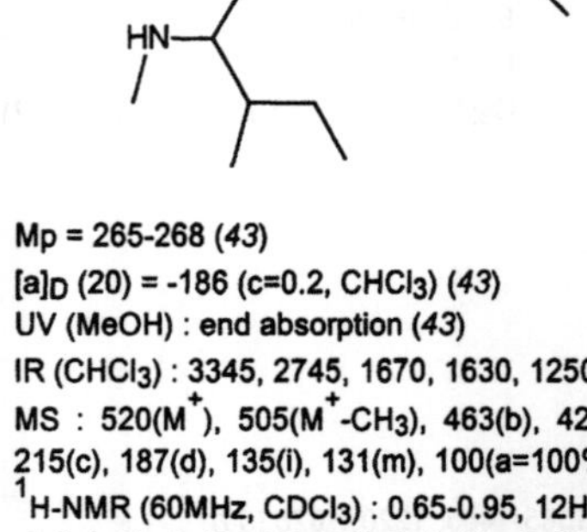

A = N-Me-Ileu
B = β-OH-Phe
C = Leu
$C_{30}H_{40}N_4O_4$, 520.3051 (MS) (*43*)

Mp = 265-268 (*43*)
[a]D (20) = -186 (c=0.2, $CHCl_3$) (*43*)
UV (MeOH) : end absorption (*43*)
IR ($CHCl_3$) : 3345, 2745, 1670, 1630, 1250, 1030 (*43*)
MS : 520(M^+), 505(M^+-CH_3), 463(b), 421(g), 378(j), 337(e), 274(h), 244(k), 224(f), 216(l), 215(c), 187(d), 135(i), 131(m), 100(a=100%), 91, 86(q), 72(p), 71(n), 58(o) (*43*)
^{1}H-NMR (60MHz, $CDCl_3$) : 0.65-0.95, 12H (4xC-Me), 2.18s, 3H (NH-Me), 6.35d, 8.0 Hz, 1H (1 olefinic-H), 6.40-7.50, 13H (9xAr-H + 3xNH + 1 olefinic-H) (*43*)
Derivatives : *N*-Acetyl-nummularine-D (*43*)
N-Methyl-dihydro-nummularine-D = Dihydro-integerrenine (*43*)
Sources : *Zizyphus nummularia* (Rhamnaceae)-root bark (*43*), stem bark (*66*)

85. Sativanine-A

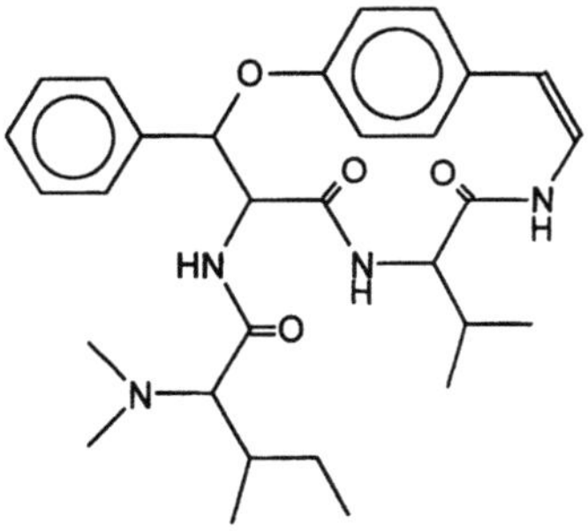

A = N,N-diMe-Ileu
B = β-OH-Phe
C = Val
$C_{30}H_{40}N_4O_4$, 520.3050 (*50*) or 520.3055 (*67*) (MS)

Mp = 80 (*50, 67*)
UV (MeOH) : 250sh., 280sh. (*50, 67*)
IR ($CHCl_3$) : 3370, 2990-2820, 2775, 1675, 1618, 1590, 1490, 1225, 1040 (*50*)
MS : 520(M^+), 505(M^+-CH_3), 491(M^+-C_2H_5), 463(b), 407(g), 364(j), 323(e), 260(h), 230(k), 229(c), 224(f), 202(l), 201(d), 135(i), 131(m), 114(a=100%), 103(m-CO), 85(n), 72(o/q) (*67*)
MS (*50*)
Sources : <u>Rhamnaceae</u>
Zizyphus sativa-stem bark (*50, 67*)
Z. spina-christi-stem bark (*106*)

86. Nummularine-E

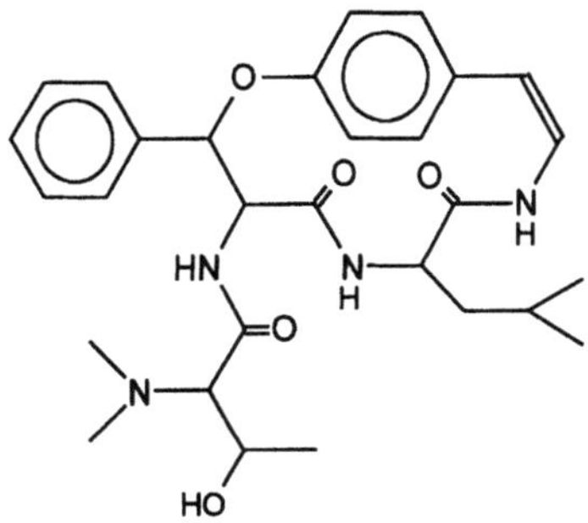

A = N,N-diMe-Thr
B = β-OH-Phe
C = Leu
$C_{29}H_{38}N_4O_5$, 522.2834 (MS) (*43*)

Mp = 276-279 (*43, 124*)
$[a]_D$ (20) = +12 (c=0.02, MeOH) (*43*)
UV (MeOH) : end absorption (*43*)
IR (KBr) : 3380, 3260, 2775, 1680, 1625, 1220, 1030 (*43*)
IR ($CHCl_3$) (*124*)
MS : 522(M^+), 507(M^+-CH_3), 478(b+H^+), 338(e+H^+), 274(h), 230(c+H^+), 224(f), 216(l), 135(i), 131(m), 102(a=100%), 86(q), 58(o), 44(CH_3-CHO^+) (*43*)
MS (*124*)
^{1}H-NMR (90MHz, CF_3COOD) : 0.77br.s, 6H (2xMe-Leu), 1.11d, 6.0 Hz, 3H (Me-Thr), 1.96s, 3H (N-Me), 2.72s, 3H (N-Me), 6.06d, 8.0 Hz, 1H (1 olefinic-H), 6.77-7.88m, 13H (9xAr-H + 3xNH + 1 olefinic-H) (*124*)
^{1}H-NMR (60MHz, DMSO-d_6) (*43*)
^{1}H-NMR (60MHz, CF_3COOD) (*43*)
Derivatives : Dihydro-nummularine-E (*43*)
O-Acetyl-nummularine-E (*43*)
Sources : <u>Rhamnaceae</u>
Zizyphus hysodrica-stem bark (*124*)
Z. nummularia-root bark (*43*), stem bark (*66*)

87. Nummularine-G

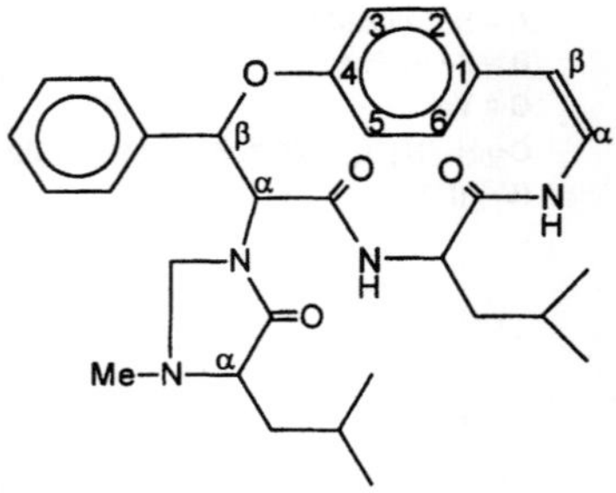

A = N-Me-Leu
B = *erythro*-β-OH-Phe[2]
C = Leu
$C_{31}H_{40}N_4O_4$, 532.3042 (MS) (*66*)

Mp = 174-175 (*66*)
[a]$_D$ (20) = -133 (c=0.085, MeOH) (*66*)
UV (MeOH) : 250sh., 280sh. (*66*)
IR ($CHCl_3$) : 3300, 2950-2880, 2800, 1675, 1610, 1585-1485,1220, 1025 (*66*)
MS : 532(M^+), 517(M^+-CH_3), 489(M^+-C_3H_7), 475(b), 378(j), 337(e), 274(h), 244(k), 227(c), 224(f), 216(l), 201(d), 155(a'), 135(i=100%), 131(m), 86(q) (*66*)
^{1}H-NMR (90MHz, $CDCl_3$) : 0.52d, 7.0 Hz, 3H (C-Me), 0.73d, 7.0 Hz, 3H (C-Me), 0.82d, 7.0 Hz, 6H (2xC-Me), 2.10s, 3H (N-Me), 2.43m, 1H (α-H-N-Me-Leu), 3.18dd, 5.4, 1.8 Hz, 1H (-N-CH_2-N), 3.85dd, 5.4, 0.9 Hz, 1H (-N-CH_2-N), 4.14m, 1H (α-H-Leu), 5.12d, 8.5 Hz, 1H (α-H-β-OH-Phe), 5.82d, 8.0 Hz, 1H (NH), 6.01d, 8.5 Hz, 1H (β-H-β-OH-Phe), 6.33-6.84m, 3H [2xolefinic-H + 1NH : +D_2O : 6.47d, 7.8 Hz, 1H (β-H-Sty), 6.66d, 7.8 Hz, 1H (α-H-Sty)], 7.13d, 7.5 Hz, 2H (3-H-Sty + 5-H-Sty), 7.22-7.62m, 7H (7xAr-H) (*66*)
Derivatives : Dihydro-nummularine-G (*66*)
Sources : *Zizyphus nummularia* (Rhamnaceae)-stem bark (*66*)

88. Discarine-C

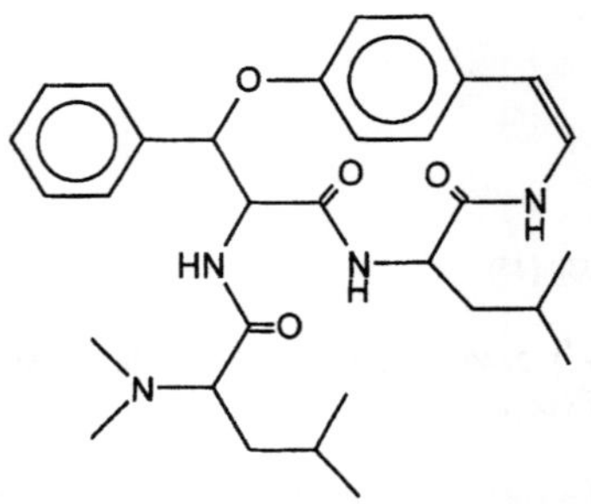

A = N,N-diMe-Leu
B = β-OH-Phe
C = Leu
$C_{31}H_{42}N_4O_4$, 534

UV : 252sh. 280sh. (*105*)
IR (KBr) : 3285, 2790, 1680, 1630, 1230 (*105*)
MS : 534(M^+), 421 (g), 135(i), 131(m), 114(a=100%), 86(q), 72(o) (*105*)
^{1}H-NMR ($CDCl_3$) : 0.68, 0.75 and 0.85, 4xd, 6.0 Hz, 12H (2xMe-Leu + 2xC-Me-N,N-diMe-Leu), 1.65s, 6H ($N-Me_2$), 7.00-7.30m, 9H (9xAr-H) (*105*)
Sources : *Discaria febrifuga* (Rhamnaceae)-stem bark (*105*)

89. Myrianthine-A

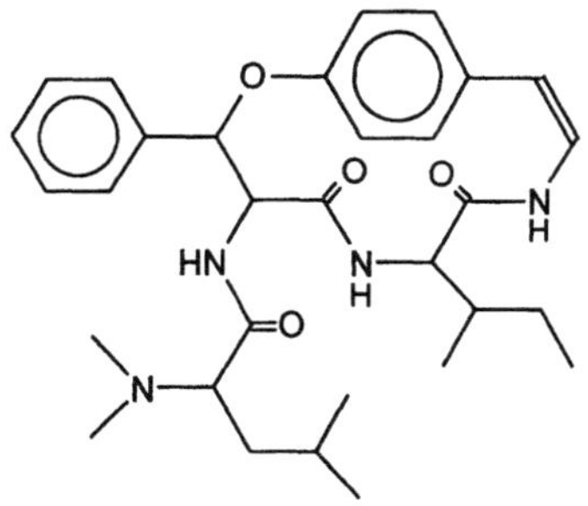

A = N,N-diMe-Leu
B = β-OH-Phe
C = Ileu
$C_{31}H_{42}N_4O_4$, 534

Mp = 286 (*100*)
[a]$_D$ (20) = -263 (c=1.0, $CHCl_3$) (*100*)
UV (EtOH) : 250 (3.79) (*100*)
IR : 3280, 1640, 1240 (*100*)
MS : 534(M^+), 477(b), 419(g-$2H^+$), 378(j), 337(e), 274(h), 244(k), 224(f), 216(l), 135(i) 131(m), 114(a), 86(p/q) (*100*)
^{1}H-NMR (60MHz, $CDCl_3$) : 0.60-1.00, 12H (4xC-Me), 1.76s, 6H (N-Me_2), 6.90-7.60, 9H (9xAr-H) (*100*)
Sources : *Myrianthus arboreus* (Urticaceae)-leaves (*100*)

90. Integerrenine

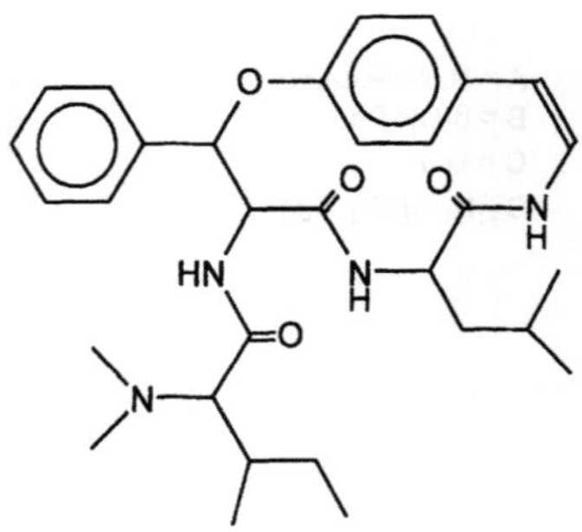

A = N,N-diMe-Ileu
B = β-OH-Phe
C = Leu
$C_{31}H_{42}N_4O_4$, 534.3200 (MS) (*45*)

Mp = 259 (*45*), 278-280 (*17, 26*)
[a]$_D$ (20) = -228 (c=0.2, $CHCl_3$) (*26*)
UV (MeOH) : 252sh. (3.70), 280 (3.13) (*17, 26*)
CD (Dioxane) : +1.40 (286.5), -21.50 (239) (*26*)
IR (KBr) : 3275, 2785, 1680, 1630, 1226 (*26*)
IR (KBr) (*17*)
MS : 534(M^+), 519(M^+-CH_3), 505(M^+-C_2H_5), 491(M^+-C_3H_7), 477(b), 337(e), 244(k), 229(c), 224(f), 216(l), 201(d), 135(i), 131(m), 114(a=100%), 97(r), 86(p/q), 85(n), 57($C_4H_9^+$) (*45*)
MS (*26, 28*)
^{1}H-NMR (270MHz, $CDCl_3$) : 0.36d, 6.7 Hz, 3H (CH-Me-Ileu), 0.78d, 6.5 Hz, 3H (Me-Leu), 0.85d, 6.5 Hz, 3H (Me-Leu), 0.86t, 7.3 Hz, 3H (CH_2-Me-Ileu) (*45*)
^{1}H-NMR (60MHz, $CDCl_3$) (*26*)
^{1}H-NMR ($CDCl_3$) (*17*)
^{1}H-NMR (60MHz, TFA) (*26*)
Derivatives : Dihydro-integerrenine (*26*)
Sources : Rhamnaceae
Ceanothus integerrimus-roots (*26*)
C. integerrimus var. *californicus*-root bark (*45*)
C. integerrimus var. *integerrimus*-root bark (*45*)
Zizyphus nummularia-root bark (*43*)
Sterculiaceae
Melochia pyramidata-leaves (*17*)

91. Nummularine-M

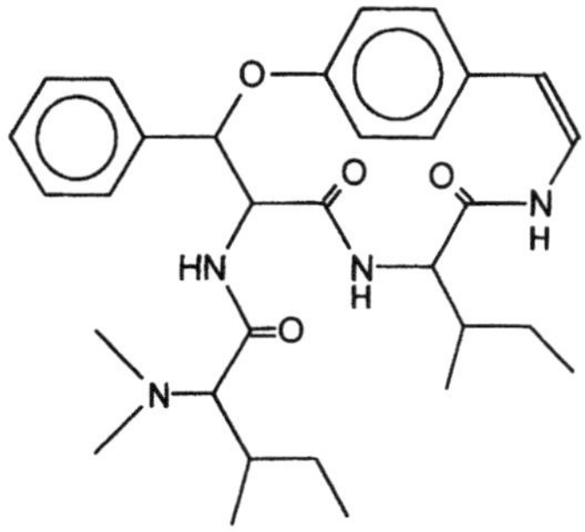

A = N,N-diMe-Ileu
B = β-OH-Phe
C = Ileu
$C_{31}H_{42}N_4O_4$, 534.3190 (MS) (*82*)

Mp = 263-265 (*82*)
[a]D = -46.66 (c=0.1, $CHCl_3$) (*82*)
UV (MeOH) : 250sh., 280sh. (*82*)
IR (KBr) : 3300, 2790, 1685, 1635, 1240 (*82*)
MS : 534(M^+), 477(b), 378(j), 337(e), 274(h), 244(k), 229(c), 224(f), 216(l), 201(d), 135(i), 131(m), 114(a=100%) (*82*)
Sources : *Zizyphus nummularia* (Rhamnaceae)-stem bark (*82*)

92. Canthiumine

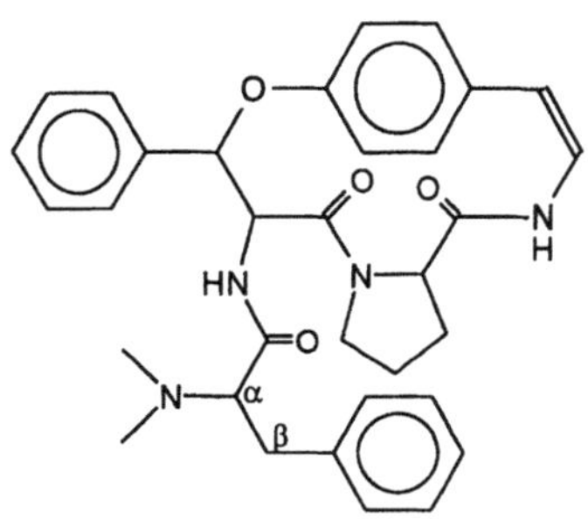

A = N,N-diMe-L-Phe[1] (*31*)
B = L-*erythro*-β-OH-Phe[1,2] (*12, 31*)
C = L-Pro[1] (*31*)
$C_{33}H_{36}N_4O_4$, 552

Mp = 232-233 (*31*)
[a]D = -254 (c=1.0, $CHCl_3$) (*31*)
UV : 250 (*31*)
IR (KBr) : 3400, 3315, 2790, 1705-1665, 1625, 1535, 1515, 1235 (*31*)
MS : 552(M^+), 461(b), 403(g-2H^+), 362(j), 258(h), 229(c), 228(k), 224(f), 201(d), 200(l), 148(a), 135(i), 131(m), 70 (*31*)
^{1}H-NMR ($CDCl_3$) : 1.67s, 6H (N-Me_2), 2.32dd, 14.0, 7.0 Hz, 1H (β-H-N,N-diMe-Phe), 2.71dd, 14.0, 6.0 Hz, 1H (β-H-N,N-diMe-Phe), 3.00t, 7.0, 6.0 Hz, 1H (α-H-N,N-diMe-Phe), 5.27dd, 10.0, 8.0 Hz, 1H (α-H-β-OH-Phe), 5.93d, 8.0 Hz, 1H (β-H-β-OH-Phe), 6.30d, 11.0 Hz, 1H (β-H-Sty), 6.37d, 7.0 Hz, 1H (NH-Sty), 6.72dd, 11.0, 7.0 Hz, 1H (α-H-Sty), 6.90-7.70m, 14H (14xAr-H), 7.00d, 10.0 Hz, 1H (NH-β-OH-Phe) (*31*)
Derivatives : Dihydro-canthiumine (*31*)
Canthiumene (=neutral compound) (*31*)
Sources : *Canthium euryoides* (Rubiaceae)-? plant part (*31*)

93. Integerresine

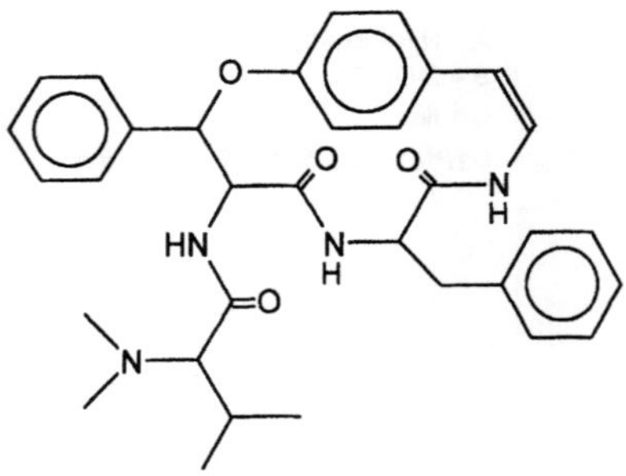

A = N,N-diMe-Val
B = β-OH-Phe
C = Phe
$C_{33}H_{38}N_4O_4$, 554.2878 (MS) (*26*)

Mp = 285 (*26*)
[a]D (20) = -164 (c=0.2, $CHCl_3$) (*26*)
UV (MeOH) : 252sh. (3.77), 280 (3.22) (*26*)
CD (Dioxane) : +1.50 (285), -19.0 (238) (*26*)
IR (KBr) : 3295, 2795, 1675, 1630, 1232 (*26*)
MS : 554(M^+), 539(M^+-CH_3), 511(b), 455(g), 412(j), 371(e), 308(h), 278(k), 250(l), 229(c), 224(f), 201(d), 135(i), 131(m), 120(q), 103(m-CO), 100(a=100%), 91, 85(a-Me^+/n) (*26*)
MS (*28*)
^{1}H-NMR (TFA) : 0.95d, 7.0 Hz, 6H (Me-N,N-diMe-Val), 2.20 and 2.81, 2xd, 4.5 Hz, 6H (N-Me_2) (*26*)
Derivatives : Dihydro-integerresine (*26*)
Sources : *Ceanothus integerrimus* (Rhamnaceae)-roots (*26*)

94. Aralionine-B

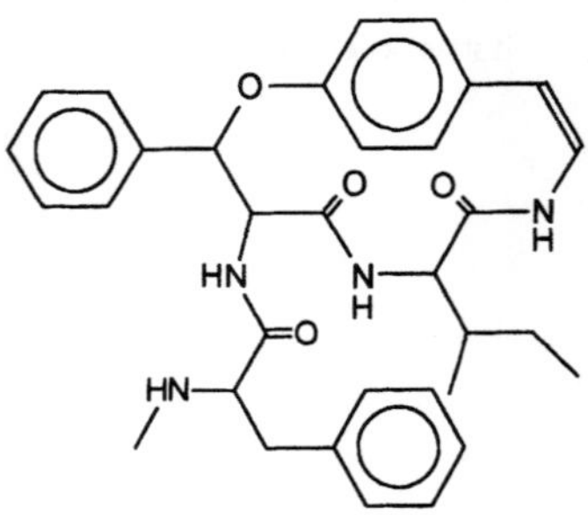

A = N-Me-Phe
B = β-OH-Phe
C = Ileu
$C_{33}H_{38}N_4O_4$, 554.2867 (MS) (*32*)

Mp = 103-105 (*32*)
[a]D (20) = -73 (c=0.1, MeOH) (*32*)
UV (MeOH) : 250 (3.30) (*32*)
CD (MeOH) : +2.02 (268), -7.26 (232) (*32*)
IR (KBr) : 3285, 2795, 1660, 1620, 1235, 1030 (*32*)
MS : 554(M^+), 463(b), 421(g), 274(h), 224(f), 215(c), 187(d), 135(i), 134(a=100%), 131(m), 86(q) (*32*)
^{1}H-NMR (60MHz, $CDCl_3$) : 0.70-1.80m, 9H (9 x aliphatic-H), 2.30s, 3H (NH-Me), 2.65d, 6.0 Hz, 2H (β-H-N-Me-Phe), 6.00-8.10m, 17H (14xAr-H + 3xNH) (*32*)
Sources : *Araliothamnus vaginatus* (Rhamnaceae)-leaves (*32*), stem bark (*117*)

95. AM-2

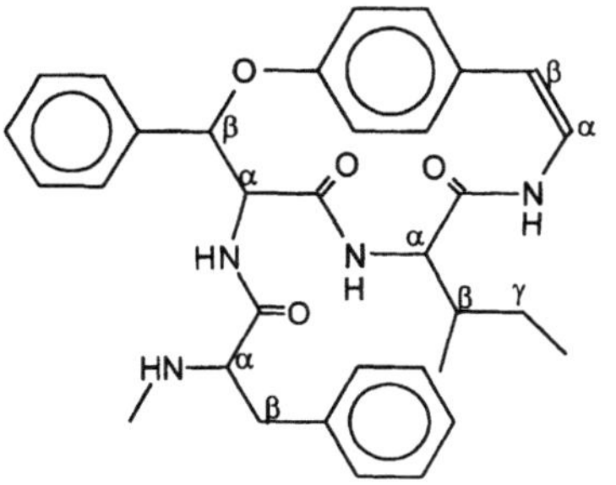

A = N-Me-L-Phe[1] (*120*)
B = *erythro*-β-OH-Phe[2]
C = L-Ileu[1] (*120*)
$C_{33}H_{38}N_4O_4$, 554

AM-2 may be Aralionine-B or its diastereoisomer. This relationship remains unresolved (*120*)
Mp = 257-258 (*120*)
UV (MeOH) : end absorption (*120*)
IR (KBr) : 3000, 1665, 1640, 1530, 1240, 1050, 700 (*120*)
MS : 554(M^+), 463(b), 378(j), 337(e), 300(y), 243(k-H^+), 224(f), 216(l), 215(c), 189(f-H^+), 187(d), 161, 148, 135(i), 134(a=100%), 132(m+H^+), 131(m), 120, 119(a-CH_3^+), 91, 86(q), 69 (*120*)
MS (CI/CH_4) : 555 (M+H^+=100%) (*120*)
^{1}H-NMR (400MHz, $CDCl_3$) : 0.70d, 7.6 Hz, 3H (CH-Me-Ileu), 0.84t, 7.8 Hz, 3H (CH_2-Me-Ileu), 0.98m, 1H (γ-H-Ileu), 1.26m, 1H (β-H-Ileu), 1.60dd, 15.3, 11.6 Hz, 1H (β-H-N-Me-Phe), 2.19m, 1H (γ-H-Ileu), 2.78dd, 15.3, 4.0 Hz, 1H (β-H-N-Me-Phe), 2.91dd, 11.6, 4.0 Hz, 1H (α-H-N-Me-Phe), 4.07dd, 9.0, 3.0 Hz, 1H (α-H-Ileu), 4.69dd, 9.6, 7.0 Hz, 1H (α-H-β-OH-Phe), 6.18d, 7.0 Hz, 1H (β-H-β-OH-Phe), 6.30d, 7.0 Hz, 1H-D_2O exchangeable (NH-Ileu), 6.41d, 8.3 Hz, 1H (β-H-Sty), 6.59d, 11.0 Hz, 1H-D_2O exchangeable (NH-Sty), 6.73dd, 11.0, 8.3 Hz, 1H (α-H-Sty), 6.97-7.53m, 14H (14xAr-H), 7.33d, 9.6 Hz, 1H-D_2O exchangeable (NH-β-OH-Phe) (*120*)
Sources : *Antidesma montana* (Euphorbiaceae)-leaves and terminal branches (*120*)

96. Crenatine-A

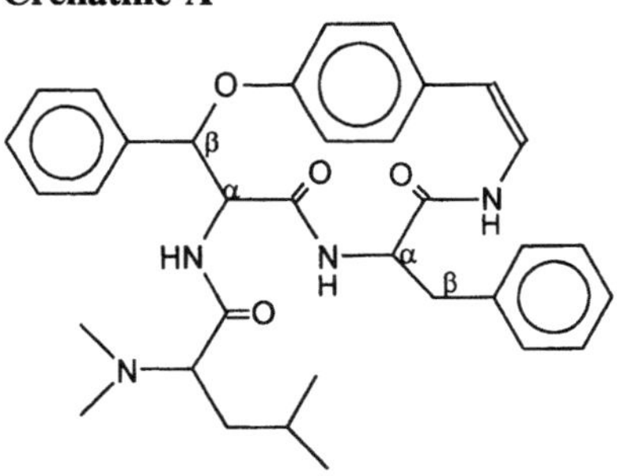

A = N,N-diMe-Leu
B = *erythro*-β-OH-Phe[2]
C = Phe
$C_{34}H_{40}N_4O_4$, 568.3022 (MS) (*125*)

Mp = 223 (*125*)
$[a]_D$ (20) = -292.58 (c=0.1, $CHCl_3$) (*125*)
IR ($CHCl_3$) : 3400, 3300, 1685, 1615, 1605, 1515 (*125*)
MS : 568(M^+), 553(M^+-CH_3), 511(b), 455(g), 412(j), 371(e), 308(h), 278(k), 250(l), 229(c), 224(f), 201(d), 135(i), 131(m), 120(q), 114(a=100%), 72(o) (*125*)
^{1}H-NMR (100MHz, $CDCl_3$) : 0.78d, 7.0 Hz, 3H (Me-N,N-diMe-Leu), 0.83d, 7.0 Hz, 3H (Me-N,N-diMe-Leu), 1.57s, 6H (N-Me_2), 2.60dd, 13.5, 9.0 Hz, 1H (β-H-Phe), 3.25dd, 13.5, 4.0 Hz, 1H (β-H-Phe), 4.50m, 9.0, 4.0 Hz, 1H (α-H-Phe), 4.50dd, 6.5 Hz, 1H (α-H-β-OH-Phe) (*125*)
Sources : *Discaria crenata* (Rhamnaceae)-leaves and stems (*97, 125*)

97. Discarine-D

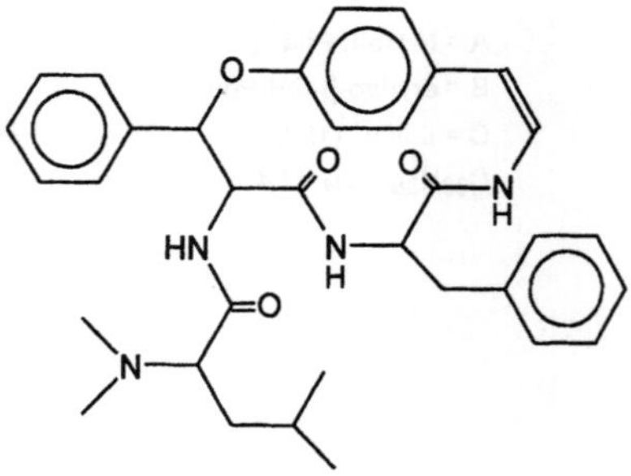

A = N,N-diMe-Leu
B = β-OH-Phe
C = Phe
$C_{34}H_{40}N_4O_4$, 568

Mp = 212 (*105*)
[a]D (20) = -148 (c=0.1, $CHCl_3$) (*105*)
UV : end absorption (*105*)
IR (KBr) : 3285, 2790, 1620, 1230 (*105*)
MS : 568(M^+), 135(i), 131(m), 120(q), 114(a=100%), 72(o) (*105*)
^{1}H-NMR ($CDCl_3$) : 0.77d, 6.0 Hz, 3H (Me-N,N-diMe-Leu), 0.89d, 6.0 Hz, 3H (Me-N,N-diMe-Leu), 1.56s, 6H (N-Me_2), 6.70m, 2H (2xolefinic-H), 7.00-7.30m, 14H, (14xAr-H) (*105*)
Sources : *Discaria febrifuga* (Rhamnaceae)-stem bark (*105*)

98. Deoxo-aralionine-A = Deoxy-aralionine-C = 5-Benzyl-8-*N*,*N*-dimethyl-isoleucyl-9-phenyl-phencyclopeptine

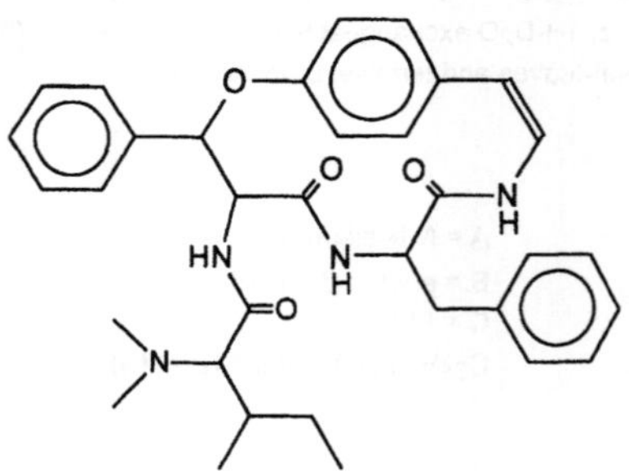

A = N,N-diMe-Ileu
B = β-OH-Phe
C = Phe
$C_{34}H_{40}N_4O_4$, 568

Mp = >350 (*45*)
MS : 568(M^+), 511(b), 412(j), 371(e), 308(h), 278(k), 250(l), 229(c), 224(f), 201(d), 135(i), 131(m), 120(q), 114(a=100%), 98, 91, 85(n), 72(o) (*45*)
^{1}H-NMR (270MHz, $CDCl_3$) : 0.18d, 6.9 Hz, 3H (CH-Me-N,N-diMe-Ileu), 0.80t, 6.9 Hz, 3H (CH_2-Me-N,N-diMe-Ileu) (*45*)
Sources : Rhamnaceae
Ceanothus integerrimus var. *californicus*-root bark (*45*)
C. integerrimus var. *integerrimus*-root bark (*45*)

99. Ceanothine-E

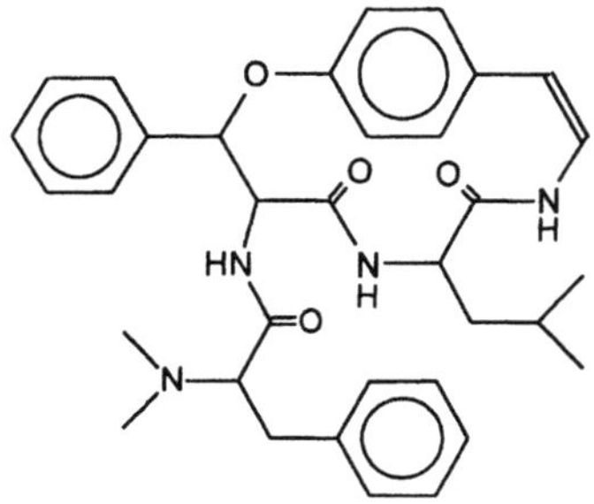

A = N,N-diMe-Phe
B = β-OH-Phe
C = Leu
$C_{34}H_{40}N_4O_4$, 568

Mp = 238-239 *(96)*
[a]D = -285 ($CHCl_3$) *(96)*
MS : 568(M^+), 553(M^+-CH_3), 477(b), 421(g), 378(j), 337(e), 274(h), 244(k), 229(c), 224(f), 216(l), 201(d), 148(a=100%), 135(i), 131(m), 91, 86(q) *(96)*
Sources : *Ceanothus americanus* (Rhamnaceae)-root bark *(96)*

100. Adouetine-Y

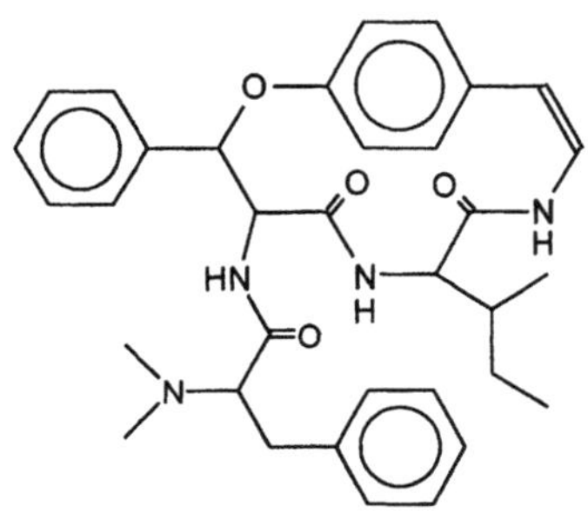

A = N,N-diMe-Phe
B = β-OH-Phe
C = Ileu
$C_{34}H_{40}N_4O_4$, 568.69 *(108)*

Adouetine-Y reported in *(9)* is a mixture of Adouetines-Y+Y' *(108)*
Mp = 272-274 *(9)*, 287-292 *(96, 108)*
[a]D (20) = -230 (c=1.0, $CHCl_3$/MeOH 9:1) *(108)*
[a]D *(9, 96)*
UV : 250 (3.83) *(108)*
IR (KBr) : 3280, 3250, 1680, 1635, 1530, 1240 *(108)*
IR (CCl_4) *(9)*
MS : 568(M^+), 477(b), 421(g), 378(j), 337(e), 274(h), 229(c), 224(f), 216(l), 201(d), 148(a), 135(i), 131(m), 86(q) *(96)*
MS *(108)*
^{1}H-NMR (60MHz, TFA) : 0.70-1.10, 6H (2xC-Me), 2.40d, 5.0 Hz, 3H (N-Me), 2.82d, 5.0 Hz, 3H (N-Me), 7.00-7.70m, 14H (14xAr-H) *(108)*
Sources : Rhamnaceae
Ceanothus americanus-root bark *(96)*
Sterculiaceae
Waltheria americana-whole plant *(9, 108)*

101. *N*-Desmethyl-integerrine = 5-β-Indolyl-8-*N*-methyl-valyl-9-phenyl-phencyclopeptine

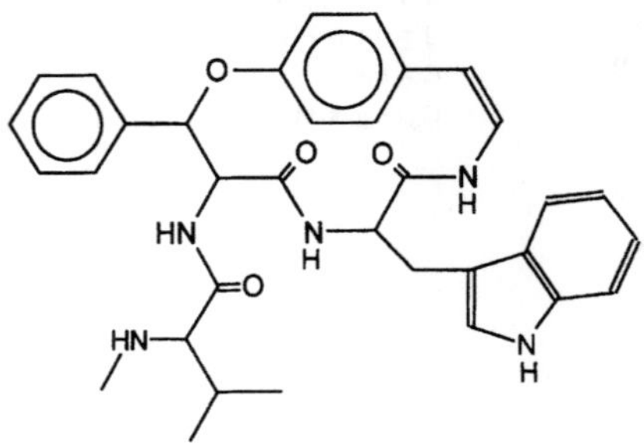

A = N-Me-Val
B = β-OH-Phe
C = Trp
$C_{34}H_{37}N_5O_4$, 579.2788 (MS) (*45*)

Mp = >350 (*45*)
MS : 579(M^+), 536(b), 494(g), 451(j), 410(e), 347(h), 317(k), 289(l), 224(f), 215(c), 187(d), 170(r), 159(q), 135(i), 131(m), 130, 117, 86(a=100%), 72(p), 58(o) (*45*)
^{1}H-NMR (270MHz, $CDCl_3$) : 0.27d, 6.9 Hz, 3H (Me-N-Me-Val), 0.57d, 6.9 Hz, 3H (Me-N-Me-Val) (*45*)
Sources : *Ceanothus integerrimus* var. *integerrimus* (Rhamnaceae)-root bark (*45*)

102. Aralionine-A

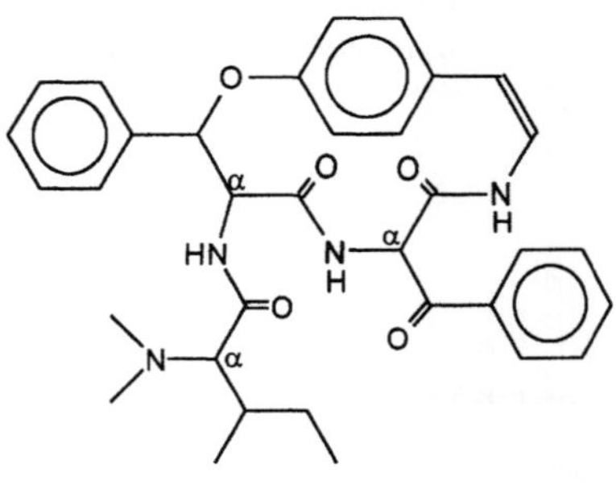

A = N,N-diMe-Ileu
B = *erythro*-β-OH-Phe[1] (*12*)
C = 2-benzoyl-Gly
$C_{34}H_{38}N_4O_5$, 582.2834 (MS) (*30*)

Mp = 165-167 (*30*)
$[a]_D$ (20) = +82 (c=0.2, MeOH) (*30*)
UV (MeOH) : 246 (4.28) (*30*)
CD (MeOH) : -0.38 (322), -2.98 (283), -2.50 (277), -4.25 (255), +9.25 (236), -7.50 (217) (*30*)
IR ($CHCl_3$) : 3385, 2780, 1685-1675, 1620, 1220 (*30*)
MS : 582(M^+), 525(b), 469(g), 467(g-$2H^+$), 426(j), 421(M^+-161), 385(e), 378(M-161-$C_3H_7^+$), 364(M-161-$C_4H_9^+$), 322(h), 229(c), 224(f), 201(d), 161, 135(i), 134(q), 131(m), 114(a=100%), 105 (*30*)
^{1}H-NMR (60MHz, $CDCl_3$) : 0.59d, 7.0 Hz, 3H (CH-Me-N,N-diMe-Ileu), 0.85m, 3H (CH_2-Me-N,N-diMe-Ileu), 2.40s, 6H (N-Me_2), 2.85d, 4.5 Hz, 1H (α-H-N,N-diMe-Ileu), 4.80d, 8.0 Hz, 1H (α-H-β-OH-Phe), 5.81d, 6.5 Hz, 1H (α-H-2-benzoyl-Gly), 6.80-7.20m, 12H (12xAr-H), 7.90-8.10m, 2H (2xAr-H-ortho-2-benzoyl-Gly) (*30*)
Derivatives : Tetrahydro-aralionine-A (*30*)
Desbenzoyl-aralionine-A (*30*)
Sources : *Araliothamnus vaginatus* (Rhamnaceae)-leaves (*30*), stem bark (*117*)

103. Aralionine-C

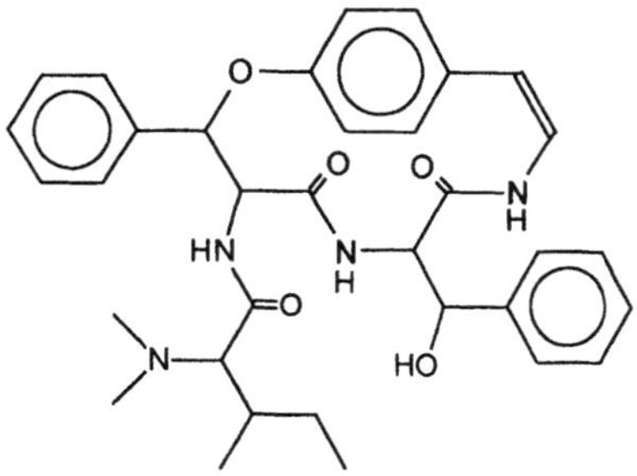

A = N,N-diMe-Ileu
B = *erythro*-β-OH-Phe[1,2] (*117*)
C = *threo*-β-OH-Phe[1] (*117*)
$C_{34}H_{40}N_4O_5$, 584.3011 (MS) (*117*)

Mp = 95-97 (*117*)
[a]D (20) = -17 (c=0.015, MeOH) (*117*)
UV (MeOH) : end absorption (*117*)
IR (KBr) : 3500, 3300, 2780, 1675, 1620, 1220, 1040 (*117*)
MS : 584(M^+), 527(b), 478(M^+-106), 229(c), 224(f), 201(d), 135(i), 131(m), 114(a=100%), 106, 105, 77 (*117*)
^{1}H-NMR ($CDCl_3$) : 0.41d, 7.0 Hz, 3H (CH-Me-N,N-diMe-Ileu), 0.81t, 7.0 Hz, 3H (CH_2-Me-N,N-diMe-Ileu), 2.15s, 6H (N-Me_2), 2.61d, 5.0 Hz, 1H (α-H-N,N-diMe-Ileu), [+D_2O : 4.61d, 8.5 Hz, 1H (α-H-β-aryloxy-Phe), 4.63d, 4.8 Hz, 1H (α-H-β-OH-Phe)], 5.10d, 4.8 Hz, 1H (β-H-β-OH-Phe), 6.00d, 8.5 Hz, 1H (β-H-β-aryloxy-Phe), [6.26-6.62m, 5H (2xolefinic-H + 2xNH + CH) : +D_2O : 6.38d, 8.0 Hz, 1H (α-H-Sty), 6.58d, 8.0 Hz, 1H (β-H-Sty)], 6.90-7.70m, 15H (14xAr-H + 1 NH) (*117*)
Derivatives : Dihydro-aralionine-C = Tetrahydro-aralionine-A (*117*)
Sources : *Araliothamnus vaginatus* (Rhamnaceae)-stem bark (*117*)

104. Integerrine

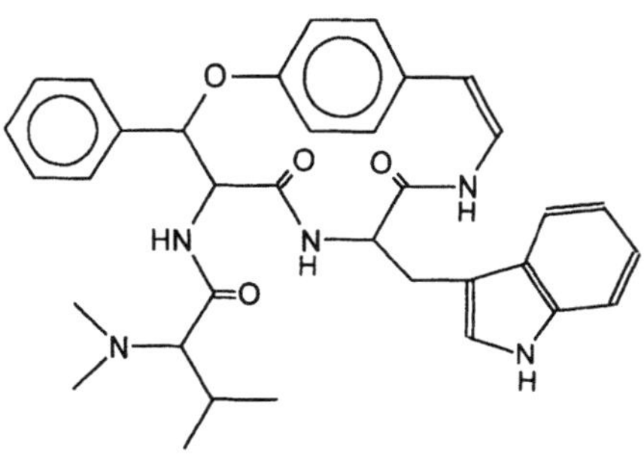

A = N,N-diMe-Val
B = β-OH-Phe
C = Trp
$C_{35}H_{39}N_5O_4$, 593.2924 (MS) (*45*)

Mp = 246 (*45*), 258 (*126*)
MS : 593(M^+), 578(M^+-CH_3), 550(b), 494(g), 451(j), 410(e), 347(h), 317(k), 304(v), 289(l), 229(c), 224(f), 201(d), 170(r), 159(q), 135(i), 131(m), 130, 103(m-CO), 100(a=100%), 85(a-Me^+/n) (*126*)
MS (*28*, *45*)
^{1}H-NMR (270MHz, $CDCl_3$) : 0.16d, 6.8 Hz, 3H (Me-N-Val), 0.70d, 6.8 Hz, 3H (Me-N-Val) (*45*)
Sources : <u>Rhamnaceae</u>
Ceanothus integerrimus-roots (*126*)
C. integerrimus var. integerrimus-root bark (*45*)

4(14)-Amphibine-F-Type Cyclopeptide Alkaloids

105. Nummularine-F

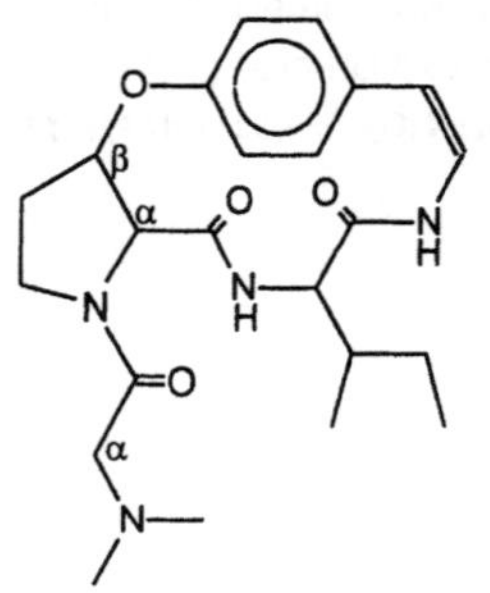

A = N,N-diMe-Gly
B = β-OH-Pro
C = Ileu
$C_{23}H_{32}N_4O_4$, 428.2429 (MS) (*43*)

Mp = 120 (*43*)
[a]$_D$ (20) = -204 (c=0.2, MeOH) (*43*)
UV (MeOH) : end absorption (*43*)
IR ($CHCl_3$) : 3380, 2790, 1675, 1620, 1225, 1030 (*43*)
MS : 428(M^+), 413(M^+-CH_3), 371(f), 356(w'-CO), 344(h), 343(i), 342(j), 274(u), 229(r), 209(p), 203(s), 186(t), 181(q), 155(k), 135(i'), 96(v), 86(q'), 68(w), 58(a=100%) (*43*)
^{1}H-NMR (60MHz, $CDCl_3$) : 0.80-0.95 complex pattern, 6H (2xC-Me), 2.21s, 6H (N-Me_2), 3.05s, 2H (α-H-N,N-diMe-Gly), 4.21d, 6.0 Hz, 1H (α-H-β-OH-Pro), 5.45m, 1H (β-H-β-OH-Pro), 6.40-7.40m, 7H (4xAr-H + 2xNH + 1 olefinic-H) (*43*)
Derivatives : Dihydro-nummularine-F (*43*)
Sources : *Zizyphus nummularia* (Rhamnaceae)-root bark (*43*), stem bark (*66*)

106. Spinanine-A

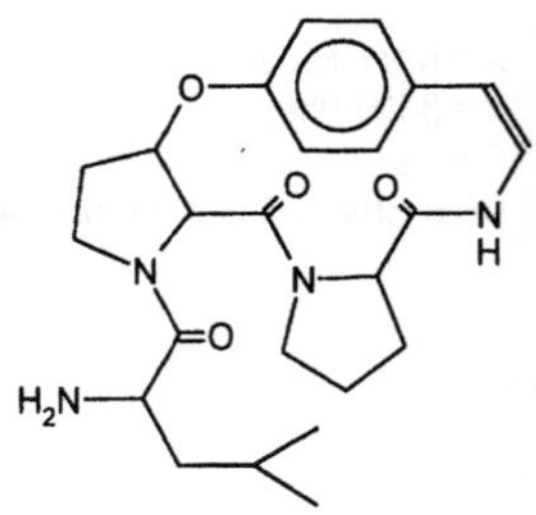

A = Leu
B = β-OH-Pro
C = Pro
$C_{24}H_{32}N_4O_4$, 440.2418 (MS) (*88*)

Mp = 175-176 (*88*)
[a]$_D$ = -121 (c=0.1, MeOH) (*88*)
UV (MeOH) : strong end absorption : 250sh., 280sh. (*88*)
IR (KBr) : 3360, 3010-2840, 1680, 1650, 1620, 1220, 1050 (*88*)
MS : 440(M^+), 383(b), 355(f), 354(g), 328(h), 327(i), 326(j), 258(u), 229(r), 203(s), 193(p), 186(t), 183(k), 165(q), 135(i'), 134(i'-H^+), 86(a=100%), 70(q'), 68(w) (*88*)
Sources : *Zizyphus spina-christi* (Rhamnaceae)-stem bark (*88*)

107. Zizyphine-G

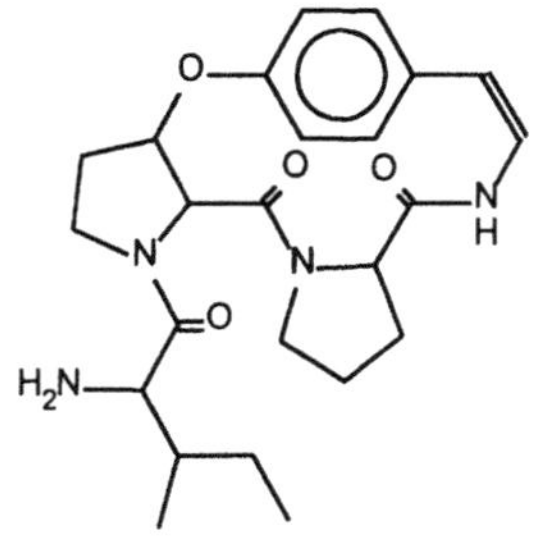

A = Ileu
B = *trans*-β-OH-Pro[1] (*40*)
C = Pro
$C_{24}H_{32}N_4O_4$, 440

Mp = 130 (*40*)
[a]D (20) = -185 (c=0.19, MeOH) (*40*)
MS : 440(M^+), 411(M^+-C_2H_5), 383(b), 355(f), 354(g), 328(h), 327(i), 326(j), 258(u), 229(r), 203(s), 193(p), 186(t), 183(k), 165(q), 135(i'), 96(v), 86(a=100%), 68(w) (*40*)
^{1}H-NMR ($CDCl_3$) : 0.96, 2xd, 6H (2xC-Me), 6.10-6.70m, 6H (4xAr-H + 2xolefinic-H) (*40*)
Sources : *Zizyphus oenoplia* (Rhamnaceae)-stem bark (*40*)

108. Mauritine-C

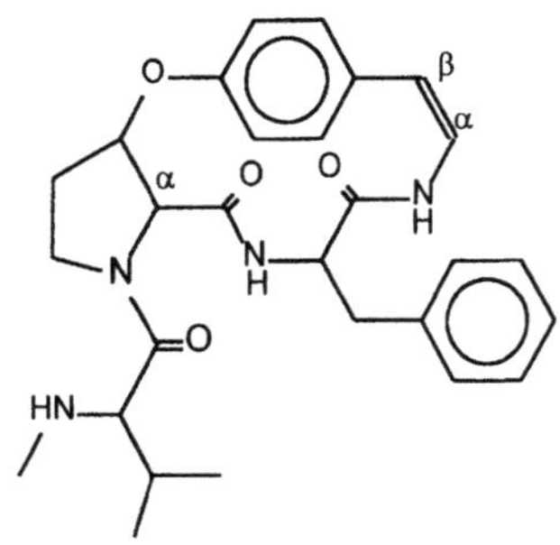

A = N-Me-Val
B = *trans*-β-OH-Pro[1] (*42*)
C = Phe
$C_{28}H_{34}N_4O_4$, 490.2575 (MS) (*42*)

Mp = Powder (*42*)
[a]D (20) = -224 (c=0.11, MeOH) (*42*)
UV (MeOH) : 252sh., 280sh. (*42*)
IR ($CHCl_3$) : 3360, 2982-2850, 2774, 1682, 1616, 1590, 1490, 1228, 1022 (*42*)
MS : 490(M^+), 447(b'), 405(f), 404(g), 378(h), 377(i), 376(j), 308(u), 243(p), 229(r), 215(q), 203(s), 186(t), 183(k), 135(i'), 120(q'), 96(v), 86(a=100%), 68(w) (*42*)
^{1}H-NMR (60 and 90MHz, $CDCl_3$) : 0.85, 2xd, 6H (2xC-Me-N-Me-Val), 2.35s, 3H (NH-Me), 4.25d, 5.5 Hz, 1H (α-H-β-OH-Pro), 6.28d, 8.0 Hz, 1H (β-H-Sty), 6.70q, 1H (α-H-Sty), 6.30-7.40m, 11H (9xAr-H + 2xNH) (*42*)
Derivatives : Dihydro-mauritine-C (*42*)
N-Methyl-dihydro-mauritine-C (*42*)
N-Formyl-mauritine-C (*52*)
Sources : Rhamnaceae
Zizyphus mauritiana-stem bark (*42, 117*)
Z. nummularia-root bark (*94*)
Z. spina-christi-stem bark (*106, 123*)

109. Amphibine-F

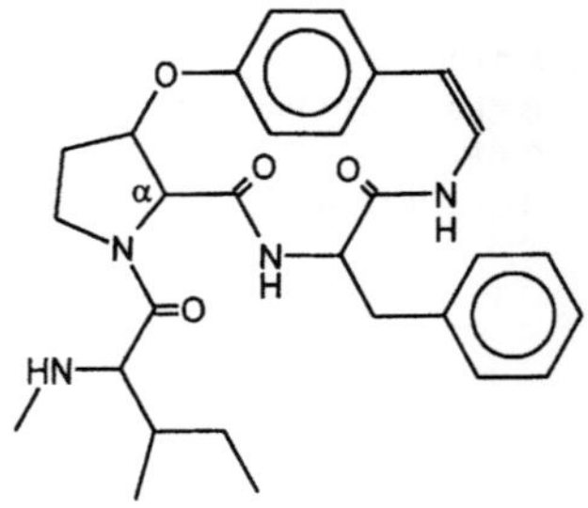

A = N-Me-Ileu
B = *trans*-β-OH-Pro[1] (*41*)
C = Phe
$C_{29}H_{36}N_4O_4$, 504.2701 (MS) (*41*)

Mp = amorphous powder (*41*)
$[a]_D$ (20) = -171 (c=0.26, $CHCl_3$) (*41*)
UV (MeOH) : end absorption : 250sh., 280sh. (*41*)
IR ($CHCl_3$) : 3380, 2780, 1686, 1620, 1220, 1030 (*41*)
MS : 504(M^+), 447(b), 405(f), 404(g), 378(h), 377(i), 376(j), 360(b'-R), 343(M^+-161), 308(u), 282(b'-$C_3H_7^+$-R_1), 243(p), 229(r), 225, 215(q), 197(k), 186(t), 165(o), 135(i'), 134(i'-H^+),120(q'), 100(a=100%), 91, 71(n'), 69(l-H^+), 68(w), 56 (*41*)
^{1}H-NMR (60 and 90MHz, $CDCl_3$) : 0.60-1.00m, 6H (2xC-Me), 2.50d, 3H (NH-Me), 4.31d, 6.0 Hz, 1H (α-H-β-OH-Pro), 6.20-6.80m, 2H (2xolefinic-H), 6.90-7.50m, 11H (9xAr-H + 2xNH) (*41*)
Derivatives : Dihydro-amphibine-F (*41*)
N-Methyl-dihydro-amphibine-F (*41*)
Sources : Rhamnaceae
Zizyphus amphibia-stem bark (*41*, *89*)
Z. mauritiana-stem bark (*42*)
Z. spina-christi-stem bark (*123*)

110. Lotusine-D (= *N*-Desmethyl-lotusine-A)

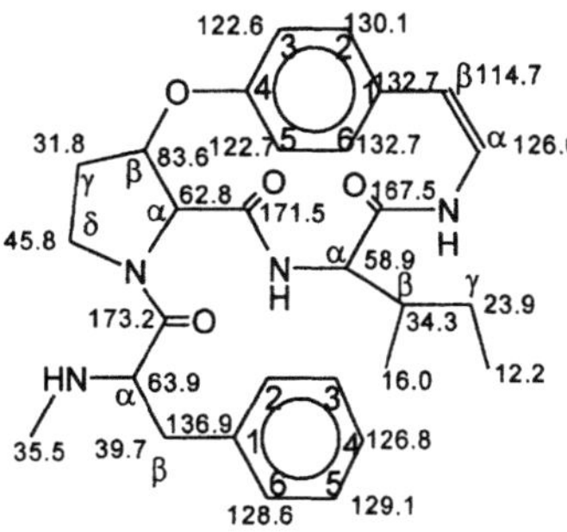

A = N-Me-Phe
B = β-OH-Pro
C = Ileu
$C_{29}H_{36}N_4O_4$, 504

$[a]_D$ = -187 (c=0.5, $CHCl_3$) *(58)*
UV (MeOH) : 211 *(58)*
IR ($CHCl_3$) : 3396, 2931, 1688, 1626, 1227, 1037 *(58)*
MS : 504(M^+), 475(M^+-C_2H_5), 448(w'-CH_3^+), 414(b+H^+), 413(b), 385(b-CO), 382(b'), 371(f), 354(b'-CO), 300(y'), 272(y'-CO), 256(w'-2xCO-161), 229(r), 209(p), 201(r-CO), 186(t), 185(t-H^+), 174, 165(o), 139(n), 135(i'), 134(a=100%), 119(a-CH_3^+), 105, 97(r'), 96(v), 91, 86(q'), 84, 77, 69(I-H^+), 68(w), 65, 57($C_4H_9^+$) *(58)*
^{1}H-NMR (300MHz, $CDCl_3$) : 0.86d, 7.0 Hz, 3H (CH-Me-Ileu), 0.90t, 7.3 Hz, 3H (CH_2-Me-Ileu), 1.15m, 1H (γ-H-Ileu), 1.31m, 1H (γ-H-Ileu), 2.05m, 1H (γ-H-β-OH-Pro), 2.20m, 1H (β-H-Ileu), 2.34br.s, 3H (NH-Me), 2.44ddd, 12.2, 7.1, 5.2 Hz, 1H (γ-H-β-OH-Pro), 2.59m, 1H (δ-H-β-OH-Pro), 2.67dd, 13.4, 7.0 Hz, 1H (β-H-N-Me-Phe), 2.90dd, 13.4, 7.0 Hz, 1H (β-H-N-Me-Phe), 3.56t, 7.0 Hz, 1H (α-H-N-Me-Phe), 3.75dd, 11.3, 8.2 Hz, 1H (δ-H-β-OH-Pro), 4.21dd, 8.5, 3.2 Hz, 1H (α-H-Ileu), 4.32d, 5.3 Hz, 1H (α-H-β-OH-Pro), 5.44ddd, 9.8, 7.1, 5.3 Hz, 1H (β-H-β-OH-Pro), 6.34d, 7.7 Hz, 1H (β-H-Sty), 6.57d, 10.5 Hz, 1H (NH-Sty), 6.75d, 8.5 Hz, 1H (NH-Ileu), 6.75dd, 10.5, 7.7 Hz, 1H (α-H-Sty), 7.06m, 2H (3-H- + 5-H-N-Me-Phe), 7.07m, 1H (6-H-Sty), 7.11m, 1H (2-H-Sty), 7.14m, 1H (3-H-Sty), 7.23m, 1H (4-H-N-Me-Phe), 7.25m, 2H (2-H- + 6-H-N-Me-Phe), 7.27m, 1H (5-H-Sty) *(58)*
^{13}C-NMR (75MHz, $CDCl_3$) : see figure *(58)*
COSY, HMBC and HMQC-NMR spectra *(58)*
Sources : *Zizyphus lotus* (Rhamnaceae)-root bark *(58)*

111. Lotusine-A

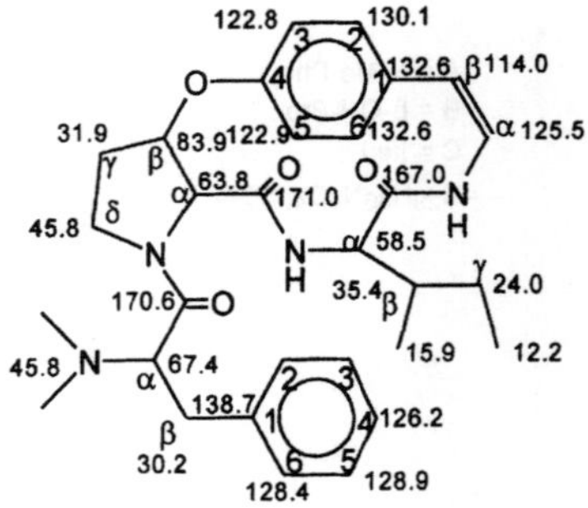

A = N,N-diMe-Phe
B = β-OH-Pro
C = Ileu
$C_{30}H_{38}N_4O_4$, 518

$[a]_D$ = -215 (c=1.0, $CHCl_3$) (*58*)
UV (MeOH) : 208, 250sh. (*58*)
IR ($CHCl_3$) : 3390, 2950, 2790, 1640, 1615, 1220, 1040 (*58*)
MS : 518(M^+), 517($M-H^+$), 475($M^+-C_3H_7$), 427(b), 322, 279, 274(u), 229(r), 209(p), 203(s), 190(u-CO-R_1), 186(t), 181(q), 167(k-77), 153(n), 149($a+H^+$), 148(a=100%), 135(i'), 134($i'-H^+$), 133($a-CH_3^+$), 105, 96(v), 91, 86(q'), 83($l-H^+$), 72($q'-Me+H^+$), 69, 68(w), 57($C_4H_9^+$) (*58*)
^{1}H-NMR (300MHz, $CDCl_3$) : 0.62d, 7.0 Hz, 3H (CH-Me-Ileu), 0.88t, 7.4 Hz, 3H (CH_2-Me-Ileu), 1.13m, 1H (γ-H-Ileu), 1.28m, 1H (γ-H-Ileu), 2.00-2.13m, 2H (β-H-Ileu + γ-H-β-OH-Pro), 2.39br.s, 6H (N-Me_2), 2.48ddd, 12.2, 7.2, 5.1 Hz, 1H (γ-H-β-OH-Pro), 2.90dd, 13.8, 4.0 Hz, 1H (β-H-N,N-diMe-Phe), 2.99-3.07m, 2H (β-H-N,N-diMe-Phe + δ-H-β-OH-Pro), 3.51dd, 9.1, 4.0 Hz, 1H (α-H-N,N-diMe-Phe), 4.14dd, 8.6, 3.0 Hz, 1H (α-H-Ileu), 4.21dd, 10.8, 8.0 Hz, 1H (δ-H-β-OH-Pro), 4.26d, 5.3 Hz, 1H (α-H-β-OH-Pro), 5.52ddd, 9.8, 7.2, 5.3 Hz, 1H (β-H-β-OH-Pro), 6.29d, 7.8 Hz, 1H (β-H-Sty), 6.51d, 10.6 Hz, 1H (NH-Sty), 6.70d, 8.6 Hz, 1H (NH-Ileu), 6.75dd, 10.6, 7.8 Hz, 1H (α-H-Sty), 7.07m, 2H (3-H- + 5-H-N,N-diMe-Phe), 7.08m, 1H (6-H-Sty), 7.10m, 1H (3-H-Sty), 7.11m, 1H (2-H-Sty), 7.15m, 1H (4-H-N,N-diMe-Phe), 7.19m, 2H (2-H- + 6-H-N,N-diMe-Phe), 7.28m, 1H (5-H-Sty) (*58*)
^{13}C-NMR (75MHz, $CDCl_3$) : see figure (*58*)
COSY, HMBC and HMQC-NMR spectra (*58*)
Sources : *Zizyphus lotus* (Rhamnaceae)-root bark (*58*)

112. Amphibine-G

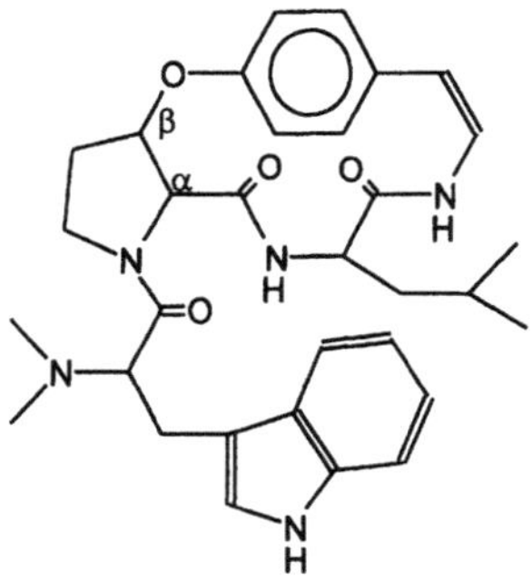

A = N,N-diMe-Trp
B = *trans*-β-OH-Pro[1] (*41*)
C = Leu
$C_{32}H_{39}N_5O_4$, 557.3031 (MS) (*41*)

Mp = 182-185 (*41*)
[a]$_D$ (20) = -218 (c=0.24, $CHCl_3$) (*41*)
UV (MeOH) : 273 (3.80), 282 (3.79), 290 (3.71) (*41*)
CD (EtOH) : +1.20 (290), +1.50 (280), +1.80 (265), -12.50 (237), +5.30 (223) (*41*)
IR ($CHCl_3$) : 3450, 3380, 2780, 1680, 1620, 1225, 1030 (*41*)
MS : 557(M^+), 514(M^+-C_3H_7), 427(b), 371(f), 344(h), 343(i), 342(j), 294(w'-161-R_1), 284(k), 274(u), 229(r), 209(p), 203(s), 188(a+H^+), 187(a=100%), 186(t), 181(q), 170, 135(i'), 134(i'-H^+), 130, 96(v), 91, 86(q'), 84(l), 68(w) (*41*)
^{1}H-NMR (60 and 90MHz, $CDCl_3$) : 0.50-1.30m, 6H (2xC-Me), 2.50s, 6H (N-Me_2), 4.30d 6.0 Hz, 1H (α-H-β-OH-Pro), 5.40m, 1H (β-H-β-OH-Pro), 6.30-6.90m, 4H (2xolefinic-H + 2xNH), 7.00-7.70m, 10H (9xAr-H + NH-Trp) (*41*)
Derivatives : Dihydro-amphibine-G (*41*)
Sources : *Zizyphus amphibia* (Rhamnaceae)-stem bark (*41*, *89*)

4(14)-Pandamine-Type Cyclopeptide Alkaloids

113. Discarine-H

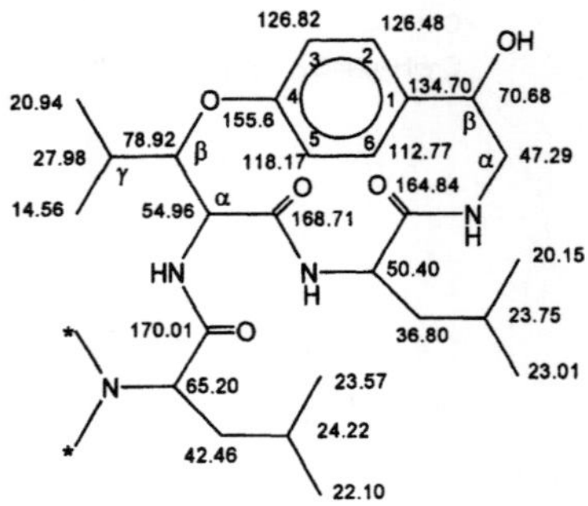

A = N,N-diMe-Leu
B = β-OH-Leu
C = Leu
$C_{28}H_{46}N_4O_5$, 518.69 (*127*)

Mp : 232 (*127*)
[a]D (20) = -266 (MeOH) (*127*)
UV (MeOH) : 230 (3.78), 276 (3.09) (*127*)
IR (KBr) : 3500-3300, 3070, 2970-2880, 2880, 1630, 1260, 1080 (*127*)
MS : 518(M^+), 461(b), 274(h), 190(f), 114(a), 97(m), 72(o) (*127*)
^{1}H-NMR (400MHz, DMSO-d_6) : 0.63d, 6.2 Hz, 3H (Me-Leu), 0.64d, 6.2 Hz, 3H (Me-Leu), 0.72d, 6.4 Hz, 3H (Me-N,N-diMe-Leu), 0.75d, 6.4 Hz, 3H (Me-N,N-diMe-Leu), 0.86d, 6.8 Hz, 3H (Me-β-OH-Leu), 0.99d, 6.8 Hz, 3H (Me-β-OH-Leu), 2.75s, 6H (N-Me_2), 3.78d, 4.0 Hz, 1H (α-H-Phe-Et), 4.58dd, 10.0, 8.0 Hz, 1H (α-H-β-OH-Leu), 4.70dd, 8.0, 2.0 Hz, 1H (β-H-β-OH-Leu), 4.91d, 4.0 Hz, 1H (β-H-Phe-Et), 5.30s, 1H (OH), 6.60-7.30m, 4H (4xAr-H) (*127*)
^{13}C-NMR (100MHz, DMSO-d_6) : see figure (*127*)
* Signals obscured by the solvent (DMSO-d_6).
Derivatives : *O*-Acetyl-discarine-H (*127*)
Sources : *Discaria febrifuga* (Rhamnaceae)-root bark (*127*)

114. Discarine-L

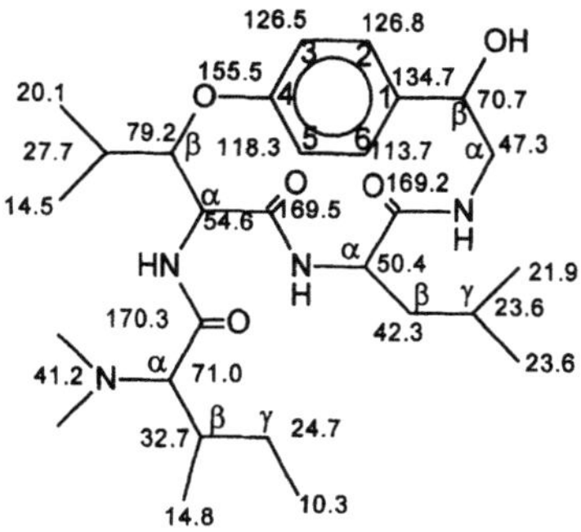

A = N,N-diMe-Ileu
B = β-OH-Leu
C = Leu
$C_{28}H_{46}N_4O_5$, 518

Mp : amorphous powder (*128*)
[a]D = -30 (c=0.5, MeOH) (*128*)
IR (KBr) : 3400-3200, 3040, 2980-2880, 2800, 1680-1630, 1230 (*128*)
MS : 518(M^+), 461(b), 387(g), 344(j), 303(e), 274(h), 210(k), 195(c), 190(f), 182(l), 135(i), 114(a=100%), 97(m), 86(p/q) (*128*)
^{1}H-NMR (400MHz, DMSO-d_6) : 0.60d, 7.0 Hz, 3H (CH-Me-N,N-diMe-Ileu), 0.68d, 7.0 Hz, 3H (Me-Leu), 0.71d, 7.0 Hz, 3H (Me-Leu), 0.78t, 7.0 Hz, 3H (CH_2-Me-N,N-diMe-Ileu), 0.89d, 7.0 Hz, 3H (Me-β-OH-Leu), 1.02d, 7.0 Hz, 3H (Me-β-OH-Leu), 1.05m, 1H (γ-H-N,N-diMe-Ileu), 1.24m, 1H (γ-H-Leu), 1.25m, 1H (β-H-Leu), 1.52m, 1H (γ-H-N,N-diMe-Ileu), 1.71m, 2H (β-H-Leu + β-H-N,N-diMe-Ileu), 2.20m, 1H (γ-H-β-OH-Leu), 2.68d, 1H (α-H-N,N-diMe-Ileu), 2.30m, 1H (α-H-Phe-Et), 4.02m, 1H (α-H-Phe-Et), 4.20m, 1H (α-H-Leu), 4.35dd, 10.0, 9.0 Hz, 1H (α-H-β-OH-Leu), 4.69dd, 9.0, 2.0, 1H (β-H-β-OH-Leu), 4.90d, 4.0 Hz, 1H (β-H-Phe-Et), 5.30br., 1H (OH), 6.72d, 9.0 Hz, 1H (NH-Leu), 6.78dd, 8.0, 2.0 Hz, 1H (2-H-Phe-Et), 6.85, 2xdd, 8.0, 2.0 Hz, 2H (3-H- + 6-H-Phe-Et), 7.24dd, 8.0, 2.0 Hz, 1H (5-H-Phe-Et), 7.54d, 10.0 Hz, 1H (NH-Phe-Et), 8.17d, 10.0 Hz, 1H (NH-β-OH-Leu) (*128*)
^{13}C-NMR (75MHz, DMSO-d_6) : see figure (*128*)
COSY, DEPT and Spin-echo NMR spectra (*128*)
Sources : *Discaria febrifuga* (Rhamnaceae)-root bark (*128*)

115. Pandaminine

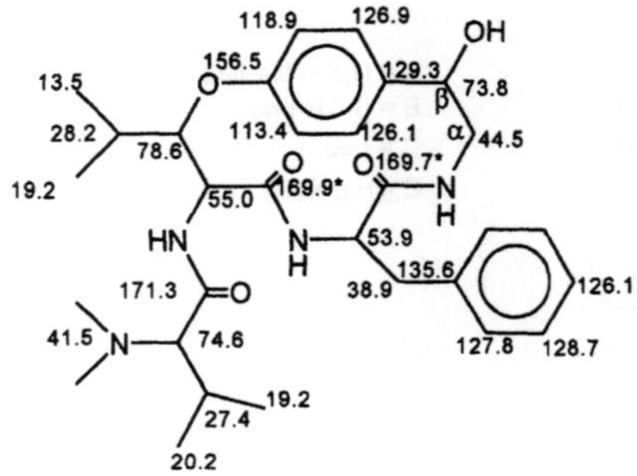

A = N,N-diMe-L-Val[1] (*11*)
B = L-*erythro*-β-OH-Leu[1,2] (*11, 16*)
C = L-Phe[1] (*11*)
$C_{30}H_{42}N_4O_5$, 538

Mp : 272 (*11*)
$[a]_D$ = -117 (c=0.5, $CHCl_3$) (*11*)
IR (KBr) : 3300, 3050, 2880, 2830, 2780, 1650, 1540, 1250, 1090 (*11*)
^{1}H-NMR (60MHz, TFA) : 0.80-1.50, 12H (4xC-Me), 3.09s, 3H (N-Me), 3.17s, 3H (N-Me), 5.40d, 1H (OH), 6.70-7.40m, 9H (9xAr-H) (*11*)
^{13}C-NMR (15MHz, $CDCl_3/CD_3OD$ 2:1 v/v) : see figure (*16*)
Derivatives : *O*-Acetyl-pandaminine (*11*)
Sources : *Panda oleosa* (Pandaceae)-root bark (*11, 24*)

116. Pandamine

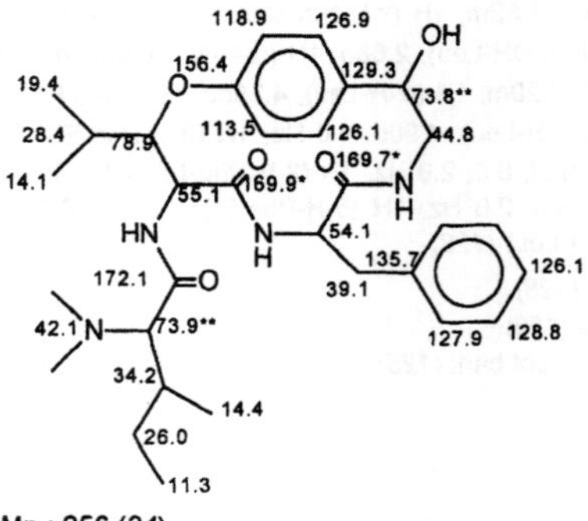

A = N,N-diMe-L-Ileu[1] (*11*)
B = L-*erythro*-β-OH-Leu[1,2] (*11, 16*)
C = L-Phe[1] (*11*)
$C_{31}H_{44}N_4O_5$, 552.69 [*11, 24*)

Mp : 256 (*24*)
$[a]_D$ = -103 (c=0.5, $CHCl_3$) (*24*)
UV (MeOH) : 230 (3.59), 277 (1.80) (*11*)
UV (MeOH) (*24*)
IR (KBr) : 3300, 3050, 2880, 2830, 2780, 1650, 1540, 1250, 1090 (*11*)
IR (KBr) (*24*)
MS : 552(M^+), 550($M-2H^+$), 509($M^+-C_3H_7$), 495(b), 491($M^+-C_3H_7-H_2O$), 477(b''), 120(q), 114(a) (*11*)
^{1}H-NMR (60MHz, TFA) : 0.80-1.30, 12H (4xC-Me), 3.05s, 3H (N-Me), 3.13s, 3H (N-Me), 5.33d, 1H (OH), 6.60-7.30m, 9H (9xAr-H) (*11, 24*)
^{13}C-NMR (15MHz, $CDCl_3/CD_3OD$ 2:1 v/v) : see figure (*16*)
* and ** Values may be interchanged
Derivatives : *O*-Acetyl-pandamine (*11, 24*)
Pandaminone=Oxo-pandamine (*11, 24*)
Desoxo-pandamine (*11*)
Chloro-pandamine (*11*)
Sources : *Panda oleosa* (Pandaceae)-root bark (*11, 24*)

117. Sanjoinine-G1

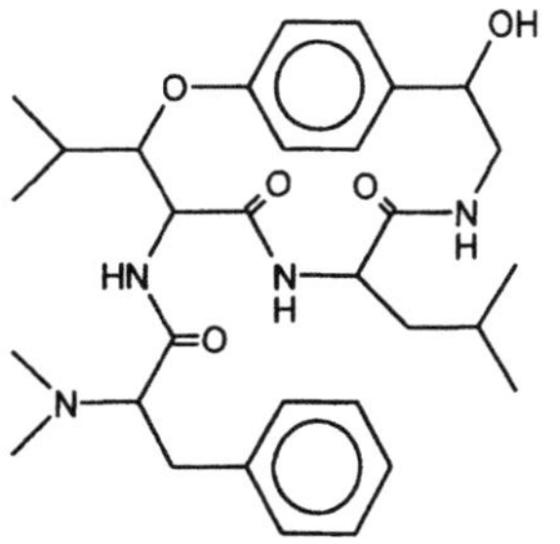

A = N,N-diMe-Phe
B = β-OH-Leu
C = Leu
$C_{31}H_{44}N_4O_5$, 552

Mp : 236-238 (*70*)
[a]D = -68.6 (*70*)
Sources : *Zizyphus vulgaris* var. *spinosus* (Rhamnaceae)-seeds (*70*)

118. Discarine-G

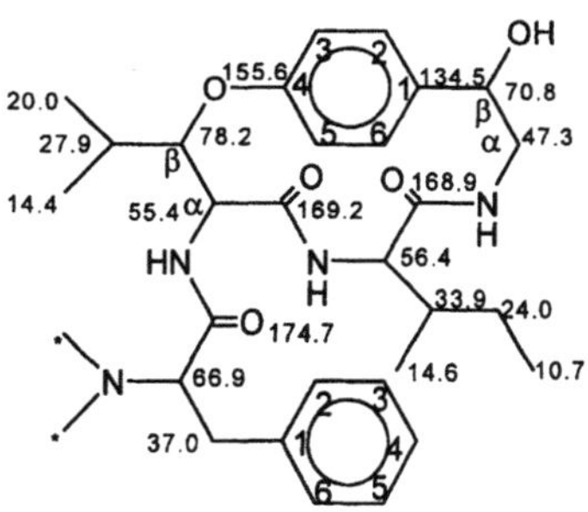

A = N,N-diMe-Phe
B = β-OH-Leu
C = Ileu
$C_{31}H_{44}N_4O_5$, 552

Mp : 257 (*129*)
[a]D (20) = -366 (c=1.0, MeOH) (*129*)
UV : 230 (3.78), 274 (3.14) (*129*)
IR : 3500-3300, 3060, 2960-2880, 2790, 1630, 1260, 1080 (*129*)
MS : 552(M^+), 537(M^+-CH_3), 509(M^+-C_3H_7), 495(M^+-C_4H_9), 462(b+H^+), 461(b), 387(g), 344(j), 303(e), 274(h), 210(k), 195(c), 190(f), 182(l), 167(d), 148(a=100%), 135(i), 105, 97(m), 91 (*129*)
^{1}H-NMR (400MHz, CD_3OD) : 0.47d, 6.6 Hz, 3H (Me-β-OH-Leu), 0.58t, 6.6 Hz, 3H (CH_2-Me-Ileu), 0.72d, 6.6 Hz, 3H (CH-Me-Ileu), 0.88d, 6.6 Hz, 3H (Me-β-OH-Leu), 2.25s, 6H (N-Me_2), 3.33m, 2H (α-H-Phe-Et), 4.12d, 8.9 Hz, 1H (α-H-β-OH-Leu), 4.54dd, 8.9, 2.0 Hz, 1H (β-H-β-OH-Leu), 4.84d, 3.1 Hz, 1H (β-H-Phe-Et), 5.26s, 1H (OH), 6.60-7.10m, 9H (9xAr-H) (*129*)
^{13}C-NMR (100MHz, DMSO-d_6) : see figure (*129*)
C-2,3,5,6-Phe-Et and C-1,2,3,4,5,6-N,N-diMe-Phe (10xC) : 117.9-128.7
* Signals obscured by the solvent (DMSO-d_6).
Derivatives : *O*-Acetyl-discarine-G (*129*)
Sources : *Discaria febrifuga* (Rhamnaceae)-root bark (*129*)

119. Sanjoinine-D (= *O*-Methyl-sanjoinine-G1)

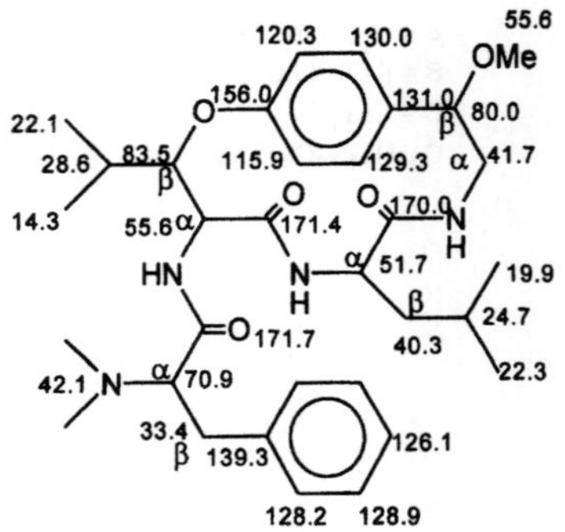

A = N,N-diMe-Phe
B = β-OH-Leu
C = Leu
$C_{32}H_{46}N_4O_5$, 566.3437 (MS) (*63*)

Mp : 256-258 (*63, 70*)
[a]D (26) = -53.6 (c=0.25, $CHCl_3$) (*63, 70*)
UV (MeOH) : 231 (3.82), 279.5 (2.82) (*63*)
IR (KBr) : 1230 (*63*)
MS : 566(M^+), 551(M^+-CH_3), 534(M^+-MeOH), 475(b), 443(b"), 222(f'), 210(k), 195(c), 190(f), 189(f-H^+), 182(l), 167(i'), 148(a=100%), 135(i), 97(m), 86(q) (*63*)
^{1}H-NMR (80MHz, $CDCl_3$) : 0.85d, 6.2 Hz, 6H (2xMe-Leu), 1.05d, 7.4 Hz, 3H (Me-β-OH-Leu), 1.22d, 4.9 Hz, 3H (Me-β-OH-Leu), 2.48s, 6H (N-Me_2), 3.35s, 3H (OMe), 3.13-3.48m, 3H (α-H-+2xβ-H-N,N-diMe-Phe), 3.95m, 1H (α-H-Leu), 4.30dd, 9.0, 1.6 Hz, 1H (β-H-β-OH-Leu), 5.30m, 2H (2xα-H-Phe-Et), 5.51d, 10.0 Hz, 1H (β-H-Phe-Et), 5.93d, 9.3 Hz, 1H (NH), 6.73-7.32m, 10H (3-H- + 6-H-Phe-Et), 6.93s, 1H (2-H-Trp), 6.97m, 1H (5-H-Trp), 7.06m, 1H (9xAr-H + NH), 7.76d, 10.0 Hz, 1H (NH) (*63*)
^{13}C-NMR (20MHz, $CDCl_3$+CD_3OD) : see figure (*63*)
Sources : *Zizyphus vulgaris* var. *spinosus* (Rhamnaceae)-seeds (*63, 70*)

120. Discarine-K

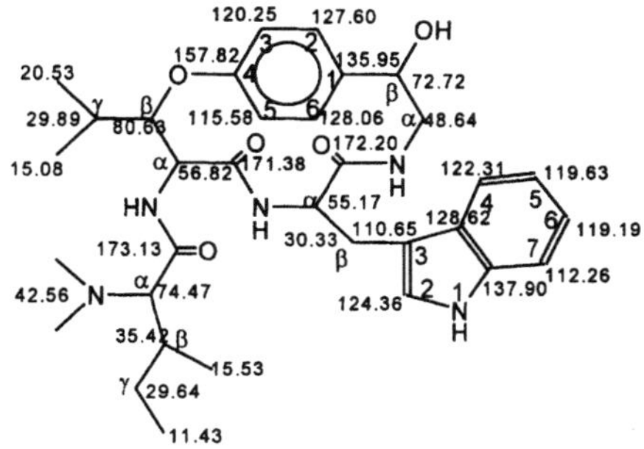

A = N,N-diMe-Ileu
B = β-OH-Leu
C = Trp
$C_{33}H_{45}N_5O_5$, 591

Mp : 237 (*55*)
[a]D (20) = -62 (MeOH) (*55*)
UV (MeOH) : 230 (3.93), 273 (3.63), 280 (3.65), 289 (3.55) (*55*)
IR (KBr) : 3400-3200, 1660-1630, 1230, 1080 (*55*)
MS : 591(M^+), 534(b), 516(b''), 462($M-R_2+H^+$), 460(g), 448($M-R_2-Me+2H^+$), 421(M^+-r), 417(j), 347(h), 283(k), 255(l), 247(t), 190(f), 186(s), 170(r), 159(q), 135(i), 130, 114(a=100%), 97(m), 86(p) (*55*)
^{1}H-NMR (400MHz, CD_3OD) : 0.78d, 6.9 Hz, 3H (CH-Me-N,N-diMe-Ileu), 0.89t, 7.2 Hz, 3H (CH_2-Me-N,N-diMe-Ileu), 1.03d, 6.7 Hz, 3H (Me-β-OH-Leu), 1.11m, 1H (γ-H-N,N-diMe-Ileu), 1.15d, 6.7 Hz, 3H (Me-β-OH-Leu), 1.60m, 1H (γ-H-N,N-diMe-Ileu), 1.84m, 1H (β-H-N,N-diMe-Ileu), 2.22m, 1H (γ-H-β-OH-Leu), 2.26s, 6H (N-Me_2), 2.62m, 1H (β-H-Trp), 2.69m, 2H (α-H-N,N-diMe-Ileu + α-H-Phe-Et), 2.94dd, 13.4, 10.0 Hz, 1H (β-H-Trp), 4.03dd, 12.1, 3.3 Hz, 1H (α-H-Phe-Et), 4.28dd, 10.1, 5.4 Hz, 1H (α-H-Trp), 4.49d, 8.7 Hz, 1H (α-H-β-OH-Leu), 4.86dd, 1H (β-H-β-OH-Leu), 4.88dd, 1H (β-H-Phe-Et), 6.78dd, 8.7, 2.0 Hz, 1H (5-H-Phe-Et), 6.91m, 2H (3-H- + 6-H-Phe-Et), 6.93s, 1H (2-H-Trp), 6.97m, 1H (5-H-Trp), 7.06m, 1H (6-H-Trp), 7.28m, 2H (2-H-Phe + 7-H-Trp), 7.43d, 6.7 Hz, 1H (4-H-Trp) (*55*)
^{13}C-NMR (100MHz, DMSO-d_6) : see figure (*55*)
COSY and Spin-echo NMR spectra (*55*)
Sources : *Discaria febrifuga* (Rhamnaceae)-roots (*55*)

5(14)-Amphibine-B-Type Cyclopeptide Alkaloids

121. Mauritine-F (= *N*-Desmethyl-mauritine-A)

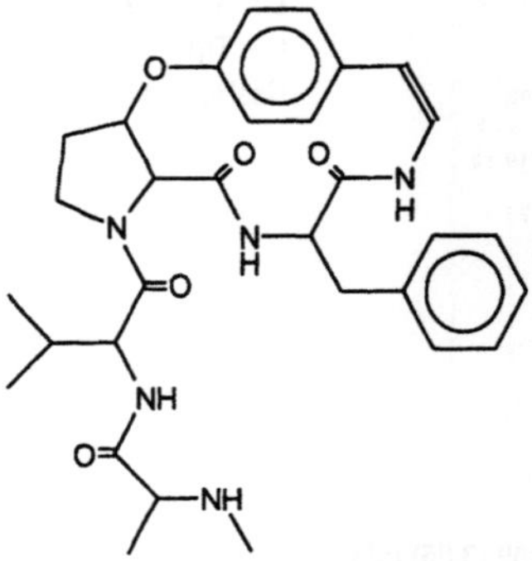

A = N-Me-Ala
B = *trans*-β-OH-Pro[1] (*42*)
C = Phe
E = Val
$C_{31}H_{39}N_5O_5$, 561.2943 (MS) (*42*)

Mp : 222-225 (*42*)
[a]$_D$ (20) = -285 (c=0.15, MeOH) (*42*)
UV (MeOH) : 250sh., 280sh. (*42*)
IR ($CHCl_3$) : 3375, 2990-2860, 2790, 1683, 1620, 1595, 1491, 1220, 1030 (*42*)
MS : 561(M^+), 546(b), 504(c), 502(d), 461(e), 405(f), 404(g), 378(h), 337(i), 376(j), 308(u), 254(k), 243(p), 229(r), 221(o), 215(q), 203(s), 186(t), 185(l), 157(m), 135(l'), 120(q'), 96(v), 72(q''), 68(w), 58(a=100%) (*42*)
^{1}H-NMR (60 and 90MHz, $CDCl_3$) : 0.85d, 6H (2xCH-Me), 1.35d, 3H (CH-Me), 2.46s, 3H (NH-Me), 6.20-7.95m, (Ar-H +NH + olefinic-H) (*42*)
Derivatives : Dihydro-mauritine-F (*42*)
N-Methyl-dihydro-mauritine-F = Dihydro-mauritine-A (*42*)
Sources : Rhamnaceae
Zizyphus mauritiana-stem bark (*42*, *117*)
Z. nummularia-root bark (*43*)

122. Mauritine-A

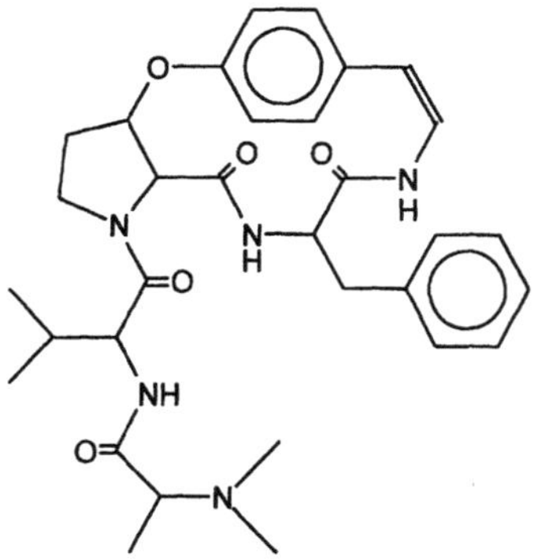

A = N,N-diMe-L-Ala[1] (*23, 35*)
B = *trans*-β-OH-Pro[1] (*23, 35*)
C = L-Phe[1] (*23, 35*)
E = L-Val[1] (*23, 35*)
$C_{32}H_{41}N_5O_5$, 575

Mp : 102-104 [*35, 68, 123*)
[a]D (20) = -315 (c=0.33, MeOH) (*35, 123*)
UV : 220 (*35*)
MS : 308(u), 243(p), 229(r), 221(o), 215(q), 199(l), 186(t), 171(m), 135(i'), 96(v), 72(a/q'') (*35*)
X-rays (*23*)
Sources : Rhamnaceae
Zizyphus jujuba-stem bark (*84*)
Z. mauritiana-stem bark (*35, 42, 117*)
Z. nummularia-stem bark (*68*)
Z. spina-christi-stem bark (*123*)

123. Mauritine-H

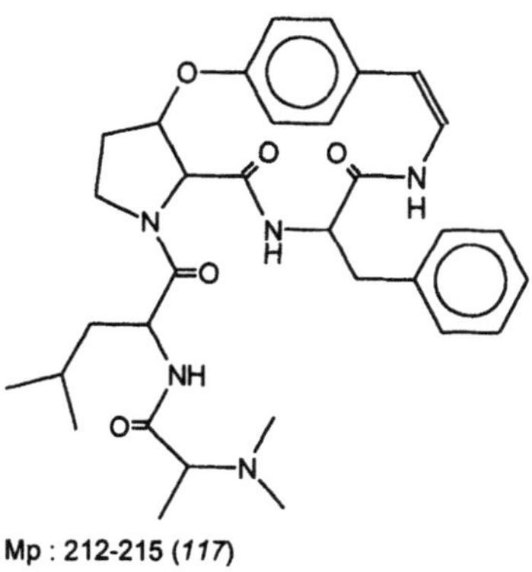

A = N,N-diMe-Ala
B = β-OH-Pro
C = Phe
E = Leu
$C_{33}H_{43}N_5O_5$, 589.3271 (MS) (*117*)

Mp : 212-215 (*117*)
[a]D (20) = -169 (c=0.013, MeOH) (*117*)
UV (MeOH) : end absorption (*117*)
IR (KBr) : 3490, 2790, 1680, 1620, 1215, 1090 (*117*)
MS : 589(M^+), 574(b), 518(c), 516(d), 475(e), 405(f), 404(g), 378(h), 377(i), 376(j), 308(u), 282(k), 243(p), 235(o), 229(r), 215(q), 213(l), 209(n), 203(s), 186(t), 185(m), 135(i'), 120(q'), 96(v), 91, 72(a=100%), 68(w) (*117*)
^{1}H-NMR ($CDCl_3$) : 0.79d, 7.0 Hz, 3H (Me-Leu), 0.89d, 7.0 Hz, 3H (Me-Leu), 1.26d, 7.0 Hz, 3H (Me-N,N-diMe-Ala), 2.36s, 6H (N-Me_2) (*117*)
Derivatives : Dihydro-mauritine-H (*117*)
Sources : *Zizyphus mauritiana* (Rhamnaceae)-stem bark (*117*)

124. Mauritine-E

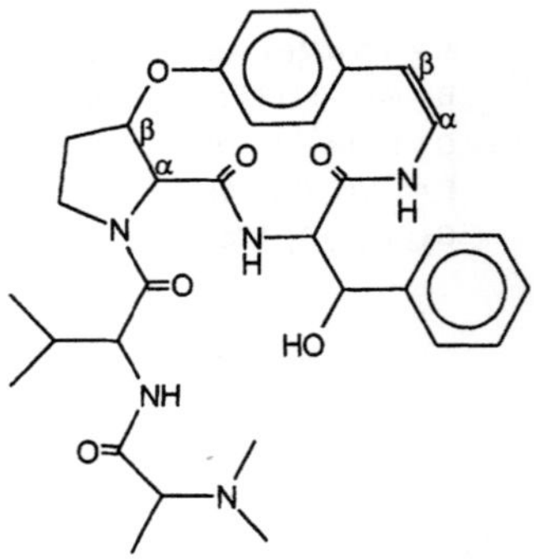

A = N,N-diMe-Ala
B = *trans*-β-OH-Pro[1] (*42*)
C = *threo*-β-OH-Phe[1] (*42*)
E = Val
$C_{32}H_{41}N_5O_6$, 591.7 (*42*)

Mp : amorphous (*42*)
$[a]_D$ (20) = -243 (c=0.11, MeOH) (*42*)
UV (MeOH) : 250sh., 280sh. (*42*)
IR ($CHCl_3$) : 3400, 3369, 2991-2861, 2780, 1690, 1624, 1600, 1496, 1230, 1030 (*42*)
MS : 485(M^+-106), 470(b), 414(c), 412(d), 371(e), 315(f), 314(g), 288(h), 287(i), 286(j), 268(k), 229(r), 221(o), 203(s), 199(l), 195(n), 186(t), 171(m), 136(q'), 135(i'), 106, 105, 96(v), 77, 72(a=100%), 68(w) (*42*)
^{1}H-NMR (60 and 90MHz, $CDCl_3$) : 0.90d, 6H (2xCH-Me), 1.25d, 3H (CH-Me), 2.30s, 6H ($N-Me_2$), 3.98d, 6.5 Hz, 1H (α-H-β-OH-Pro), 6.29d, 8.0 Hz, 1H (β-H-Sty), ~6.70q, 1H (α-H-Sty), 6.40-7.80m, (Ar-H +NH) (*42*)
Derivatives : Dihydro-mauritine-E (*42*)
Sources : *Zizyphus mauritiana* (Rhamnaceae)-stem bark (*42, 117*)

125. Mauritine-D

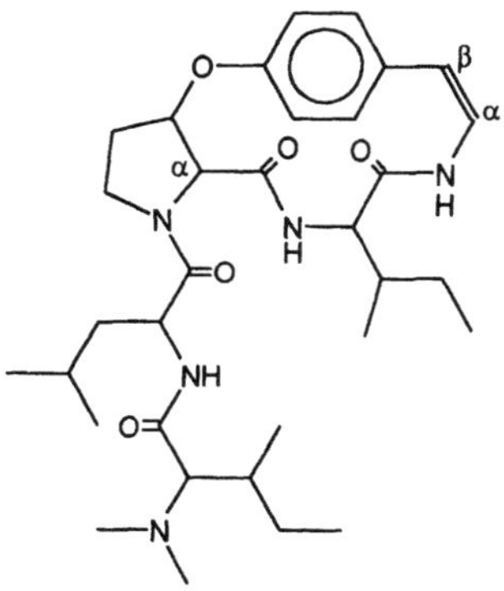

A = N,N-diMe-Ileu
B = *trans*-β-OH-Pro[1] (*42*)
C = Ileu
E = Leu
$C_{33}H_{51}N_5O_5$, 597.3876 (MS) (*42*)

Mp : amorphous (*42*)
[a]D (20) = -256 (c=0.1, $CHCl_3$) (*80*), -259 (c=0.16, MeOH) (*42*)
UV (MeOH) : 250sh., 280sh. (*42, 80*)
IR ($CHCl_3$) : 3400, 3010-2870, 2790, 1690, 1630, 1600, 1500, 1214, 1025 (*42, 80*)
MS : 597(M^+), 540(b), 484(c), 482(d), 441(e), 371(f), 370(g), 344(h), 343(i), 342(j), 324(k), 274(u), 255(l), 235(o), 229(r), 227(m), 209(n/p), 203(s), 186(t), 181(q), 135(i'), 114(a=100%), 96(v), 86(p'/q'/q"), 68(w) (*42, 80*)
^{1}H-NMR (60 and 90MHz, $CDCl_3$) : 0.60-1.00m, 18H (6xC-Me), 2.30s, 6H (N-Me_2), 4.24d, 5.8 Hz, 1H (α-H-β-OH-Pro), 6.25d, 8.0 Hz, 1H (β-H-Sty), ~6.70q, 1H (α-H-Sty), 6.30-7.30m, 7H (4xAr-H + 3xNH) (*42*)
Derivatives : Dihydro-mauritine-D (*42*)
Sources : Rhamnaceae
Zizyphus mauritiana-stem bark (*42, 117*)
Z. nummularia-stem bark (*80*)
Z. xylopyra-stem bark (*85*)

126. Hysodricanine-A

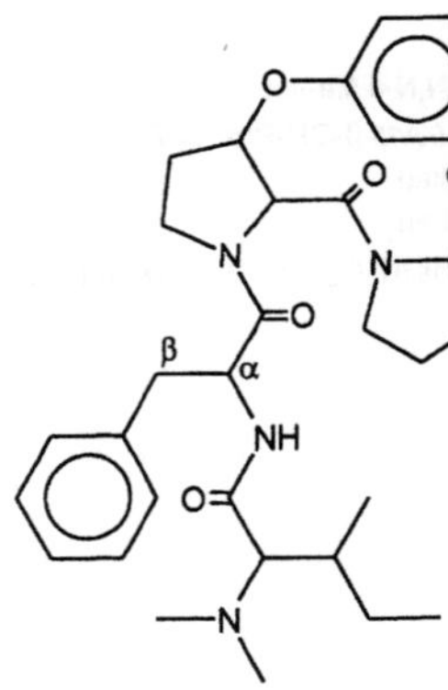

A = N,N-diMe-Ileu
B = β-OH-Pro
C = Pro
E = Phe
$C_{35}H_{45}N_5O_5$, 615.3408 (MS) (*117*)

Mp : 92-96 (*117*, *124*)
[a]D (20) = -215 (c=0.05, $CHCl_3$) (*117*)
UV (MeOH) : end absorption (*117*)
IR (KBr) : 3380, 2790, 1645, 1620, 1230, 1020 (*124*)
IR (KBr) (*117*)
MS : 615(M^+), 558(b), 358(k), 328(h), 327(i), 326(j), 289(l), 269(o), 261(m), 258(u), 243(n), 232(l-R), 229(r), 203(s), 193(p), 186(t), 181(q), 165(q), 135(i'), 120(q''), 114(a=100%), 96(v), 68(w) (*117*)
MS (*124*)
^{1}H-NMR (90MHz, $CDCl_3$) : 0.66d, 6.5 Hz, 3H (CH-Me-N,N-diMe-Ileu), 0.84t, 6.0 Hz, 3H (CH_2-Me-N,N-diMe-Ileu), 2.17s, 6H (N-Me_2), 2.50d, 5.2 Hz, 2H (β-H-Phe), 6.31 (vinyl-H), 7.00-7.33m, 9H (9xAr-H) (*124*)
^{1}H-NMR (90MHz, $CDCl_3$+D_2O) : (*117*)
Derivatives : Dihydro-hysodricanine-A (*117*)
Sources : <u>Rhamnaceae</u>
Zizyphus hutchinsonii-stem bark (*124*)
Z. hysodrica-stem bark (*117*)

127. Lotusine-C

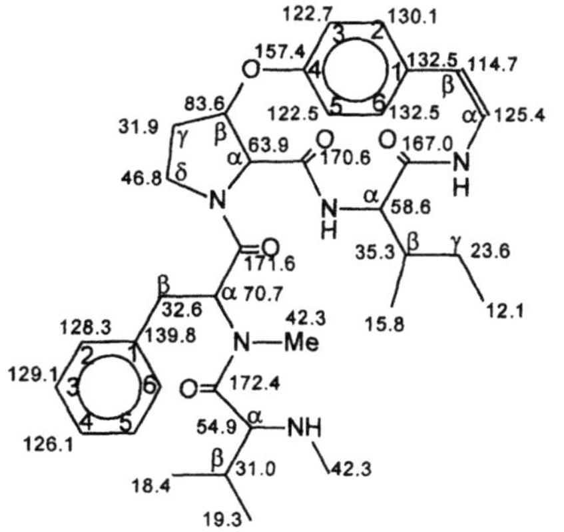

A = N-Me-Val
B = β-OH-Pro
C = Ileu
E = N-Me-Phe
$C_{35}H_{47}N_5O_5$, 617

UV (MeOH) : 204 (*75*)

IR ($CHCl_3$) : 3310, 2960, 1615, 1220, 1045 (*75*)

MS : 617(M^+), 603($M\text{-}CH_3\text{+}H^+$), 544($b'\text{+}H^+$), 542($b'\text{-}H^+$), 528($w'\text{-}CH_3^+\text{-}C_3H_7^+$), 526($M^+\text{-}R_1$), 513($w'\text{-}2xCH_3^+\text{-}C_3H_7^+$), 427($b\text{-}R_1\text{-}R_2$), 347, 344(h/k), 316(h-CO), 314(j-CO), 286($i\text{-}R_2$), 285($j\text{-}R_2$), 272($i\text{-}CO\text{-}C_3H_7^+$), 233($m\text{-}CH_3^+$), 229(r), 221, 213($n\text{-}HCO\text{-}CH_3^+$), 203(s), 183($l\text{-}R_1$), 169($l\text{-}R_1\text{-}CH_3^+$), 156($m\text{-}R_1$), 148($o\text{-}CO\text{-}R_1\text{-}CH_3^+$), 135(i'), 134(q''=100%), 105, 100, 96(v), 91, 86(a/q'), 77, 69, 68(w), 56 (*75*)

^{1}H-NMR (300MHz, $CDCl_3$) : 0.75d, 6.9 Hz, 3H (CH-Me-Ileu), 0.86t, 6.5 Hz, 3H (CH_2-Me-Ileu), 0.79-0.93m, 6H (2xC-Me-N-Me-Val), 1.01-1.13m, 1H (γ-H-Ileu), 1.19-1.29m, 1H (γ-H-Ileu), 1.81-2.01m, 1H (β-H-N-Me-Val), 2.10-2.20m, 1H (β-H-Ileu), 2.13-2.34m, 1H (γ-H-β-OH-Pro), 2.34br.s, 3H (N-Me-N-Me-Val), 2.36br.s, 3H (N-Me-N-Me-Phe), 2.53-2.61m, 1H (γ-H-β-OH-Pro), 2.92dd, 14.0, 6.4 Hz, 1H (β-H-N-Me-Phe), 3.21dd, 14.0, 6.4 Hz, 1H (β-H-N-Me-Phe), 3.36t, 6.4 Hz, 1H (α-H-N-Me-Phe), 3.43m, 1H (δ-H-β-OH-Pro), 4.17dd, 8.6, 3.0 Hz, 1H (α-H-Ileu), 4.23d, 5.4 Hz, 1H (α-H-β-OH-Pro), 4.32dd, 10.6, 8.3 Hz, 1H (δ-H-β-OH-Pro), 4.43t, 8.5 Hz, 1H (α-H-N-Me-Val), 5.55m, 1H (β-H-β-OH-Pro), 6.32d, 7.7 Hz, 1H (β-H-Sty), 6.46br.d, 8.6 Hz, 1H (NH-Ileu), 6.47br.d, 10.1 Hz, 1H (NH-Sty), 6.74dd, 10.1, 7.7 Hz, 1H (α-H-Sty), 7.09m, 1H (2-H-Sty), 7.12m, 1H (6-H-Sty), 7.14m, 1H (3-H-Sty), 7.22m, 1H (4-H-N-Me-Phe), 7.23m, 2H (3-H- + 5-H-N-Me-Phe), 7.26m, 2H (2-H- + 6-H-N-Me-Phe), 7.28m, 1H (5-H-Sty) (*75*)

^{13}C-NMR (75MHz, $CDCl_3$) : see figure (*75*)

COSY, HMBC and HMQC-NMR spectra (*75*)

Sources : *Zizyphus lotus* (Rhamnaceae)-root bark (*75*)

128. Mauritine-B

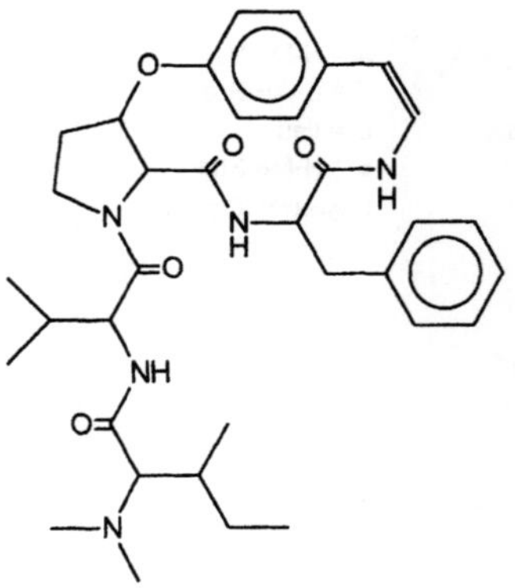

A = N,N-diMe-L-Ileu[1] (*35*)
B = *trans*-β-OH-L-Pro[1] (*35*)
C = L-Phe[1] (*35*)
E = L-Val[1] (*35*)
$C_{35}H_{47}N_5O_5$, 617

Mp : amorphous powder (*35*)
[a]D (20) = -151 (c=0.44, MeOH) (*35*)
UV : 220 (*35*)
MS : 308(u), 243(p), 241(l), 229(r), 221(o), 215(q), 213(m), 186(t), 135(i'), 120(q'), 114(a), 96(v) (*35*)
Sources : *Zizyphus mauritiana* (Rhamnaceae)-stem bark (*35, 42, 117*)

129. Lotusine-B

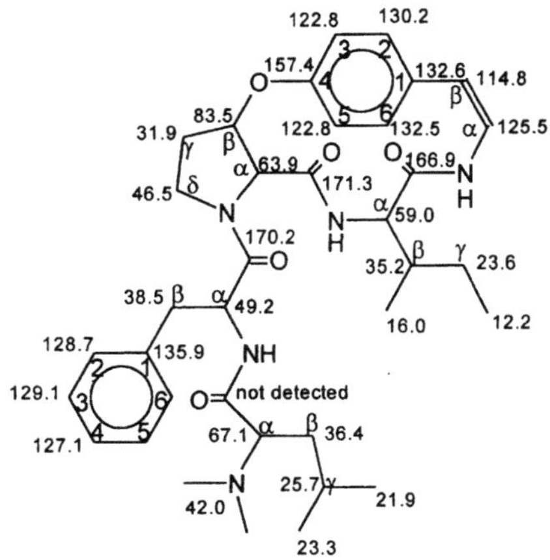

A = N,N-diMe-Leu
B = β-OH-Pro
C = Ileu
E = Phe
$C_{35}H_{49}N_5O_5$, 631

$[a]_D$ = -179 (c=0.32, $CHCl_3$) (*75*)
UV (MeOH) : 207, 245sh. (*75*)
IR ($CHCl_3$) : 3310, 2920, 1655, 1225 (*75*)
MS : 631(M^+), 574(b), 519($c+H^+$), 427($c-R_1$), 274(u), 243(n), 227(o-NCO), 209(p), 203(s), 200(n-HNCO), 190, 148($s''+H^+$), 135(i'), 134($i'-H^+$), 114(a=100%), 96(v), 91, 86(q'), 84($a-2xCH_3^+$), 72(o'), 68(w) (*75*)
^{1}H-NMR (300MHz, $CDCl_3$) : 0.82d, 6.9 Hz, 3H (CH-Me-Ileu), 0.85d, 6.4 Hz, 3H (Me-N,N-diMe-Leu), 0.87d, 6.4 Hz, 3H (Me-N,N-diMe-Leu), 0.91t, 7.2 Hz, 3H (CH_2-Me-Ileu), 1.05-1.20m, 1H (γ-H-Ileu), 1.20-1.35m, 1H (γ-H-Ileu), 2.10-2.25m, 1H (γ-H-β-OH-Pro), 2.22s, 3H ($N-Me_2$), 2.17-2.30m, 1H (β-H-Ileu), 2.45-2.60m, 1H (γ-H-β-OH-Pro), 2.86br.d, 7.0 Hz, 2H (β-H-Phe), 2.90-3.05m, 1H (δ-H-β-OH-Pro), 4.18-4.20m, 1H (δ-H-β-OH-Pro), 4.20dd, 8.6, 3.0 Hz, 1H (α-H-Ileu), 4.29d, 5.7 Hz, 1H (α-H-β-OH-Pro), 4.89q, 7.0 Hz, 1H (α-H-Phe), 5.54m, 1H (β-H-β-OH-Pro), 6.32d, 7.7 Hz, 1H (β-H-Sty), 6.50br.d, 10.5 Hz, 1H (NH-Sty), 6.56br.d, 8.6 Hz, 1H (NH-Ileu), 6.75dd, 10.5, 7.7 Hz, 1H (α-H-Sty), 7.07m, 1H (6-H-Sty), 7.10m, 1H (2-H-Sty), 7.12m, 1H (3-H-Sty), 7.15m, 2H (3-H- + 5-H-Phe), 7.19m, 1H (4-H-Phe), 7.22m, 2H (2-H- + 6-H-Phe), 7.28m, 1H (5-H-Sty) (*75*)
^{13}C-NMR (75MHz, $CDCl_3$) : see figure (*75*)
COSY, HMBC and HMQC-NMR spectra (*75*)
Sources : *Zizyphus lotus* (Rhamnaceae)-root bark (*75*)

130. Amphibine-C

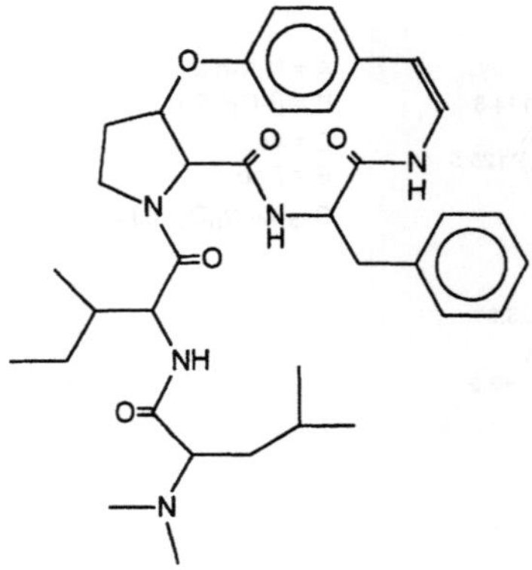

A = N,N-diMe-Leu
B = *trans*-β-OH-Pro[1] (*36*)
C = Phe
E = Ileu
$C_{36}H_{49}N_5O_5$, 631.3725 (MS) (*36*)

Mp : amorphous (*36*)
$[a]_D$ (20) = -224 (c=0.075, MeOH) (*36*)
UV (MeOH) : 252, 280 (*36*)
CD (EtOH) : +0.90 (282), -26.90 (237), -46.6 (204) (*36*)
IR ($CHCl_3$) : 3390, 2785, 1695, 1235 (*36*)
MS : 631(M^+), 602(M^+-C_2H_5), 588(M^+-C_3H_7), 574(b), 518(c), 516(d), 475(e), 446(e'), 445(e"), 405(f), 404(g), 378(h), 377(i), 376(j), 324(k), 308(u), 255(l), 243(p), 235(o), 229(r), 227(m), 215(q), 209(n), 203(s), 186(t), 135(i'), 120(q'), 114(a=100%), 96(v), 91, 86(q"), 72(o'), 68(w) (*36*)
^{1}H-NMR (60 and 90MHz, $CDCl_3$) : 0.70-1.10m, 12H (4xC-Me), 2.37s, 6H (N-Me_2), 4.20d, 5.8 Hz, 1H (α-H-β-OH-Pro), 6.20-6.80m, (olefinic-H), 7.00-7.60m, (Ar-H +NH) (*36*)
Sources : *Zizyphus amphibia* (Rhamnaceae)-stem bark (*36*, *89*)

131. Amphibine-D

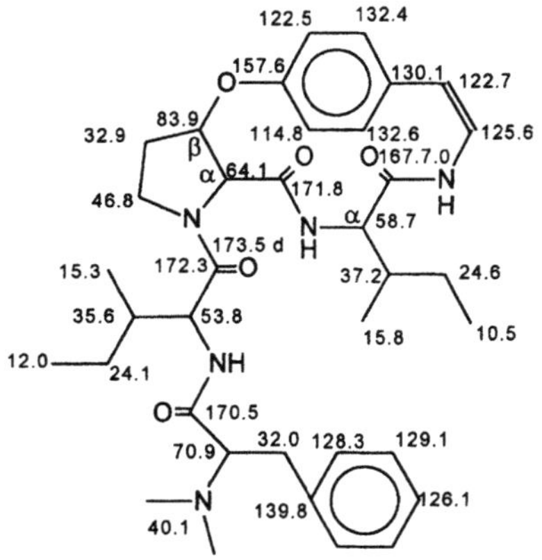

A = N,N-diMe-Phe
B = *trans*-β-OH-Pro[1] (*36*)
C = Ileu
E = Ileu
$C_{36}H_{49}N_5O_5$, 631.3706 (MS) (*36*)

Mp : amorphous (*36, 63*)
[a]D (25) = -201.5 (c=0.7, MeOH) (*63*)
[a]D (20) (*36*)
UV (MeOH) : 250sh., 280sh. (*36*)
CD (Dioxane) : +0.40 (290), -27.90 (237) (*36*)
IR ($CHCl_3$) : 3395, 2790, 1693, 1235 (*36*)
IR ($CHCl_3$) (*124*)
MS : 631(M^+), 630($M-H^+$), 602($M^+-C_2H_5$), 588($M^+-C_3H_7$), 574($M^+-C_4H_9$), 540(b), 484(c), 482(d), 441(e), 411(e''), 371(f), 370(g), 358(k), 344(h), 343(i), 342(j), 289(l), 274(u), 261(m), 235(o), 229(r), 209(n/p), 203(s), 186(t), 181(q), 148(a=100%), 135(i'), 96(v), 91, 86(q'/q''), 68(w) (*36*)
MS (*63, 124*)
^{1}H-NMR (80MHz, $CDCl_3$) : 0.70-0.86m, 12H (4xC-Me), 2.30s, 6H (N-Me_2), 4.20-4.51m, 2H (α-H-Ileu + α-H-β-OH-Pro), 5.51m, 1H (β-H-β-OH-Pro), 6.25-6.75m, 4H (2xolefinic-H + 2xNH), 7.12-7.26m, 10H (9xAr-H + 1xNH) (*63*)
^{1}H-NMR (90MHz, $CDCl_3$) : (*36, 124*)
^{13}C-NMR (20MHz, $CDCl_3$) : see figure (*63*)
^{13}C-NMR (20/25MHz, $CDCl_3$) (*92*)
Sources : Rhamnaceae
Zizyphus amphibia-stem bark (*36, 89*)
Z. juazeiro-stem bark (*124*)
Z. mauritiana-stem bark (*35, 42*)
Z. rugosa-stem bark (*124*)
Z. vulgaris var. *spinosus*-seeds (*63*)

132. Amphibine-B

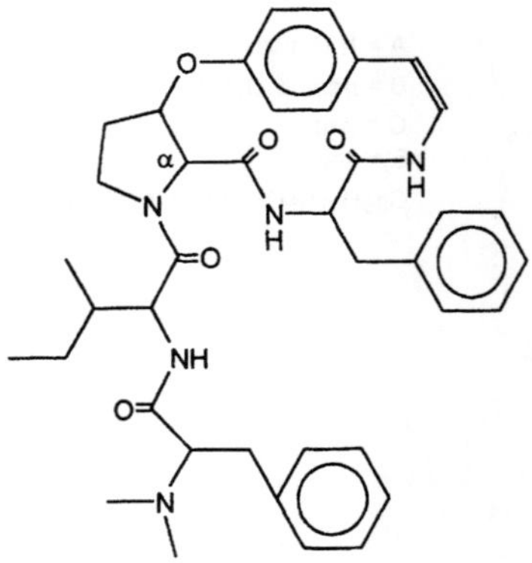

A = N,N-diMe-Phe
B = *trans*-β-OH-Pro[1] (*36*)
C = Phe
E = Ileu
$C_{39}H_{47}N_5O_5$, 665.3548 (MS) (*36*)

Mp : amorphous powder (*36*)
[a]D (20) = -181 (c=0.08, MeOH) (*36*)
UV (MeOH) : 250sh., 280sh. (*36*)
CD (EtOH) : +0.40 (284), -6.60 (234), -13.00 (203) (*36*)
IR ($CHCl_3$) : 3390, 2780, 1685, 1235 (*36*)
MS : 665(M^+), 663($M-2H^+$), 622($M^+-C_3H_7$), 574(b), 518(c), 516(d), 475(e), 446(e'), 445(e''), 405(f), 404(g), 378(h), 377(i), 376(j), 358(k), 308(u), 289(l), 261(m), 243(p), 235(o), 229(r), 215(q), 209(n), 203(s), 186(t), 148(a=100%), 135(i'), 120(q'), 96(v), 91, 68(w) (*36*)
^{1}H-NMR (60 and 90MHz, $CDCl_3$) : 0.60-1.00m, 6H (2xC-Me), 2.40s, 6H ($N-Me_2$), 4.14d, 6.0 Hz, 1H (α-H-β-OH-Pro), 6.20-6.80m, 2H (2xolefinic-H), 7.00-7.50m, 17H (14xAr-H + 3xNH) (*36*)
Derivatives : Dihydro-amphibine-B (*36*)
Sources : Rhamnaceae
Zizyphus amphibia-stem bark (*36, 89*)
Z. mauritiana-stem bark (*42*)

133. Amphibine-E

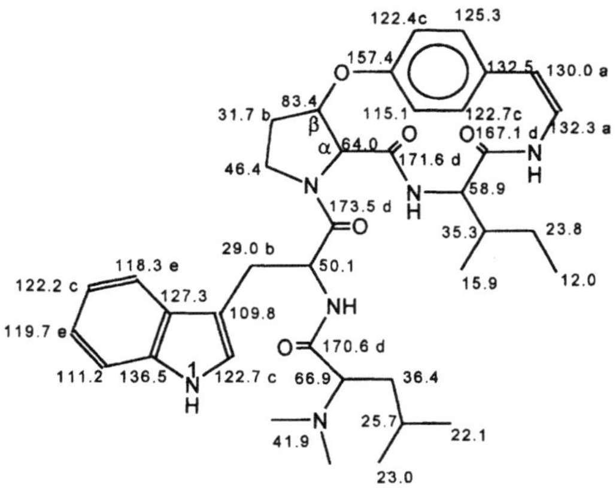

A = N,N-diMe-Leu
B = *trans*-β-OH-Pro[1] (*36*)
C = Ileu
E = Trp
$C_{38}H_{50}N_6O_5$, 670.3801 (MS)(*36*)

Mp : amorphous powder (*36*)
[a]D (20) = -175 (c=0.14, MeOH) (*36*)
UV (MeOH) : 220 (4.46), 270 (3.78), 281 (3.77), 290 (3.68) (*36*)
CD (EtOH) : +1.90 (285), -26.50 (237), -83.10 (203) (*36*)
IR ($CHCl_3$) : 3390, 2780, 1691, 1235 (*36*)
MS : 670(M^+), 641(M^+-C_2H_5), 627(M^+-C_3H_7), 613(b), 557(c), 555(d), 514(e), 371(f), 344(h), 343(i), 342(j), 328(l), 300(m), 274(u), 209(p), 203(s), 181(q), 170(r''), 135(i'), 130, 114(a=100%), 86(q'), 72(o'), 68(w) (*36*)
^{1}H-NMR (60 and 90MHz, $CDCl_3$) : 0.60-1.00m, 12H (4xC-Me), 2.22s, 6H (N-Me_2), 4.28d, 5.3 Hz, 1H (α-H-β-OH-Pro), 6.20-6.80m, 2H (2xolefinic-H), 7.00-7.80m, 12H (9xAr-H + 3xNH), 8.40s, 1H (1-NH-Trp) (*36*)
^{13}C-NMR (20/25MHz, $CDCl_3$) : see figure (*92*)
a, b, c, d = values may be interchanged.
Sources : <u>Rhamnaceae</u>
Zizyphus amphibia-stem bark (*36*, *89*)
Z. mauritiana-stem bark (*42*)
Z. spina-christi-stem bark (*123*)

5(14)-Scutianine-A-Type Cyclopeptide Alkaloids

134. Lasiodine-B

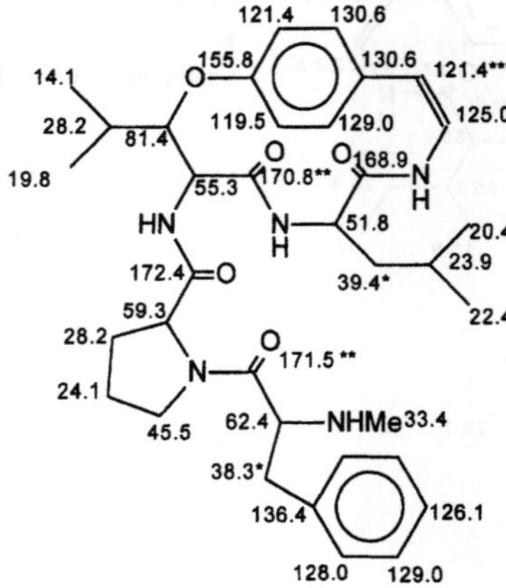

A = N-Me-L-Phe[1] (*62*)
B = L-*erythro*-β-OH-Leu[1,2] (*12-14, 16, 62*)
C = L-Leu[1] (*62*)
E = L-Pro[1] (*62*)
$C_{35}H_{47}N_5O_5$, 617

Mp : 221 (*62*)
$[a]_D$ (20) = -301 (c=1.0, $CHCl_3$/MeOH 1:1 v/v) (*62*)
UV (EtOH) : 250 (3.90) (*62*)
IR (Nujol) : 3260, 1665, 1630, 1505, 1240 (*62*)
MS : 617(M^+), 526(b), 385(g-2H^+), 344(j), 278(c), 274(h), 252(d), 189(f-H^+), 135(i), 134(a), 97(m), 86(q), 70(q') (*62*)
^{1}H-NMR (60MHz, $CDCl_3$+tr. CD_3OD) : 0.60-1.40m, 12H (4xC-Me), 2.29s, 6H (N-Me_2), 6.50-7.60m, 13H (9xAr-H + 4xNH) (*62*)
^{13}C-NMR (15MHz, $CDCl_3$/CD_3OD 2:1 v/v) : see figure (*16*)
*, ** and *** Values may be interchanged.
Derivatives : *N*-Acetyl-lasiodine-B (*14*)
Lasiodine-B imido aldehyde (*14*)
Sources : *Lasiodiscus marmoratus* (Rhamnaceae)-leaves (*62*)

135. Scutianine-F (= *N*-Desmethyl-scutianine-A)

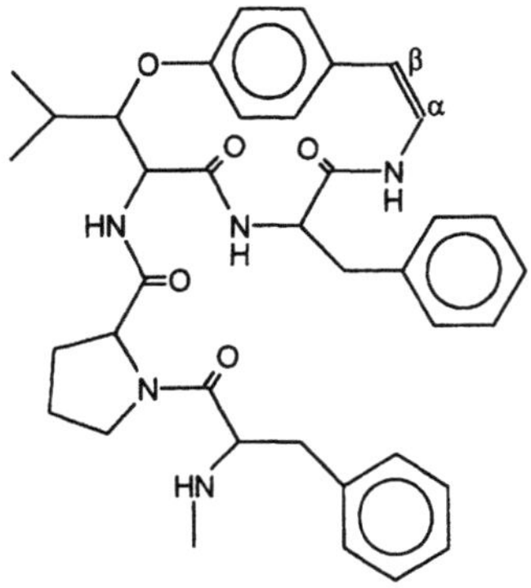

A = N-Me-L-Phe
B = β-OH-Leu
C = Phe
E = Pro
$C_{38}H_{45}N_5O_5$, 651.3416 (MS) (*117*)

Mp : 208 (*117*)
[a]D (20) = -132 (c=0.02, MeOH) (*117*)
IR (KBr) : 3370, 2780, 1680, 1620, 1230, 1040 (*117*)
MS : 651(M^+), 560(b), 518(c'), 461(y), 421(g), 419(g-2H^+), 378(j), 371(z), 308(h), 278(c), 252(d), 189(f-H^+), 135(i), 134(a=100%), 120(q) (*117*)
^{1}H-NMR ($CDCl_3$) : 0.85d, 6.5 Hz, 3H (Me-β-OH-Leu), 1.20d, 6.5 Hz, 3H (Me-β-OH-Leu), 2.25s, 3H (NH-Me), [+D_2O : 6.30d, 8.0 Hz, 1H (α-H-Sty), 6.57d, 8.0 Hz, 1H (β-H-Sty)] (*117*)
Derivatives : Dihydro-*N*-Methyl-scutianine-F = Dihydro-scutianine-A (*117*)
Sources : *Scutia buxifolia* (Rhamnaceae)-stem bark (*117*)

136. Scutianine-A

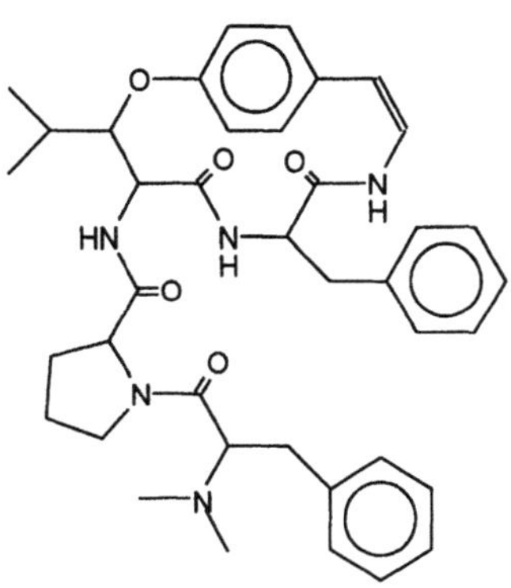

A = N,N-diMe-L-Phe[1] (*64*)
B = β-OH-Leu
C = L-Phe[1] (*64*)
E = L-Pro[1] (*64*)
$C_{39}H_{47}N_5O_5$, 665.3552 (MS) (*25*)

Mp : 186-187 (*25*)
[a]D (20) = -399 (c=0.15, $CHCl_3$) (*25*)
UV (MeOH) : 252sh. (3.78), 277 (3.22) (*25*)
IR ($CHCl_3$) : 3390, 2790, 1690, 1640, 1625, 1500, 1235, 865, 690 (*25*)
MS : 665(M^+), 574(b), 518(c'), 475(y), 421(g), 419(g-2H^+), 387, 385(z), 378(j), 308(h), 292(c), 266(d), 189(f-H^+), 148(a=100%), 135(i), 120(q) (*25*)
^{1}H-NMR (60MHz) : 0.80d, 7.0 Hz, 3H (Me-β-OH-Leu), 1.20d, 7.0 Hz, 3H (Me-β-OH-Leu), 2.07s, 6H (N-Me_2), 6.40d, 7.5 Hz, 1H (olefinic-H), 6.62d, 7.5 Hz, 1H (olefinic-H) (*25*)
Derivatives : Dihydro-scutianine-A (*25*)
Scutianine-A amido aldehyde (*25*)
Sources : *Scutia buxifolia* (Rhamnaceae)-stem bark (*18, 19, 25, 117*)

137. Feretine (= *N*-Desmethyl-adouetine-Z)

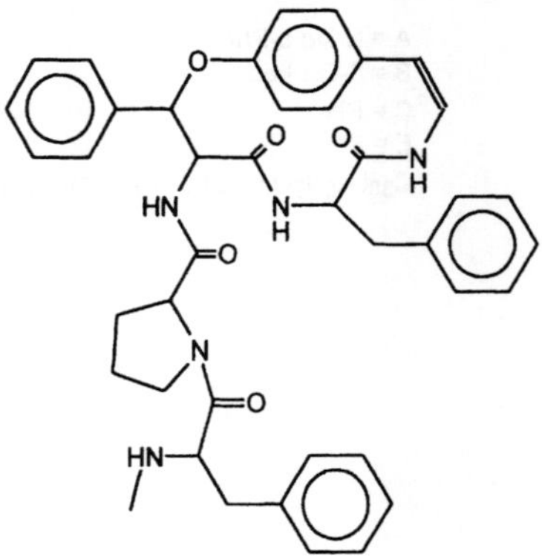

A = N-Me-Phe
B = β-OH-Phe
C = Phe
E = Pro
$C_{41}H_{43}N_5O_5$, 685

Mp : 123 (*130*)
$[a]_D$ = -139 (c=1.0, $CHCl_3$) (*130*)
UV (EtOH) : 230 (*130*)
IR (KBr) : 3300, 1650, 1510, 1230 (*130*)
MS : 685(M^+), 594(b), 308(h), 286(d), 224(f), 135(i), 134(a), 131(m), 120(q), 70(q') (*130*)
^{1}H-NMR ($CDCl_3$) : 1.95s, 3H (NH-Me), 6.80-7.60m, (Ar-H + NH) (*130*)
Sources : *Feretia apodanthera* (Rubiaceae)-leaves (*130*)

138. Adouetine-Z

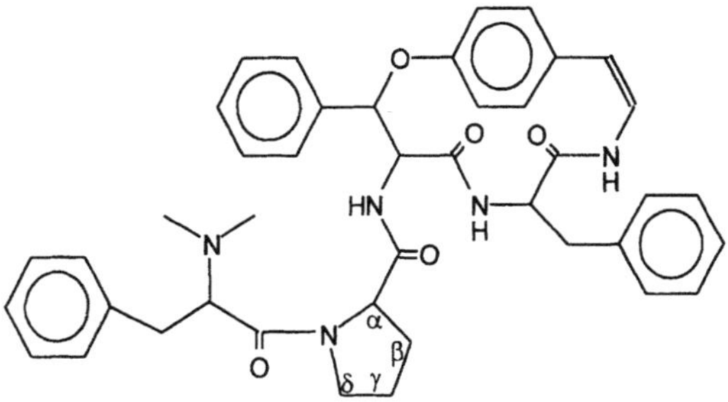

A = N,N-diMe-Phe
B = β-OH-Phe
C = Phe
E = *cis/trans*-Pro[1] (*17*)
$C_{42}H_{45}N_5O_5$, 699.8574 (MS) (*17*)

Adouetine-Z is not a pure compound as originally assumed but a mixture of *cis* and *trans* isomers (*17*)
Mp : amorphous (*17*), 135-140 (*130*), 140-145 (*9, 108*)
[a]D (20) = -184 (c=1.0, $CHCl_3$) (*9, 108*)
[a]D (*17, 130*)
UV (EtOH) : 250 (3.94) (*108*)
IR (KBr) : 3280, 2785, 1630, 1230 (*17*)
IR (CCl_4) : (*9, 108*)
MS : 699(M^+), 608(b), 455(g), 453(g-$2H^+$), 412(j), 326(c), 308(h), 300(d), 224(f), 148(a=100%), 135(i), 131(m), 120(q), 70(q') (*108*)
MS (*17, 130*)
^{1}H-NMR (60MHz, CCl_4) : 1.45s, 3H, 1.95, s, 3H (N-Me), 2.31s, 3H (N-Me), 7.00-7.50, large band, 15-17H (Ar-H), [+TFA : 2.95d, 5.0 Hz, 3H (N-Me), 3.27d, 5.0 Hz, 3H (N-Me)] (*9, 108*)
^{1}H-NMR ($CDCl_3$) : (*130*)
^{13}C-NMR : γ-C-Pro : 21.6 ppm (*trans*), 23.8 ppm (*cis*) (*17*)
Derivatives : Dihydro-adouetine-Z (*9, 17, 108*)
Sources : Rubiaceae
Feretia apodanthera-leaves (*130*)
Sterculiaceae
Melochia pyramidata-leaves (*17*)
Waltheria americana-whole plant (*9, 108*)

139. Hymenocardine

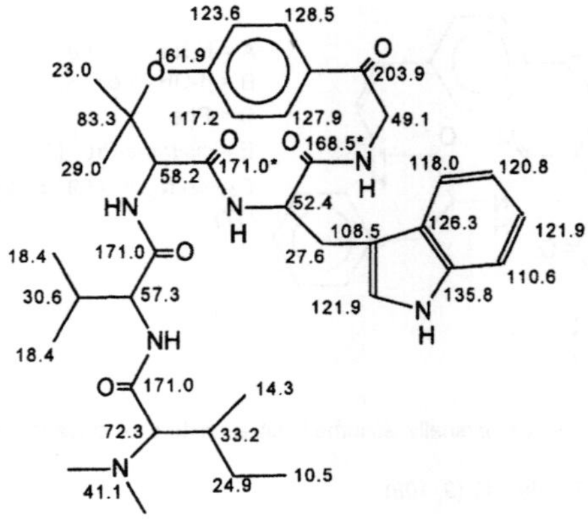

A = N,N-diMe-L-Ileu[1] (*29*)
B = β-OH-Val
C = L-Trp[1] (*29*)
E = L-Val[1] (*29*)
$C_{37}H_{50}N_6O_6$, 674

Mp : 261 (*29*, *131*)
[a]$_D$ (20) = -124 (c=1.0, $CHCl_3$ or $CHCl_3$/MeOH 9:1 v/v) (*29*, *131*)
UV (EtOH) : 222 (4.66), 263 (4.17), 289 (4.02) (*29*, *131*)
IR (Nujol) : 3280, 1700, 1635, 1520 (*29*, *131*)
MS : 674(M^+), 617(b), 559(d'), 462(g), 460(g-$2H^+$), 393(w), 338(x), 320(v), 170(r), 159(q), 135(i), 130, 121(i'), 114(a=100%), 85(n), 72(o/q') (*29*)
^{1}H-NMR (60MHz, $CDCl_3$+tr. CD_3OD) : 0.70-1.10, 12H (4xC-Me), 1.48s, 3H (Me-β-OH-Val), 1.94s, 3H (Me-β-OH-Val), [2.30s, 6H (N-Me_2), +TFA : 3.18s, 6H (N-Me_2)], 6.67-7.95, 13H (Ar-H + NH) (*29*)
^{1}H-NMR (60MHz, $CDCl_3$) : (*131*)
^{13}C-NMR (15MHz, $CDCl_3$/CD_3OD 2:1 v/v) : see figure (*16*)
Derivatives : Hymenocardinols (two epimeric alcohols) (*29*)
Sources : *Hymenocardia acida* (Euphorbiaceae)-root bark (*29*, *131*)

4(15)-Mucronine-A-Type Cyclopeptide Alkaloids

140. Abyssinine-C (= *N*-Desmethyl-abyssinine-B)

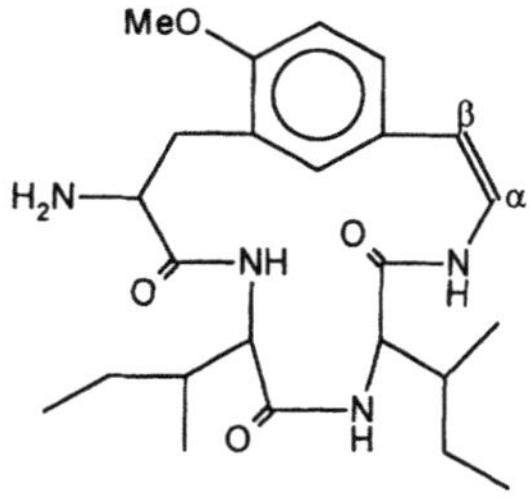

A = Ala
B = Ileu
C = Ileu
$C_{24}H_{36}N_4O_4$, 444.2741 (MS) (*132*)

Mp : amorphous powder (*132*)
[a]$_D$ (20) = +144 (c=0.12, $CHCl_3$) (*132*)
-15 (c=0.13, MeOH) (*132*)
UV (MeOH) : 220sh., 278 (4.22), 315sh. (*132*)
IR ($CHCl_3$) : 3390, 3290-3230, 2820, 1670, 1645, 1240, 1020 (*132*)
MS : 444(M^+=100%), 163(k) (*132*)
^{1}H-NMR (60 and 90MHz) : 0.75-1.15m, 12H (4xC-Me), 2.15s, 2H (NH_2-Ala), 3.85s, 3H (OMe), 5.60d, 10.0 Hz, 1H (α-H-Sty), 6.70-7.40m, 3H (3xAr-H), 6.75d, 10.0 Hz, 1H (β-H-Sty), 8.25-8.60m, 2H (2xNH-CO), 9.66d, 9.0 Hz, 1H (NH-CO) (*132*)
Derivatives: *N,N*-Dimethyl-dihydro-abyssinine-C = *N*-Methyl-dihydro-abyssinine-B (*132*)
N-Acetyl-abyssinine-C (*132*)
N-Tri-deuterio-acetyl-abyssinine-C (*132*)
N-Acetyl-abyssinine-C-aldehyde (*132*)
Sources : *Zizyphus abyssinica* (Rhamnaceae)-stem bark (*132*)

141. Abyssinine-B

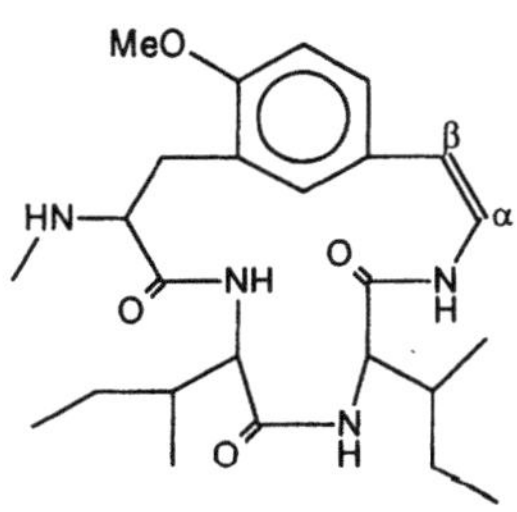

A = N-Me-Ala
B = Ileu
C = Ileu
$C_{25}H_{38}N_4O_4$, 458.2879 (MS) (*132*)

Mp : 229-230 (*132*)
[a]$_D$ (20) = +151 (c=0.16, $CHCl_3$) (*132*)
UV (MeOH) : 220sh., 277 (4.28), 315sh. (*132*)
IR ($CHCl_3$) : 3400, 3350-3230, 2830, 2790, 1690, 1675, 1655, 1255, 1030 (*132*)
MS : 458(M^+=100%), 163(k) (*132*)
^{1}H-NMR (60 and 90MHz) : 0.75-1.15m, 12H (4xC-Me), 2.30s, 1H (NH-N-Me-Ala), 2.50s, 3H (NH-Me), 3.85s, 3H (OMe), 5.60d, 10.0 Hz, 1H (α-H-Sty), 6.65-7.35m, 3H (3xAr-H), 6.75d, 10.0 Hz, 1H (β-H-Sty), 8.25d, 8.0 Hz, 1H (NH-CO), 8.48d, 11.0 Hz, 1H (NH-CO), 9.60d, 8.0 Hz, 1H (NH-CO) (*132*)
Derivatives : *N*-Methyl-dihydro-abyssinine-B (*132*)
N-Acetyl-abyssinine-B (*132*)
N-Acetyl-abyssinine-B-aldehyde (*132*)
Sources : Rhamnaceae
Zizyphus abyssinica-stem bark (*132*)
Z. oenoplia-stem bark (*87*)

142. Abyssinine-A (= *N*-Desmethyl-mucronine-C)

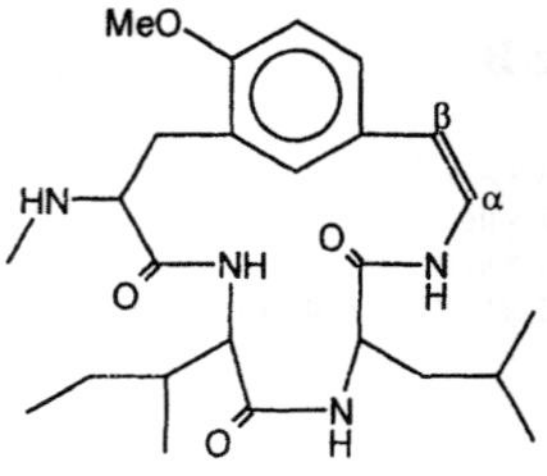

A = N-Me-Ala
B = Ileu
C = Leu
$C_{25}H_{38}N_4O_4$, 458.2896 (MS) (*132*)

Mp : 237-239 (*132*)
[a]$_D$ (20) = +160 (c=0.22, $CHCl_3$) (*132*)
-58 (c=0.1, MeOH) (*132*)
UV (MeOH) : 220sh., 276 (4.14) (*132*)
CD (EtOH) : +1.14 (308), -8.73 (258), +17.09 (228), -16.33 (206) (*132*)
IR ($CHCl_3$) : 3400, 3370, 3290-3230, 2830, 2785, 1690, 1655, 1240, 1030 (*132*)
MS : 458(M^+=100%), 443(M-CH_3^+), 430(M^+-CO), 415(a), 400(a'), 386(a''), 302(c), 259(c-$C_3H_7^+$), 231(d), 205(e), 163(k), 86(m/n) (*132*)
^{1}H-NMR (60 and 90MHz) : 0.70-1.10m, 12H (4xC-Me), 2.50s, 3H (NH-Me), 3.85s, 3H (OMe), 5.60d, 10.0 Hz, 1H (α-H-Sty), 6.70-7.30m, 3H (3xAr-H), 6.72d, 10.0 Hz, 1H (β-H-Sty), 8.15d, 7.0 Hz, 1H (NH-CO), 8.50d, 12.0 Hz, 1H (NH-CO), 9.45d, 8.0 Hz, 1H (NH-CO) (*132*)
Derivatives : Dihydro-abyssinine-A (*132*)
N-Methyl-dihydro-abyssinine-A (*132*)
N-Acetyl-abyssinine-A (*132*)
N-Acetyl-abyssinine-A-aldehyde (*132*)
Sources : Rhamnaceae
Zizyphus abyssinica-stem bark (*132*)
Z. oenoplia-stem bark (*87*)

143. Zizyphine-E (= *N*-Desmethyl-zizyphine-D)

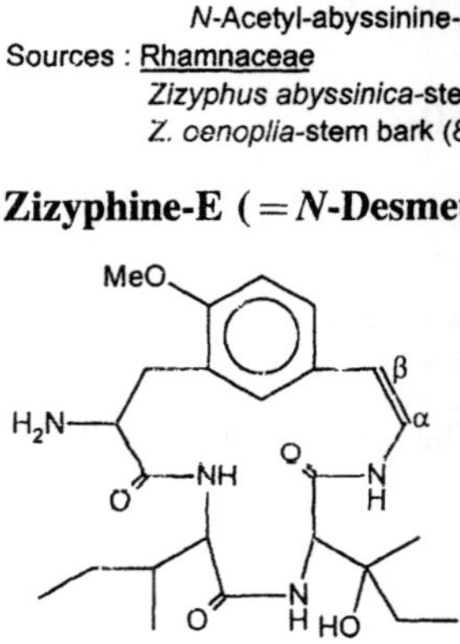

A = Ala
B = Ileu
C = β-OH-Ileu
$C_{24}H_{36}N_4O_5$, 460.2683 (MS) (*87*)

Mp : amorphous (*87*)
[a]$_D$ (20) = +150±2 (c=0.10, $CHCl_3$) (*87*)
-111±2 (c=0.10, MeOH) (*87*)
UV (MeOH) : 279 (4.20) (*87*)
IR ($CHCl_3$) : 3300, 2820, 1673, 1647, 1602, 1250, 1020, 704 (*87*)
MS : 460(M^+), 388(M-$C_4H_8O^+$), 345(a-$C_4H_8O^+$), 330(a'-$C_4H_8O^+$), 316(a''-$C_4H_8O^+$), 72($C_4H_8O^+$), 57($C_4H_9^+$), 43($C_3H_7^+$) (*87*)
^{1}H-NMR (60 and 90MHz, $CDCl_3$) : 0.80-1.20m, 9H (3xC-Me), 1.31s, 3H (C-Me), 1.59m, 2H (-CH_2-), 3.89s, 3H (OMe), 4.56d, 8.3 Hz, 1H (α-H-amino acid), 5.67dd, 11.4, 9.5 Hz, 1H (α-H-Sty), 6.89d, 9.5 Hz, 1H (β-H-Sty), 6.58-7.20m, 3H (3xAr-H), 8.32d, 7.5 Hz, 1H (NH), 8.50d, 11.4 Hz, 1H (NH-Sty), 9.90d, 8.9 Hz, 1H (NH) (*87*)
Derivatives : *N*-Acetyl-zizyphine-E (*87*)
N,O-Diacetyl-zizyphine-E (*87*)
Sources : *Zizyphus oenoplia* (Rhamnaceae)-stem bark (*87*)

144. Mucronine-C

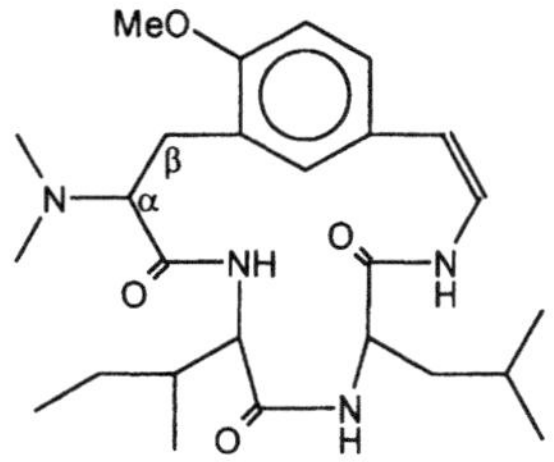

A = N,N-diMe-Ala
B = Ileu
C = Leu
$C_{26}H_{40}N_4O_4$, 472.3050 (MS) (*133*)

Mp : 257 (*133*)
[a]D (20) = -39.4 (c=0.09, $CHCl_3$) (*133*)
UV (MeOH) : 223sh. (4.11), 273 (4.22) (*133*)
CD (Dioxane) : +0.493 (312), -11.76 (272), +12.38 (232) (*133*)
IR (KBr) : 3410, 2820, 2775, 1680, 1665, 1635, 1240, 1025 (*133*)
MS : 472(M^+), 444(M^+-CO), 429(a), 414(a'), 400(a''), 387(M-CO-$C_4H_9^+$), 386(a-$C_3H_7^+$), 373(a-$C_4H_9^+$+H^+), 359(b), 316(c), 273(c-$C_3H_7^+$), 245(d), 219(e), 86(m/n) (*133*)
^{1}H-NMR (60 and 90MHz, $CDCl_3$) : 0.80-1.10, 12H (4xC-Me), 2.45s, 6H (N-Me_2), 3.05d, 6.0 Hz, 2H (β-H-N,N-diMe-Ala), 3.85s, 3H (OMe), 5.70d, 9.0 Hz, 1H (olefinic-H), 6.70-7.20m, 5H (3xAr-H + 1xNH + 1xolefinic-H), 7.55 and 8.40, 2H (2xNH) (*133*)
Sources : Rhamnaceae
Zizyphus abyssinica-stem bark (*133*)
Z. mucronata-stem bark (*133*)

145. Mucronine-F (= *N*-Desmethyl-mucronine-E)

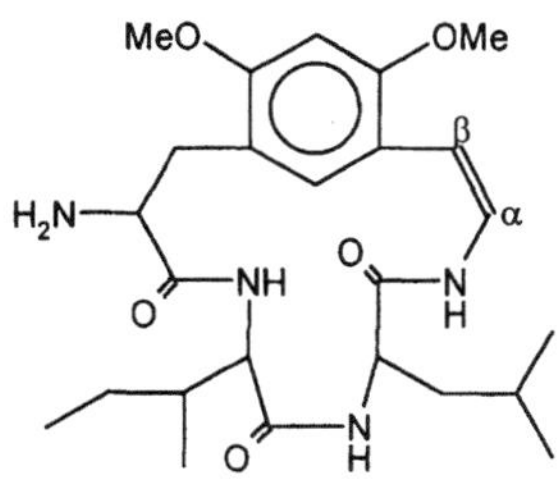

A = Ala
B = Ileu
C = Leu
$C_{25}H_{38}N_4O_5$, 474

Mp : 208-214 (*132*)
[a]D (20) = +17.4 (c=0.092, MeOH) (*132*)
UV (MeOH) : 220sh. (4.19), 274 (4.21), 305sh. (3.89) (*132*)
IR (KBr) : 2860, 1680, 1660, 1640, 1260, 1030 (*132*)
MS : 474(M^+=100%), 193(k) (*132*)
^{1}H-NMR (60 and 90MHz) : 0.80-1.10m, 12H (4xC-Me), 2.25s, 2H (NH_2-Ala), 3.82s, 3H (OMe), 3.86s, 3H (OMe), 5.75d, 10.0 Hz, 1H (α-H-Sty), 6.45s, 1H (Ar-H), 6.70d, 10.0 Hz, 1H (β-H-Sty), 7.02s, 1H (Ar-H), 8.05-8.60m, 2H (2xNH-CO), 9.42d, 8.5 Hz, 1H (NH-CO) (*132*)
Derivatives : Dihydro-mucronine-F (*132*)
N,N-Dimethyl-dihydro-mucronine-F = *N*-Methyl-dihydro-mucronine-E (*132*)
N-Acetyl-mucronine-F (*132*)
N-Acetyl-mucronine-F-aldehyde (*132*)
Sources : *Zizyphus mucronata* (Rhamnaceae)-stem bark (*132*)

146. Mucronine-G

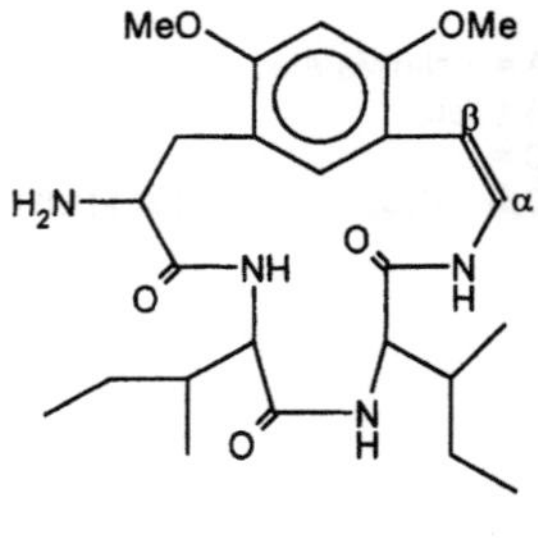

A = Ala
B = Ileu
C = Ileu
$C_{25}H_{38}N_4O_5$, 474

$[a]_D$ (20) = -50 (c=0.084, MeOH) (*132*)
UV (MeOH) : 220sh. (4.16), 274 (4.23), 305sh. (3.90) (*132*)
IR (KBr) : 2860, 1680, 1660, 1640, 1260, 1030 (*132*)
MS : 474(M^+=100%), 193(k) (*132*)
^{1}H-NMR (60 and 90MHz) : 0.80-1.10m, 12H (4xC-Me), 2.02s, 2H (NH_2-Ala), 3.82s, 3H (OMe), 3.86s, 3H (OMe), 5.75d, 10.0 Hz, 1H (α-H-Sty), 6.46s, 1H (Ar-H), 6.80d, 10.0 Hz, 1H (β-H-Sty), 6.98s, 1H (Ar-H), 8.15-8.55m, 2H (2xNH-CO), 9.62d, 9.0 Hz, 1H (NH-CO) (*132*)
Derivatives : *N*-Acetyl-mucronine-G (*132*)
N-Acetyl-mucronine-G-aldehyde (*132*)
Sources : *Zizyphus mucronata* (Rhamnaceae)-stem bark (*132*)

147. Zizyphine-D

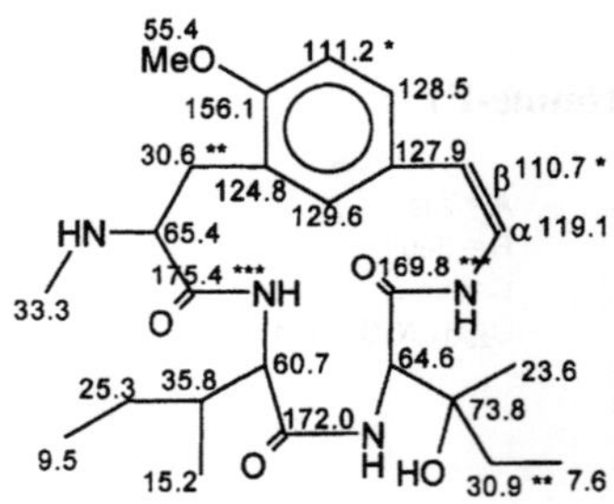

A = N-Me-Ala
B = Ileu
C = β-OH-Ileu
$C_{25}H_{38}N_4O_5$, 474.2842 (MS) (*87*)

Mp : 195 (*87*)
$[a]_D$ (20) = +236±4 (c=0.10, $CHCl_3$) (*87*)
-121±2 (c=0.10, MeOH) (*87*)
UV (MeOH) : 274 (4.16) (*87*)
IR ($CHCl_3$) : 3400, 1670, 1647, 1602, 1249, 1030, 704 (*87*)
MS : 474(M^+=100%), 402(M-$C_4H_8O^+$), 359(a-$C_4H_8O^+$), 344(a'-$C_4H_8O^+$), 330(a''-$C_4H_8O^+$), 317(c-H^+), 163(k), 148(l), 86(m), 72($C_4H_8O^+$), 57($C_4H_9^+$), 43($C_3H_7^+$=100%) (*87*)
^{1}H-NMR (60 and 90MHz, $CDCl_3$) : 0.80-1.20m, 9H (3xC-Me), 1.32s, 3H (C-Me), 1.60m, 2H (-CH_2-), 2.53s, 3H (NH-Me), 3.89s, 3H (OMe), 4.51d, 8.3 Hz, 1H (α-H-amino acid), 5.67dd, 11.4, 9.5 Hz, 1H (α-H-Sty), 6.60d, 9.5 Hz, 1H (β-H-Sty), 6.58-7.33m, 3H (3xAr-H), 8.25d, 7.0 Hz, 1H (NH), 8.50d, 11.4 Hz, 1H (NH-Sty), 9.80br.d, 8.3 Hz, 1H (NH) (*87*)
^{13}C-NMR (20/25MHz, $CDCl_3$) : see figure (*87*)
*, **, *** = values may be interchanged
Derivatives : *N*-Acetyl-zizyphine-D (*87*)
N,*O*-Diacetyl-zizyphine-D (*87*)
Zizyphine-D imido-aldehyde (*87*)
Sources : *Zizyphus oenoplia* (Rhamnaceae)-stem bark (*87*)

148. Mucronine-H (= *N*-Desmethyl-mucronine-B)

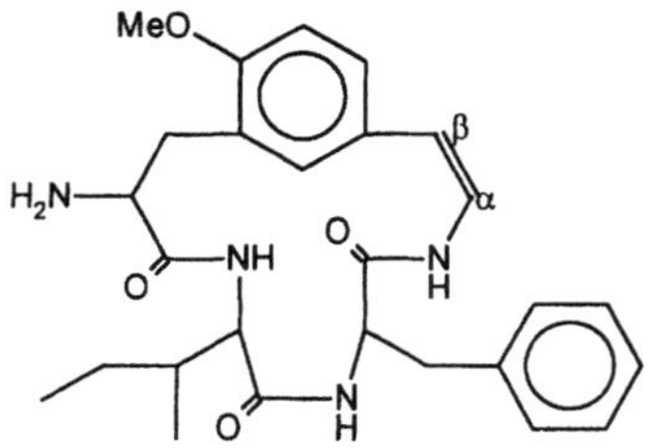

A = Ala
B = Ileu
C = Phe
$C_{27}H_{34}N_4O_4$, 478

[a]D (20) = +5 (c=0.1, MeOH) (*132*)
UV (MeOH) : 276 (4.30) (*132*)
IR (KBr) : 3410, 2825, 1680, 1670, 1650, 1250, 1030 (*132*)
MS : 478(M^+=100%), 163(k) (*132*)
^{1}H-NMR (60 and 90MHz) : 0.28d, 7.0 Hz, 3H (CH-Me-Ileu), 0.80t, 3H (CH_2-Me-Ileu), 3.80s, 3H (OMe), 5.75d, 10.0 Hz, 1H (α-H-Sty), 6.85-7.40m, 3H (3xAr-H), 6.88d, 10.0 Hz, 1H (β-H-Sty), 7.38s, 5H (5xAr-H) (*132*)
Derivatives : *N*-Acetyl-mucronine-H (*132*)
N,N-Dimethyl-dihydro-mucronine-H (*132*)
Sources : *Zizyphus mucronata* (Rhamnaceae)-stem bark (*132*)

149. Mucronine-E (= 4-Methoxy-abyssinine-A)

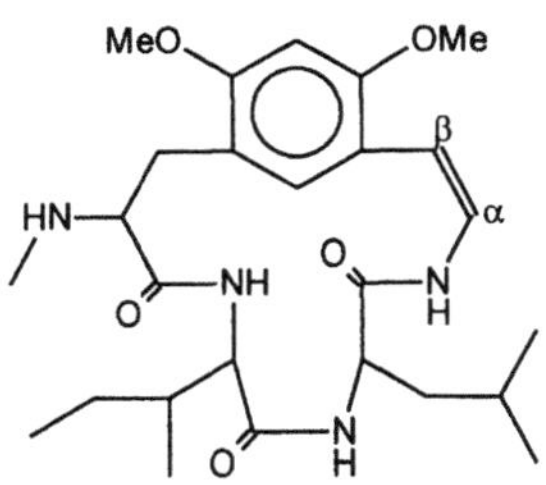

A = N-Me-Ala
B = Ileu
C = Leu
$C_{26}H_{40}N_4O_5$, 488.2996 (MS) (*132*)

Mp : 232-234 (*132*)
[a]D (20) = -89 (c=0.084, MeOH) (*132*)
UV (MeOH) : 220 (4.23), 273 (4.25), 297sh. (3.94) (*132*)
CD (Dioxane) : +3.27 (317), +1.26 (287), -4.18 (261.5), +14.60 (231) (*132*)
IR ($CHCl_3$) : 3410, 3370, 3290-3230, 2830, 2800, 1675, 1650, 1270, 1040 (*132*)
MS : 488(M^+=100%), 193(k) (*132*)
^{1}H-NMR (60 and 90MHz) : 0.80-1.10m, 12H (4xC-Me), 2.37s, 1H (Me-NH-N-Me-Ala), 2.46s, 3H (NH-Me), 3.79s, 3H (OMe), 3.85s, 3H (OMe), 5.72d, 10.0 Hz, 1H (α-H-Sty), 6.42s, 1H (Ar-H), 6.70d, 10.0 Hz, 1H (β-H-Sty), 6.89s, 1H (Ar-H), 8.05d, 7.5 Hz, 1H (NH-CO), 8.40d, 11.0 Hz, 1H (NH-CO), 9.32d, 8.0 Hz, 1H (NH-CO) (*132*)
Derivatives : Dihydro-mucronine-E (*132*)
N-Methyl-dihydro-mucronine-E (*132*)
N-Acetyl-mucronine-E (*132*)
N-Acetyl-mucronine-E-aldehyde (*132*)
N-Tri-deuterio-acetyl-mucronine-E (*132*)
Sources : *Zizyphus mucronata* (Rhamnaceae)-stem bark (*132*)

150. Mucronine-B (=*N*-Desmethyl-mucronine-A)

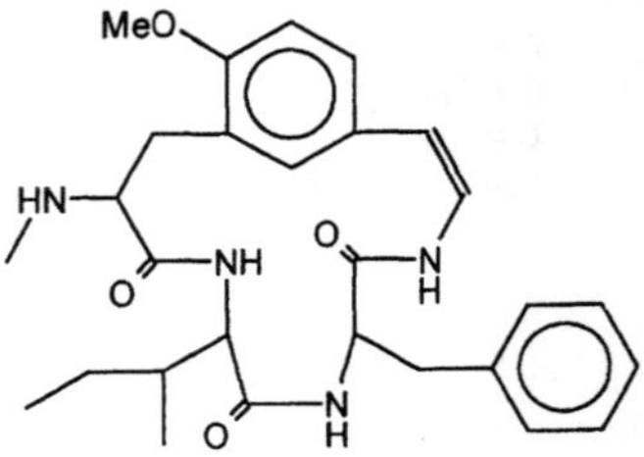

A = N-Me-Ala
B = Ileu
C = Phe
$C_{28}H_{36}N_4O_4$, 492.2737 (MS) (*133*)

Mp : 222-224 (*133*)
$[a]_D$ (25) = +175 (c=0.2, $CHCl_3$) (*133*)
UV (MeOH) : 277 (4.24) (*133*)
CD (Dioxane) : +6.13 (284.5), -0.79 (252.5), +0.52 (241), +5.07 (227) (*133*)
IR ($CHCl_3$) : 3410, 3370, 2830, 2800, 1675, 1650, 1600, 1250 (*133*)
MS : 492(M^+), 449(a), 379(b), 336(c), 231(d), 205(e), 163(k), 120(n), 86(m) (*133*)
^{1}H-NMR (60 and 90MHz, $CDCl_3$) : 0.17d, 6.0 Hz, 3H (CH-Me-Ileu), 0.80t, 3H (CH_2-Me-Ileu), 2.38, 1H (NH-N-Me-Ala), 2.47s, 3H (NH-Me), 2.80-3.60m, 4H (2xCH_2-Φ), 3.85s, 3H (OMe), 5.64d, 10.0 Hz, 1H (olefinic-H), 7.30-8.50m, 11H (8xAr-H + 2xNH + 1xolefinic-H), 9.50m, 1H (NH) (*133*)
Derivatives : Dihydro-mucronine-B (*133*)
Sources : Rhamnaceae
Zizyphus abyssinica-stem bark (*132*)
Z. mucronata-stem bark (*133*)

151. Mucronine-A

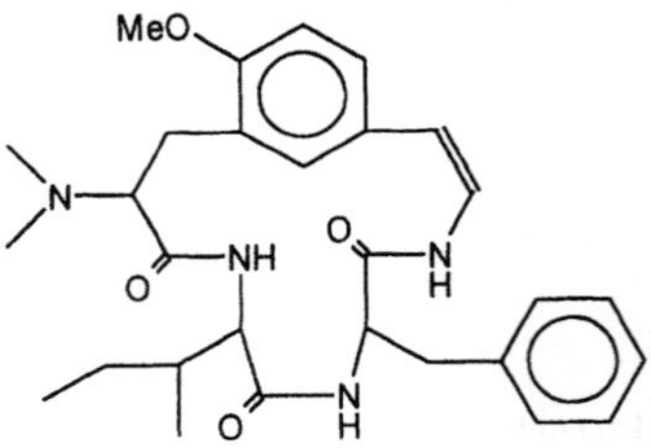

A = N,N-diMe-Ala
B = Ileu
C = Phe
$C_{29}H_{38}N_4O_4$, 506.2896 (MS) (*133*)

Mp : 235 (*133*)
$[a]_D$ (20) = -28.3 (c=0.06, $CHCl_3$) (*133*)
UV (MeOH) : 217sh. (4.37), 273 (4.24) (*133*)
CD (Dioxane) : +0.303 (313), -10.76 (273), +5.00 (235), -19.15 (216) (*133*)
IR ($CHCl_3$) : 3400, 2830, 2785, 1690, 1655, 1240, 1030 (*133*)
MS : 506(M^+), 478(M^+-CO), 463(a), 448(a'), 434(a''), 421(M-CO-$C_4H_9^+$), 393(b), 350(c), 245(d), 219(e), 120(n), 86(m) (*133*)
^{1}H-NMR (60 and 90MHz, $CDCl_3$) : 0.58d, 6.0 Hz, 3H (CH-Me-Ileu), 0.85t, 3H (CH_2-Me-Ileu), 3.00-3.30m, 4H (2xCH_2-Φ), 3.40s, 6H (N-Me_2), 3.85s, 3H (OMe), 5.75d, 10.0 Hz, 1H (olefinic-H), 6.90-7.40m, 11H (8xAr-H + 2xNH + 1 x olefinic-H), 8.30m, 1H (NH) (*133*)
Derivatives : Dihydro-mucronine-A (*133*)
Mucronine-A-amido-aldehyde (*133*)
Sources : Rhamnaceae
Zizyphus abyssinica-stem bark (*132*)
Z. mucronata-stem bark (*133*)

Linear (Open) Peptide Alkaloids

152. Sanjoinine-G2 (= Frangufoline-amido-aldehyde)

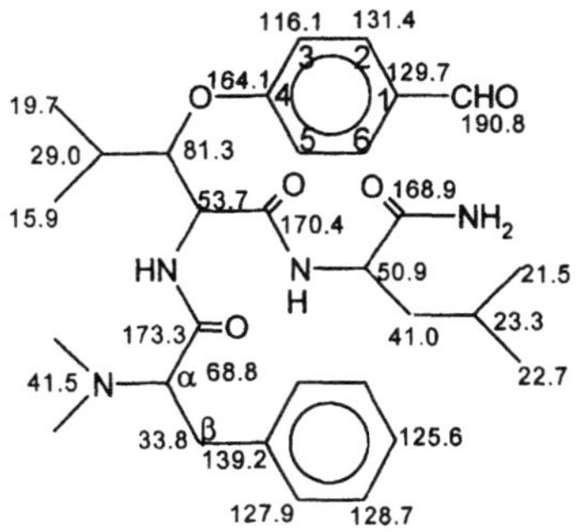

A = N,N-diMe-Phe
B = β-OH-Leu
C = Leu
$C_{30}H_{42}N_4O_5$, 538.3129 (MS) *(63)*

Mp : 182 *(63, 70)*
[a]D (26) = -79.2 (c=0.275, $CHCl_3$) *(63, 70)*
UV (MeOH) : 283.5 (3.81) *(63)*
IR (KBr) : 3290, 2770, 1685-1630, 1596, 1240 *(63)*
MS : 538(M^+), 537(M-H^+), 523(M^+-CH_3), 495(M^+-C_3H_7), 447(b), 409(M^+-leucinamide), 325(b-i'-H^+), 287(M^+-leucinamide-i'), 226(k+NH_2), 195(c), 182(l), 167(d), 148(a=100%), 121(i'), 97(m), 86(n) *(63)*
^{1}H-NMR (80MHz, $CDCl_3$) : 0.72-0.95, 12H (4xC-Me), 2.18s, 6H (NMe_2), 2.88q, 6.2 Hz, 2H (β-H-N,N-diMe-Phe), 3.25t, 6.3 Hz, 1H (α-H-N,N-diMe-Phe), 5.26 and 5.79, 1H each (NH_2), 6.51d, 8.1 Hz, 1H (NH), 6.93d, 8.6 Hz, 2H (3-H- + 5-H-benzaldehyde), 7.11, 5H (5xAr-H-N,N-diMe-Phe), 7.50, 1H (NH), 7.66d, 8.6 Hz, 2H (2-H- + 6-H-benzaldehyde), 9.72s, 1H (CHO) *(63)*
^{13}C-NMR (20MHz, DMSO-d_6) : see figure *(63)*
Sources : *Zizyphus vulgaris* var. *spinosus* (Rhamnaceae)-seeds *(63, 70)*
* Sanjoinine-G2 was also prepared from frangufoline *(63)* :
Mp=182-184, [a]D (23) = -88.3 (c=0.65, $CHCl_3$)

153. Lasiodine-A

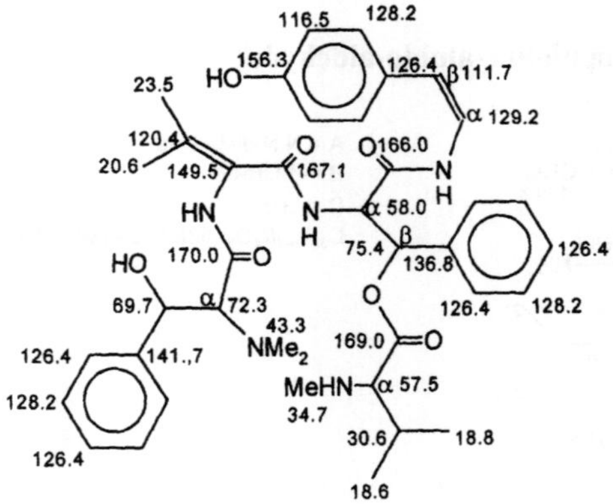

A = N,N-diMe-β-OH-Phe
B = Dide-Val
C = β-OH-Phe
E = N-Me-L-Val[1] (*62*)
$C_{39}H_{49}N_5O_7$, 699

Mp : 195 (*62*)
[a]$_D$ (20) = +38 (c=1.0, $CHCl_3$) (*62*)
UV (EtOH) : 281 (4.49) (*62*)
IR (Nujol) : 3300, 3220, 1750, 1700, 1650, 1515 (*62*)
MS : 135(i), 106, 86(a), 77, 58(o) (*62*)
^{1}H-NMR (60MHz, $CDCl_3$) : 0.69d, 6.7 Hz, 3H (Me-N-Me-Val), 0.82d, 6.7 Hz, 3H (Me-N-Me-Val), 1.80s, 6H (N-Me-N-Me-Val + Me-dide-Val), 1.93s, 3H (Me-dide-Val), 2.73s, 6H (NMe_2), 2.90d, 6.0 Hz, 1H (α-H-N-Me-Val), 3.37d, 6.5 Hz, 1H (α-H-N,N-diMe-β-OH-Phe), 5.17m, 2H (β-H-β-OH-Phe + α-H-Sty), 5.83d, 9.5 Hz, 1H (β-H-Sty), 6.60-7.50, 18H (15xAr-H + 3xNH) (*62*)
^{13}C-NMR (15MHz, $CDCl_3/CD_3OD$ 2:1 v/v) : see figure (*16*)
Derivatives : *O*-Methyl-lasiodine-A (*62*)
O-Methyl-*O*-acetyl-*N*-acetyl-lasiodine-A (*62*)
Dihydro-*N*-methyl-lasiodine-A (*62*)
Sources : *Lasiodiscus marmoratus* (Rhamnaceae)-leaves (*62*)

154. Celenamide-C

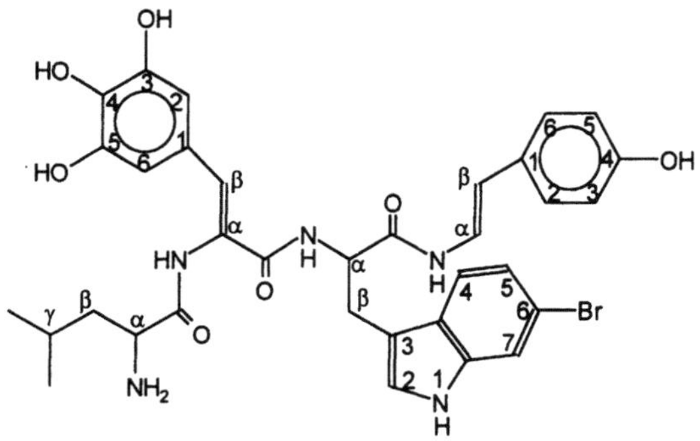

A = Leu
B = Dide-Phe
C = 6-Br-Trp
D = OH-Sty
$C_{46}H_{48}BrN_5O_{13}$, 959/957

Isolated as the pentaacetyl-derivative (*61*)
<u>Pentaacetyl-celenamide-C</u>
$[a]_D$ (25) = +14 (c=0.30, MeCOMe) (*61*)
IR ($CHCl_3$) : 3300(br.), 1780, 1660(br.) (*61*)
MS : 432(f), 390(f-Ac), 348(f-2Ac), 319(e), 306(f-3Ac), 277(e-Ac), 235(e-2Ac), 210/208(c), 193(e-3Ac), 177(d), 135(d-Ac), 128(b), 86(a), 43(CH_3CO^+=100%) (*61*)
^{1}H-NMR (270MHz, CD_3COCD_3) : 0.98d, 7.0 Hz, 3H (Me-Leu), 1.03d, 7.0 Hz, 3H (Me-Leu), 1.75m, 2H (β-H-Leu), 1.89m, 1H (γ-H-Leu), 1.99s, 3H ($NHCOCH_3$), 2.24s, 3H ($OCOCH_3$), 2.28s, 6H (2x$OCOCH_3$), 2.29s, 3H ($OCOCH_3$), 3.28dd, 15.0, 11.0 Hz, 1H (β-H-6-Br-Trp), 3.55dd, 15.0, 4.0 Hz, 1H (β-H-6-Br-Trp), 4.32m, 1H (α-H-Leu), 4.77ddd, 11.0, 9.0, 4.0 Hz, 1H (α-H-6-Br-Trp), 6.56d, 15.0 Hz, 1H (β-H-OH-Sty), 6.84s, 1H (β-H-dide-Phe), 7.04d, 9.0 Hz, 2H (3-H- + 5-H-OH-Sty), 7.15dd, 8.8, 1.4 Hz, 1H (5-H-6-Br-Trp), 7.23s, 2H (2-H- + 6-H-dide-Phe), 7.32d, 2.2 Hz, 1H (2-H-6-Br-Trp), 7.38d, 9.0 Hz, 2H (2-H- + 6-H-OH-Sty), 7.51dd, 15.0, 10.0 Hz, 1H (α-H-OH-Sty), 7.55d, 1.4 Hz, 1H (7-H-6-Br-Trp), 7.64d, 8.8 Hz, 1H (4-H-6-Br-Trp), 7.88d, 9.0 Hz, 1H (NH-6-Br-Trp), 7.94d, 5.0 Hz, 1H (NH-Leu), 9.46s, 1H (NH-dide-Phe), 9.46d, 10.0 Hz, 1H (NH-OH-Sty), 10.21s, 1H (1-H-6-Br-Trp) (*61*)
Derivatives : Pentaacetyl-celenamide-d_{15}-C (*61*)
Primary amide (product of partial acid hydrolysis) (*61*)
Sources : *Cliona celata* (sponge) (*61*)

155. Celenamide-B

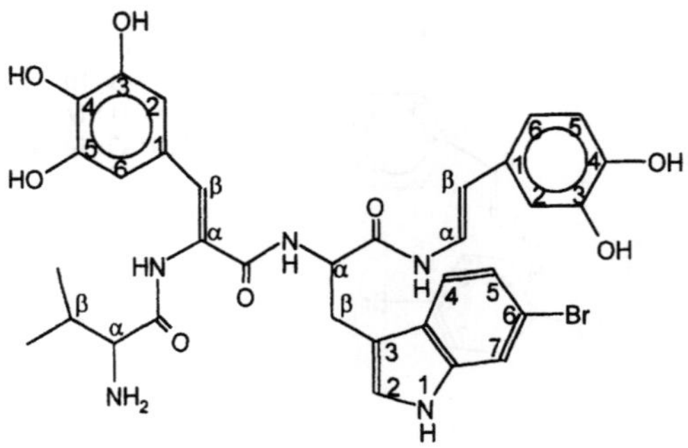

A = Val
B = Dide-Phe
C = 6-Br-Trp
D = diOH-Sty
$C_{45}H_{46}BrN_5O_{14}$, 961/959

Isolated as the hexaacetyl-derivative (*60*)
Hexaacetyl-celenamide-B
[a]$_D$ (25) = +22 (c=1.1, MeCOMe) (*60*)
UV (MeOH) : 227 (4.90), 289br. (4.78) (*60*)
IR ($CHCl_3$) : 3340(br.), 1780, 1660(br.), 1510(br.), 1380, 1200 (*60*)
MS : 585/583(k), 543/541(k-Ac), 501/499(k-2Ac), 484/482(g), 418(f), 376(f-Ac), 334(f-2Ac), 319(e), 292(f-3Ac), 277(e-Ac), 235(d/e-2Ac), 210/208(c), 193(d-Ac/e-3Ac), 165(e-3Ac-CO), 151(d-2Ac), 142(l), 130, 129(c-Br), 123, 114(b), 102, 99, 72(a), 43(CH_3CO^+=100%) (*60*)
^{1}H-NMR (270MHz, CD_3COCD_3) : 0.99d, 7.0 Hz, 3H (Me-Val), 1.01d, 7.0 Hz, 3H (Me-Val), 1.91s, 3H ($NHCOCH_3$), 2.24m, 1H (β-H-Val), 2.24s, 9H (3x$OCOCH_3$), 2.25s, 3H ($OCOCH_3$), 2.29s, 3H ($OCOCH_3$), 3.29dd, 15.4, 11.0 Hz, 1H (β-H-6-Br-Trp), 3.48dd, 15.4, 4.5 Hz, 1H (β-H-6-Br-Trp), 4.46dd, 8.2, 6.0 Hz, 1H (α-H-Val), 4.68ddd, 11.0, 10.0, 4.5 Hz, 1H (α-H-6-Br-Trp), 6.40d, 15.0 Hz, 1H (β-H-diOH-Sty), 7.15d, 9.0 Hz, 1H (5-H-diOH-Sty), 7.16s, 1H (β-H-dide-Phe), 7.18dd, 9.0, 2.0 Hz, 1H (6-H-diOH-Sty), 7.19dd, 8.8, 1.4 Hz, 1H (5-H-6-Br-Trp), 7.21d, 2.0 Hz, 1H (2-H-diOH-Sty), 7.34d, 2.2 Hz, 1H (2-H-6-Br-Trp), 7.41s, 2H (2-H- + 6-H-dide-Phe), 7.49dd, 15.0, 10.0 Hz, 1H (α-H-diOH-Sty), 7.59d, 1.4 Hz, 1H (7-H-6-Br-Trp), 7.64d, 8.8 Hz, 1H (4-H-6-Br-Trp), 7.76d, 10.0 Hz, 1H (NH-6-Br-Trp)*, 7.80d, 6.0 Hz, 1H (NH-Val)*, 9.47s, 1H (NH-dide-Phe), 9.54d, 10.0 Hz, 1H (NH-diOH-Sty), 10.19s, 1H (1-H-6-Br-Trp) (*60*)
* Assignments may be interchanged.
Derivatives : Hexaacetyl-celenamide-d_{18}-B (*60*)
Primary amide (product of partial acid hydrolysis) (*61*)
Sources : *Cliona celata* (sponge) (*60*)

156. Celenamide-A

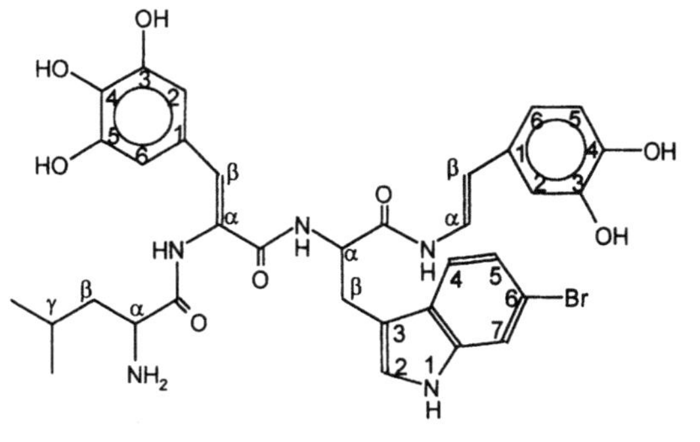

A = Leu
B = Dide-Phe
C = 6-Br-Trp
D = diOH-Sty
$C_{46}H_{48}BrN_5O_{14}$, 975/973

Isolated as the hexaacetyl-derivative *(60)*
Hexaacetyl-celenamide-A
$[a]_D$ (25) = +40 (c=1.1, MeCOMe) *(60)*
UV (MeOH) : 227 (4.86), 289br. (4.71) *(60)*
IR ($CHCl_3$) : 3300(br.), 1780, 1660(br.), 1500(br.), 1380, 1200 *(60)*
MS : 585/583(k), 543/541(k-Ac), 501/499(k-2Ac), 484/482(g), 442/440(g-Ac), 432(f), 390(f-Ac), 348(f-2Ac), 319(e), 306(f-3Ac), 277(e-Ac), 235(d/e-2Ac), 210/208(c), 193(d-Ac/e-3Ac), 165(e-3Ac-CO), 156(l), 151(d-2Ac), 130, 129(c-Br), 128(b), 123, 99, 86(a), 43(CH_3CO^+=100%) *(60)*
^{1}H-NMR (270MHz, CD_3COCD_3) : 0.97d, 7.0 Hz, 3H (Me-Leu), 0.99d, 7.0 Hz, 3H (Me-Leu), 1.73m, 2H (β-H-Leu), 1.86m, 1H (γ-H-Leu), 1.97s, 3H ($NHCOCH_3$), 2.25s, 3H ($OCOCH_3$), 2.27s, 3H ($OCOCH_3$), 2.28s, 9H ($3xOCOCH_3$), 3.26dd, 15.0, 11.0 Hz, 1H (β-H-6-Br-Trp), 3.54dd, 15.0, 4.0 Hz, 1H (β-H-6-Br-Trp), 4.32m, 1H (α-H-Leu), 4.75ddd, 11.0, 10.0, 4.0 Hz, 1H (α-H-6-Br-Trp), 6.52d, 15.0 Hz, 1H (β-H-diOH-Sty), 6.79s, 1H (β-H-dide-Phe), 7.15d, 9.0 Hz, 1H (5-H-diOH-Sty), 7.16dd, 8.8, 1.4 Hz, 1H (5-H-6-Br-Trp), 7.17s, 2H (2-H- + 6-H-dide-Phe), 7.23dd, 9.0, 2.0 Hz, 1H (6-H-diOH-Sty), 7.24d, 2.0 Hz, 1H (2-H-diOH-Sty), 7.31d, 2.2 Hz, 1H (2-H-6-Br-Trp), 7.50dd, 15.0, 10.0 Hz, 1H (α-H-diOH-Sty), 7.57d, 1.4 Hz, 1H (7-H-6-Br-Trp), 7.62d, 8.8 Hz, 1H (4-H-6-Br-Trp), 7.91d, 10.0 Hz, 1H (NH-6-Br-Trp), 7.93d, 5.0 Hz, 1H (NH-Leu), 9.42s, 1H (NH-dide-Phe), 9.50d, 10.0 Hz, 1H (NH-diOH-Sty), 10.18s, 1H (1-H-6-Br-Trp) *(60, 61)*
Derivatives : Hexaacetyl-celenamide-d_{18}-A *(60)*
Primary amide (product of partial acid hydrolysis) *(60)*
Sources : *Cliona celata* (sponge) *(60)*

157. Celenamide-D

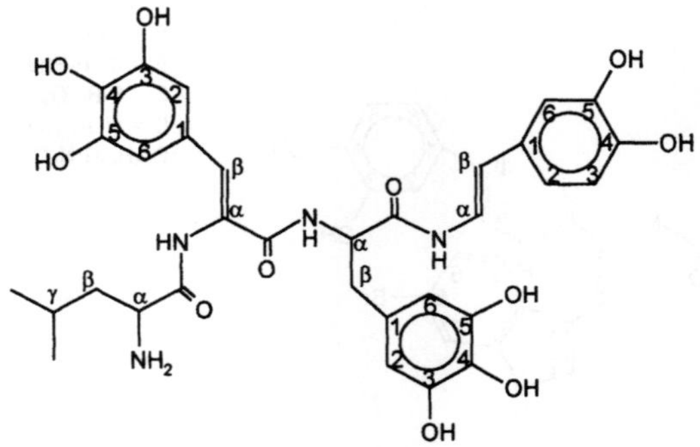

A = Leu
B = Dide-Phe
C = Dide-Phe
D = diOH-Sty
$C_{50}H_{52}N_4O_{20}$, 1028

Isolated as the nonacetyl-derivative (*61*)
Nonacetyl-celenamide-D
[a]$_D$ (25) = -25 (c=0.54, MeCOMe) (*61*)
IR ($CHCl_3$) : 3300(br.), 1780, 1680(br.) (*61*)
MS : 638(j), 596(j-Ac), 554(h/j-2Ac), 512(h-Ac/j-3-Ac), 470(h-2Ac/j-4Ac), 432(f), 428(h-3Ac/j-5Ac), 390(f-Ac), 386(h-4Ac), 348(f-2Ac), 344(h-5Ac), 319(e), 306(f-3Ac), 277(e-Ac), 235(d/e-2Ac), 193(e-3Ac), 151(d-2Ac), 128(b), 86(a), 43(CH_3CO^+=100%) (*61*)
^{1}H-NMR (270MHz, CD_3COCD_3) : 0.95d, 7.0 Hz, 3H (Me-Leu), 0.98d, 7.0 Hz, 3H (Me-Leu), 1.78m, 2H (β-H-Leu), 1.86m, 1H (γ-H-Leu), 1.90s, 3H ($NHCOCH_3$), 2.26s, 9H (3x$OCOCH_3$), 2.27s, 3H ($OCOCH_3$), 2.29s, 3H ($OCOCH_3$), 2.31s, 9H (3x$OCOCH_3$), 4.48m, 1H (α-H-Leu), 6.60d, 15.0 Hz, 1H (β-H-diOH-Sty), 7.06s, 1H (β-H-dide-Phe), 7.15d, 9.0 Hz, 1H (5-H-diOH-Sty), 7.26d, 2.0 Hz, 1H (2-H-diOH-Sty), 7.28dd, 9.0, 2.0 Hz, 1H (6-H-diOH-Sty), 7.43s, 2H (2-H- + 6-H-dide-Phe), 7.47s, 2H (2-H- + 6-H-dide-Phe), 7.59s, 1H (β-H-dide-Phe), 7.59dd, 15.0, 10.0 Hz, 1H (α-H-diOH-Sty), 7.91d, 5.0 Hz, 1H (NH-Leu), 9.18s, 1H (NH-dide-Phe), 9.65d, 10.0 Hz, 1H (NH-diOH-Sty), 9.89s, 1H (NH-dide-Phe) (*61*)
Derivatives : Nonacetyl-celenamide-d_{27}-D (*61*)
Primary amide (product of partial acid hydrolysis) (*61*)
Sources : *Cliona celata* (sponge) (*61*)

Neutral Compounds Related to Cyclopeptide Alkaloids

158. Sanjoinenine (Sanjoinine-F-like)

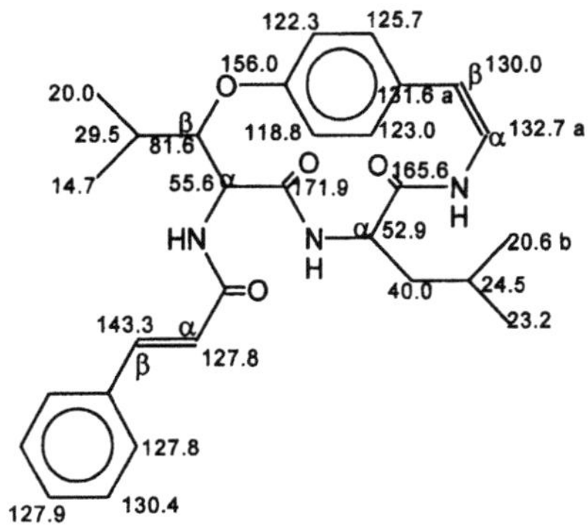

A = *trans*-cinnamic acid[1] (*59, 63*)
B = *threo*[2]- (*59*) or *erythro*[2]-(*63*)-β-OH-Leu
C = Leu
$C_{29}H_{35}N_3O_4$, 489.2601 (*63*) or 489.2665 (*59*) (MS)

Mp = colorless amorphous solid (*59*), 281-282 (*63, 70*)
[a]$_D$ (22) = -272.5 (c=1.6, C_5H_5N) (*63, 70*)
UV (MeOH) : 224.5 (4.38), 281 (4.39) (*63*)
UV (MeOH) (*59*)
IR (KBr) : 3280, 1680-1620, 1640, 1605, 1235 (*63*)
IR ($CHCl_3$) (*59*)
MS : 489(M^+), 355(x), 327(x-CO), 303(e), 300(y), 272(y-CO), 242(z+CO), 215(z+H^+), 214(z), 190(f), 189(f-H^+), 136(i+H^+), 135(i), 134(i-H^+), 132(a+H^+), 131(a=100%), 103, 86(q), 84 (*59*)
MS (*63*)
MS (FD) : 489(M^+) (*59*)
^{1}H-NMR (500MHz, $CDCl_3$) : 0.59d, 6.5 Hz, 3H (Me-Leu), 0.73d, 6.5 Hz, 3H (Me-Leu), 1.01d, 6.8 Hz, 3H (Me-β-OH-Leu), 1.28d, 6.8 Hz, 3H (Me-β-OH-Leu), 1.38d, 7.3 Hz, 2H (β-H-Leu), 4.05m, 1H (α-H-Leu), 4.69dd, 7.4, 2.0 Hz, 1H (α-H-β-OH-Leu), 5.00dd, 7.4, 2.0 Hz, 1H (β-H-β-OH-Leu), 5.90d, 7.8 Hz, 1H (NH-Leu), 5.96d, 9.6 Hz, 1H (NH-Sty), 6.30d, 15.5 Hz, 1H (α-H-*trans*-cinnamic acid), 6.38d, 7.2 Hz, 1H (NH-β-OH-Leu), 6.50m, 1H (β-H-Sty), 6.66m, 1H (α-H-Sty), 7.00-7.50m, 9H (9xAr-H), 7.61d, 15.5 Hz, 1H (β-H-*trans*-cinnamic acid) (*59*)
^{1}H-NMR (80MHz, DMSO-d_6) (*63*)
^{13}C-NMR (125MHz, $CDCl_3$) : see figure (*59*)
a, b and c : values may be interchanged.
^{13}C-NMR (20MHz, DMSO) (*63*)
Sources : Rhamnaceae
Zizyphus lotus-aerial parts (*59*)
Z. vulgaris var. *spinosus*-seeds (*63, 70*)

159. Scutianene-C (Scutianine-C-like)

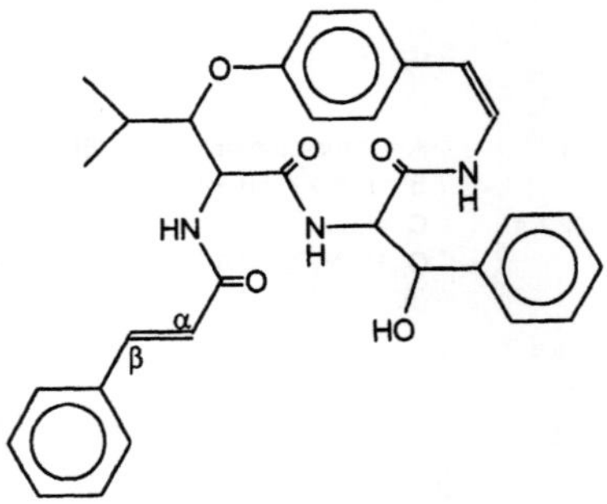

A = *trans*-cinnamic acid[1] (*64*)
B = β-OH-Leu
C = β-OH-Phe
$C_{32}H_{33}N_3O_5$, 539

Mp = 232-234 (*64*)
$[a]_D$ = +203 (c=0.12, $CHCl_3$-MeOH 3:2 v/v) (*64*)
UV : 207 (4.43), 218 (4.39), 275(4.32) (*64*)
IR : 3410, 3380, 3350, 1630 (*64*)
MS : 190(f), 189(f-H^+), 135(i), 131(a=100%), 106, 105, 103, 77 (*64*)
^{1}H-NMR (220MHz, $DMSO_6$) : 0.88d, 7.0 Hz, 3H (Me-β-OH-Leu), 1.15d, 7.0 Hz, 3H (Me-β-OH-Leu), 1.83m, 1H, 4.00t, 9.0 Hz, 1H, 4.10-4.15m, 3H, 5.51d, 4.5 Hz, 1H (OH), 5.94d, 7.0 Hz, 1H, 6.43d, 16.0 Hz, 1H (α-H-*trans*-cinnamic acid), 6.63dd, 7.0, 2.0 Hz, 1H, 6.57-7.13m, 9H, 7.23d, 9.0 Hz, 1H (NH), 7.41t, 7.0 Hz, 2H, 7.44t, 7.0 Hz, 1H, 7.69d, 7.0 Hz, 2H, 8.01d, 10.0 Hz, 1H (NH), 8.52s, 1H (NH) (*64*)
Derivatives : Tetrahydro-scutianene-C (*64*)
Sources : *Scutia buxifolia* (Rhamnaceae)-roots (*64*)

160. Lotusanine-B (Scutianine-A-like)

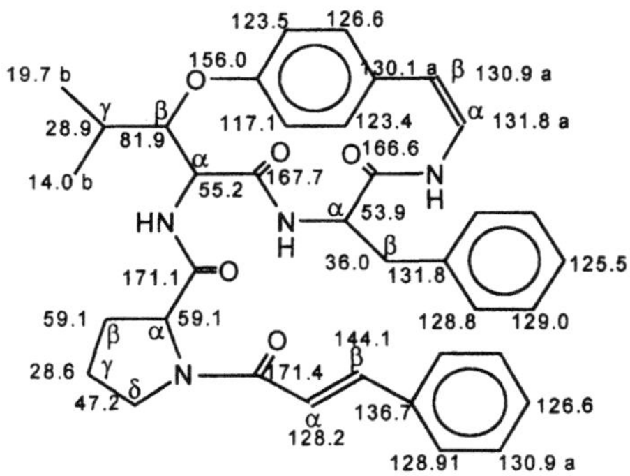

A = *trans*-cinnamic acid[1] (*59*)
B = L-*erythro*-β-OH-Leu[1] (*59*)
C = Phe
E = Pro
$C_{37}H_{40}N_4O_5$, 620.3180 (MS) (*59*)

Mp : amorphous solid (*59*)
UV (MeOH) : 244 (3.74), 280 (3.12) (*59*)
IR ($CHCl_3$) : 3450, 1660, 1610, 1100-1000 (*59*)
MS : 620(M^+), 486(x), 458(x-CO), 431(y), 403(y-CO), 337(e), 228(o), 200(n), 190(f), 189(f-H^+), 135(i), 131(a=100%), 120(q), 103, 97(m), 70(q') (*59*)
MS (FD) : 620(M^+) (*59*)
^{1}H-NMR (500MHz, $CDCl_3$) : 0.66d, 6.6 Hz, 3H (Me-β-OH-Leu), 1.12d, 6.6 Hz, 3H (Me-β-OH-Leu), 1.80m, 1H (γ-H-β-OH-Leu), 3.39-3.55m, 7H (α-H- + β-H- + γ-H- + δ-H-Pro), 4.30dd, 9.2, 6.7 Hz, 1H (α-H-Phe), 4.60m, 1H (α-H-β-OH-Leu), 4.91dd, 6.6, 1.9 Hz, 1H (β-H-β-OH-Leu), 6.02d, 7.8 Hz, 1H (β-H-Sty), 6.32d, 7.6 Hz, 1H (α-H-Sty), 6.68d, 15.4 Hz, 1H (α-H-*trans*-cinnamic acid), 7.38-7.54m, 14H (14xAr-H), 7.72d, 15.4 Hz, 1H (β-H-*trans*-cinnamic acid) (*59*)
^{13}C-NMR (125MHz, $CDCl_3$) : see figure (*59*)
a and b : values may be interchanged.
Sources : *Zizyphus lotus* (Rhamnaceae)-aerial parts (*59*)

Additionally, the isolation of fifteen unidentified cyclopeptide alkaloids has been claimed: a) from the leaves of *Zizyphus jujuba* (Rhamnaceae) five alkaloids designated as fraction 2 and subfractions I, II, III and IV (*134*), b) from the leaves of *Cocculus villosus* (Menispermaceae) four alkaloids designated as subfraction I (fraction II), subfraction II (fraction II), subfraction I (fraction IV) and subfraction II (fraction IV) (*135*), c) from the stem of *C. villosus* six alkaloids designated as subfraction I (fraction III), subfraction II (fraction III), subfraction I (fraction IV), subfraction II (fraction IV), subfraction III (fraction IV) and subfraction I (fraction VI) (*136*).

7. Synthesis

There have been quite a few attempts to synthesize cyclopeptide alkaloids but so far the results are rather meager. The appropriate literature is well reviewed in (*1*, *5*, *6*). Recent efforts are described in (*65*, *137–146*).

The general approach in the synthesis of cyclopeptide alkaloids is first to synthesize the open macrocycle or its dihydro derivative and then, depending on the nature of the ends of the macrocycle, to close the ring either by forming a peptide bond between units B and C or between units C and D (the ends are an activated carboxy group and a protected amino group which react under conditions that favor an intramolecular process) or by forming the ether linkage between units D (the end can be a phenol function) and B (the end can be a dehydro amino acid residue). The formation of the β-aryloxy amino acid, one of the most difficult problems, has recently been accomplished successfully (*146*).

Recently the total synthesis of mucronine-B (**150**) (*147*), of the 4(15)-mucronine-A-type, and zizyphine-A (**27**) (*148*), the major representative of the 5(13)-type, was achieved. Total synthesis of dihydro derivatives of cyclopeptide alkaloids was more fruitful: dihydro-zizyphine-A and -B (*149*) of the 5(13)-type, dihydro-mauritine-A (*150*) of the 5(14)-amphibine-B-type and dihydro-zizyphine-G (*151*, *152*) of the 4(14)-amphibine-F-type were synthesized. The total synthesis of the linear peptide alkaloid hexaacetyl-celenamide-A was also achieved (**156**) (*153*).

No cyclopeptide alkaloid, nor even a dihydro derivative, of the 4(13) type, the 4(14)-frangulanine-, -integerrine- or -pandamine-type or the 5(14)-scutianine-A-type has so far been synthesized.

8. Biological Activity

Almost nothing is known about the physiological role of cyclopeptide alkaloids in plants. The low natural abundance of these compounds (0.0002–1%) and the lack of practical synthetic methods hamper systematic studies on their biological properties. The following activity have been reported.

8.1. Sedative Activity

Adouetine-Z was reported to possess discrete central sedative activity possibly of central sympatholytic origin (*154*).

Frangufoline showed strong sedative activity at a dose of 3 mg/kg (the effect was monitored by measuring the prolongation of hexobarbital-induced sleeping time of mice). Additivity was observed between frangufoline and nuciferine, the latter an aporphine alkaloid also obtained from the seeds of *Zizyphus vulgaris* var. *spinosus* (*70*, *155*, *156*).

Sanjoinine-Ah1, an artifact produced by heating of sanjoinine-A (=frangufoline), showed higher activity at the same dosages than sanjoinine-A, thus verifying the classical record on heat treatment of Sanjoin (seeds of *Zizyphus vulgaris* var. *spinosus*) which appears in the Old Chinese Materia Medica and mentions that roasting of Sanjoin potentiates its hypnotic activity. Based on the descriptions found in the Old Chinese Materia Medica, Sanjoin has been frequently used as an important and reliable hypnotic or sedative agent for treatment of insomnia. Oriental herbal medicine doctors customarily use the drug after it has been roasted on a hot plate at high temperature (*70*, *155*, *156*).

8.2. Antibacterial Activity

Scutianine-A, -B, -C and -E (as the hydrochloride salts) showed some antibacterial activity against the Gram (+) *Bacillus subtilis*. Scutianine-C completely inhibits development of *B. subtilis* at a concentration of 200 μg/ml (*18*). Mauritine-A, -B and -D (hydrochloride salts) showed weak antibacterial activity against *B. subtilis* (*42*).

Rugosanine-A and -B, nummularine-B, -R and -S and amphibine-H (all 13-membered cyclopeptide alkaloids) showed antibacterial activity against the Gram-(-) bacteria *Klebsiella pneumoniae* and *Escherichia coli* (*73*). The 14-membered cyclopeptide alkaloids frangufoline and nummularine-K exhibited significant antibacterial activity against the Gram-(-) bacteria *K. pneumoniae* and *E. coli* and a Gram-(+) bacterium *Staphylococcus aureus* (*73*). Of all the compounds in this study, amphibine-H (13-membered) showed the highest antibacterial activity against *E. coli*.

Mucronine-F, -G and -H (as the hydrochloride salts) exhibited antibacterial activity against *B. subtilis* and *E. coli* (*132*).

8.3. Antifungal Activity

Scutianine-A, -B, -C, -D and -E (hydrochloride salts) (*18*), mauritine-A, -B, -C, -D and -E (hydrochloride salts) (*42*) and mucronine-F, -G and -H (hydrochloride salts) (*132*) inhibited development of *Pythium debaryanum*. Pubescine-A and abyssinine-C showed fungicidal effect against *P. debaryanum* and *Trichoderma viride* (*20*, *132*).

Frangufoline, amphibine-H, rugosanine-A and -B and nummularine-B, -K, -R and -S showed significant activity against *Aspergillus niger*, but no activity against *Candida albicans* (*73*).

9. Biosynthesis – Tissue Culture

There is only one paper dealing with tissue culture and biosynthesis of cyclopeptide alkaloids (*157*). According to this paper, calli did not accumulate alkaloids, but after feeding with appropriate amino acids, tetrapeptides were formed whose structures accorded perfectly with those of the cyclopeptide alkaloids produced in the roots of the plant. As expected, the styrylamine unit arose from tyrosine. No tetrapeptides were formed in suspension cultures or in calli that were not fed with appropriate amino acids. Interestingly, plants or plantlets regenerated *in vitro* did not produce tetrapeptides. No radiolabelling experiments were performed. It should be noted that ceanothamine-E discussed in reference (*157*) corresponds to ceanothamine-A (frangulanine).

The biosynthetic origin of the styrylamine unit in 4(14)- and 5(14)-type cyclopeptide alkaloids is obvious (tyrosine). However, in the 4(13)- and 5(13)-type cyclopeptide alkaloids the β-(2-alkoxy-5-hydroxy)styrylamine unit cannot be attributed to tyrosine. Studies on the biosynthetic origin of the styrylamine unit would be of interest.

10. Conclusions

During the last two decades great progress has been made in the area of cyclopeptide alkaloids. The use of powerful spectrometric methods like MS and NMR has resulted in much faster identification and unequivocal structure elucidation. However little attention has been paid to their role in the plant. More knowledge on their physiological role may also lead to further clues for possible applications of these compounds, *e.g.* for drug development. Also, little is known about their biosynthesis. Cell cultures of the plants producing the alkaloids could be an interesting tool for such studies, but so far no work has been done in this direction.

11. General Tables

Table 7. *Cyclopeptide Alkaloids and Related Compounds in Order of Increasing Molecular Weight, Including Type, Molecular Formula and Molecular Weight*

Compound	Type	Molecular formula	Molecular weight
Subfraction II (*69*) (**1**)	4(13)-nummularine-C	$C_{22}H_{28}N_4O_5$	428
Nummularine-F (**105**)	4(14)-amphibine-F	$C_{23}H_{32}N_4O_4$	428
Spinanine-A (**106**)	4(14)-amphibine-F	$C_{24}H_{32}N_4O_4$	440
Zizyphine-G (**107**)	4(14)-amphibine-F	$C_{24}H_{32}N_4O_4$	440
Abyssinine-C (**140**)	4(15)-mucronine-A	$C_{24}H_{36}N_4O_4$	444
Abyssinine-B (**141**)	4(15)-mucronine-A	$C_{25}H_{38}N_4O_4$	458
Abyssinine-A (**142**)	4(15)-mucronine-A	$C_{25}H_{38}N_4O_4$	458
Zizyphine-E (**143**)	4(15)-mucronine-A	$C_{24}H_{36}N_4O_5$	460
Ceanothine-D (**42**)	4(14)-frangulanine	$C_{26}H_{38}N_4O_4$	470
Ceanothine-C (**43**)	4(14)-frangulanine	$C_{26}H_{38}N_4O_4$	470
N-Desmethyl-myrianthine-C (**44**)	4(14)-frangulanine	$C_{26}H_{40}N_4O_4$	472
Mucronine-C (**144**)	4(15)-mucronine-A	$C_{26}H_{40}N_4O_4$	472
Mucronine-F (**145**)	4(15)-mucronine-A	$C_{25}H_{38}N_4O_5$	474
Mucronine-G (**146**)	4(15)-mucronine-A	$C_{25}H_{38}N_4O_5$	474
Zizyphine-D (**147**)	4(15)-mucronine-A	$C_{25}H_{38}N_4O_5$	474
Mucronine-H (**148**)	4(15)-mucronine-A	$C_{27}H_{34}N_4O_4$	478
Desbenzoyl-aralionine-A (**81**)	4(14)-integerrine	$C_{27}H_{34}N_4O_4$	478
Pubescine-A (**45**)	4(14)-frangulanine	$C_{27}H_{42}N_4O_4$	486
Melonovine-A (**46**)	4(14)-frangulanine	$C_{27}H_{42}N_4O_4$	486
Daechuine-S5 (**47**)	4(14)-frangulanine	$C_{27}H_{42}N_4O_4$	486
Myrianthine-C (**48**)	4(14)-frangulanine	$C_{27}H_{42}N_4O_4$	486
Discarine-F (**49**)	4(14)-frangulanine	$C_{27}H_{42}N_4O_4$	486
Hovenine (**50**)	4(14)-frangulanine	$C_{27}H_{42}N_4O_4$	486
Melofoline (**51**)	4(14)-frangulanine	$C_{26}H_{40}N_4O_5$	488
Mucronine-E (**149**)	4(15)-mucronine-A	$C_{26}H_{40}N_4O_5$	488
Sanjoinenine (**158**)	Neutral	$C_{29}H_{35}N_3O_4$	489
Mauritine-C (**108**)	4(14)-amphibine-F	$C_{28}H_{34}N_4O_4$	490
Mucronine-B (**150**)	4(15)-mucronine-A	$C_{28}H_{36}N_4O_4$	492
Franganine (**52**)	4(14)-frangulanine	$C_{28}H_{44}N_4O_4$	500
Adouetine-X (**53**)	4(14)-frangulanine	$C_{28}H_{44}N_4O_4$	500
Frangulanine (**54**)	4(14)-frangulanine	$C_{28}H_{44}N_4O_4$	500
Discarine-E (**55**)	4(14)-frangulanine	$C_{28}H_{44}N_4O_4$	500
Ceanothine-B (**56**)	4(14)-frangulanine	$C_{29}H_{36}N_4O_4$	504
Amphibine-F (**109**)	4(14)-amphibine-F	$C_{29}H_{36}N_4O_4$	504
Lotusine-D (**110**)	4(14)-amphibine-F	$C_{29}H_{36}N_4O_4$	504
Mucronine-A (**151**)	4(15)-mucronine-A	$C_{29}H_{38}N_4O_4$	506
Daechuine-S7 (**2**)	4(13)-nummularine-C	$C_{28}H_{42}N_4O_5$	514
Compound 2 (*71*) (**3**)	4(13)-nummularine-C	$C_{28}H_{42}N_4O_5$	514
Sativanine-G (**4**)	4(13)-nummularine-C	$C_{28}H_{42}N_4O_5$	514
Sativanine-K (**5**)	4(13)-nummularine-C	$C_{27}H_{38}N_4O_6$	514
Sativanine-B (**82**)	4(14)-integerrine	$C_{30}H_{38}N_4O_4$	518

Table 7. *(continued)*

Compound	Type	M. formula	M. weight
Lotusine-A (**111**)	4(14)-amphibine-F	$C_{30}H_{38}N_4O_4$	518
Discarine-H (**113**)	4(14)-pandamine	$C_{28}H_{46}N_4O_5$	518
Discarine-L (**114**)	4(14)-pandamine	$C_{28}H_{46}N_4O_5$	518
Nummularine-S (**6**)	4(13)-nummularine-C	$C_{29}H_{36}N_4O_5$	520
Tscheschamine (**7**)	4(13)-nummularine-C	$C_{29}H_{36}N_4O_5$	520
Lotusine-F (**8**)	4(13)-nummularine-C	$C_{29}H_{36}N_4O_5$	520
Sanjoinine-B (**57**)	4(14)-frangulanine	$C_{30}H_{40}N_4O_4$	520
N-Desmethyl-myrianthine-B (**58**)	4(14)-frangulanine	$C_{30}H_{40}N_4O_4$	520
Ceanothine-A (**59**)	4(14)-frangulanine	$C_{30}H_{40}N_4O_4$	520
5-Isobutyl-8-*N*-methyl-isoleucyl-9-phenyl-phencyclopeptine (**83**)	4(14)-integerrine	$C_{30}H_{40}N_4O_4$	520
Nummularine-D (**84**)	4(14)-integerrine	$C_{30}H_{40}N_4O_4$	520
Sativanine-A (**85**)	4(14)-integerrine	$C_{30}H_{40}N_4O_4$	520
Nummularine-E (**86**)	4(14)-integerrine	$C_{29}H_{38}N_4O_5$	522
Subfraction I (*69*) (**9**)	4(13)-nummularine-C	$C_{30}H_{36}N_4O_5$	532
Nummularine-G (**87**)	4(14)-integerrine	$C_{31}H_{40}N_4O_4$	532
Discarine-C (**88**)	4(14)-integerrine	$C_{31}H_{42}N_4O_4$	534
Myrianthine-A (**89**)	4(14)-integerrine	$C_{31}H_{42}N_4O_4$	534
Integerrenine (**90**)	4(14)-integerrine	$C_{31}H_{42}N_4O_4$	534
Nummularine-M (**91**)	4(14)-integerrine	$C_{31}H_{42}N_4O_4$	534
Daechucyclopeptide-I (**10**)	4(13)-nummularine-C	$C_{30}H_{38}N_4O_5$	534
Scutianine-C (**60**)	4(14)-frangulanine	$C_{31}H_{42}N_4O_4$	534
Frangufoline (**61**)	4(14)-frangulanine	$C_{31}H_{42}N_4O_4$	534
Adouetine-Y′ (**62**)	4(14)-frangulanine	$C_{31}H_{42}N_4O_4$	534
Lotusanine-A (**63**)	4(14)-frangulanine	$C_{31}H_{42}N_4O_4$	534
Melonovine-B (**64**)	4(14)-frangulanine	$C_{30}H_{40}N_4O_5$	536
Pandaminine (**115**)	4(14)-pandamine	$C_{30}H_{42}N_4O_5$	538
Sanjoinine-G2 (**152**)	Linear	$C_{30}H_{42}N_4O_5$	538
Scutianene-C (**159**)	Neutral	$C_{32}H_{33}N_3O_5$	539
Americine (**65**)	4(14)-frangulanine	$C_{31}H_{39}N_5O_4$	545
Nummularine-C (**11**)	4(13)-nummularine-C	$C_{31}H_{40}N_4O_5$	548
Daechuine-S6 (**12**)	4(13)-nummularine-C	$C_{31}H_{40}N_4O_5$	548
Scutianine-H (**66**)	4(14)-frangulanine	$C_{31}H_{42}N_4O_5$	550
Sanjoinine-F (**67**)	4(14)-frangulanine	$C_{31}H_{42}N_4O_5$	550
Pandamine (**116**)	4(14)-pandamine	$C_{31}H_{44}N_4O_5$	552
Sanjoinine-G1 (**117**)	4(14)-pandamine	$C_{31}H_{44}N_4O_5$	552
Discarine-G (**118**)	4(14)-pandamine	$C_{31}H_{44}N_4O_5$	552
Canthiumine (**92**)	4(14)-integerrine	$C_{33}H_{36}N_4O_4$	552
Integerresine (**93**)	4(14)-integerrine	$C_{33}H_{38}N_4O_4$	554
Aralionine-B (**94**)	4(14)-integerrine	$C_{33}H_{38}N_4O_4$	554
AM-2 (**95**)	4(14)-integerrine	$C_{33}H_{38}N_4O_4$	554
Sativanine-H (**17**)	5(13)-zizyphine-A	$C_{29}H_{43}N_5O_6$	557
Nummularine-P (**18**)	5(13)-zizyphine-A	$C_{29}H_{43}N_5O_6$	557
Sativanine-C (**19**)	5(13)-zizyphine-A	$C_{29}H_{43}N_5O_6$	557
Amphibine-G (**112**)	4(14)-amphibine-F	$C_{32}H_{39}N_5O_4$	557
N-Methyl-americine (**68**)	4(14)-frangulanine	$C_{32}H_{41}N_5O_4$	559

Table 7. *(continued)*

Compound	Type	M. formula	M. weight
Homoamericine **(69)**	4(14)-frangulanine	$C_{32}H_{41}N_5O_4$	559
Mauritine-F **(121)**	5(14)-amphibine-B	$C_{31}H_{39}N_5O_5$	561
Sanjoinine-D **(119)**	4(14)-pandamine	$C_{32}H_{46}N_4O_5$	566
Scutianine-B **(70)**	4(14)-frangulanine	$C_{34}H_{40}N_4O_4$	568
Crenatine-A **(96)**	4(14)-integerrine	$C_{34}H_{40}N_4O_4$	568
Discarine-D **(97)**	4(14)-integerrine	$C_{34}H_{40}N_4O_4$	568
Deoxo-aralionine-A **(98)**	4(14)-integerrine	$C_{34}H_{40}N_4O_4$	568
Ceanothine-E **(99)**	4(14)-integerrine	$C_{34}H_{40}N_4O_4$	568
Adouetine-Y **(100)**	4(14)-integerrine	$C_{34}H_{40}N_4O_4$	568
Sativanine-D **(20)**	5(13)-zizyphine-A	$C_{30}H_{43}N_5O_6$	569
Texensine **(71)**	4(14)-frangulanine	$C_{33}H_{43}N_5O_4$	573
Discarine-B **(72)**	4(14)-frangulanine	$C_{33}H_{43}N_5O_4$	573
Discarine-X **(73)**	4(14)-frangulanine	$C_{33}H_{43}N_5O_4$	573
Nummularine-K **(74)**	4(14)-frangulanine	$C_{33}H_{43}N_5O_4$	573
Discarine-A **(75)**	4(14)-frangulanine	$C_{33}H_{43}N_5O_4$	573
Amphibine-A **(76)**	4(14)-frangulanine	$C_{33}H_{43}N_5O_4$	573
Mauritine-A **(122)**	5(14)-amphibine-B	$C_{32}H_{41}N_5O_5$	575
N-Desmethyl-integerrine **(101)**	4(14)-integerrine	$C_{34}H_{37}N_5O_4$	579
Aralionine-A **(102)**	4(14)-integerrine	$C_{34}H_{38}N_4O_5$	582
Aralionine-C **(103)**	4(14)-integerrine	$C_{34}H_{40}N_4O_5$	584
Scutianine-D **(77)**	4(14)-frangulanine	$C_{34}H_{40}N_4O_5$	584
Scutianine-E **(78)**	4(14)-frangulanine	$C_{34}H_{40}N_4O_5$	584
Scutianine-G **(79)**	4(14)-frangulanine	$C_{34}H_{40}N_4O_5$	584
Rugosanine-A **(21)**	5(13)-zizyphine-A	$C_{30}H_{43}N_5O_7$	585
Sativanine-E **(13)**	4(13)-nummularine-C	$C_{33}H_{41}N_5O_5$	587
Nummularine-R **(14)**	4(13)-nummularine-C	$C_{33}H_{41}N_5O_5$	587
Daechuine-S10 **(15)**	4(13)-nummularine-C	$C_{33}H_{41}N_5O_5$	587
Mauritine-H **(123)**	5(14)-amphibine-B	$C_{33}H_{43}N_5O_5$	589
Mauritine-E **(124)**	5(14)-amphibine-B	$C_{32}H_{41}N_5O_6$	591
Nummularine-N **(22)**	5(13)-zizyphine-A	$C_{32}H_{41}N_5O_6$	591
Nummularine-B **(23)**	5(13)-zizyphine-A	$C_{32}H_{41}N_5O_6$	591
Discarine-K **(120)**	4(14)-pandamine	$C_{33}H_{45}N_5O_5$	591
Integerrine **(104)**	4(14)-integerrine	$C_{35}H_{39}N_5O_4$	593
Zizyphine-B **(24)**	5(13)-zizyphine-A	$C_{32}H_{47}N_5O_6$	597
Zizyphine-F **(25)**	5(13)-zizyphine-A	$C_{32}H_{47}N_5O_6$	597
Mauritine-D **(125)**	5(14)-amphibine-B	$C_{33}H_{51}N_5O_5$	597
Scutianine-J **(80)**	4(14)-frangulanine	$C_{34}H_{40}N_4O_6$	600
Amphibine-H **(26)**	5(13)-zizyphine-A	$C_{33}H_{43}N_5O_6$	605
Zizyphine-A **(27)**	5(13)-zizyphine-A	$C_{33}H_{49}N_5O_6$	611
Daechuine-S8-1 **(28)**	5(13)-zizyphine-A	$C_{33}H_{51}N_5O_6$	613
Hysodricanine-A **(126)**	5(14)-amphibine-B	$C_{35}H_{45}N_5O_5$	615
Lotusine-C **(127)**	5(14)-amphibine-B	$C_{35}H_{47}N_5O_5$	617
Mauritine-B **(128)**	5(14)-amphibine-B	$C_{35}H_{47}N_5O_5$	617
Lasiodine-B **(134)**	5(14)-scutianine-A	$C_{35}H_{47}N_5O_5$	617
Nummularine-T **(29)**	5(13)-zizyphine-A	$C_{33}H_{41}N_5O_7$	619
Lotusanine-B **(160)**	Neutral	$C_{37}H_{40}N_4O_5$	620

Table 7. *(continued)*

Compound	Type	M. formula	M. weight
Rugosanine-B (**16**)	4(13)-nummularine-C	$C_{36}H_{39}N_5O_5$	621
Daechuine-S3 (**30**)	5(13)-zizyphine-A	$C_{34}H_{53}N_5O_6$	627
Lotusine-B (**129**)	5(14)-amphibine-B	$C_{36}H_{49}N_5O_5$	631
Amphibine-C (**130**)	5(14)-amphibine-B	$C_{36}H_{49}N_5O_5$	631
Amphibine-D (**131**)	5(14)-amphibine-B	$C_{36}H_{49}N_5O_5$	631
Sativanine-F (**31**)	5(13)-zizyphine-A	$C_{34}H_{43}N_5O_7$	633
Zizyphine-C (**32**)	5(13)-zizyphine-A	$C_{36}H_{47}N_5O_6$	645
Lotusine-E (**33**)	5(13)-zizyphine-A	$C_{36}H_{49}N_5O_6$	647
Paliurine-B (**34**)	5(13)-zizyphine-A	$C_{36}H_{49}N_5O_6$	647
Nummularine-A (**35**)	5(13)-zizyphine-A	$C_{36}H_{49}N_5O_6$	647
O-Desmethyl-mucronine-D (**36**)	5(13)-zizyphine-A	$C_{36}H_{49}N_5O_6$	647
Scutianine-F (**135**)	5(14)-scutianine-A	$C_{38}H_{45}N_5O_5$	651
Mucronine-D (**37**)	5(13)-zizyphine-A	$C_{37}H_{51}N_5O_6$	661
Amphibine-B (**132**)	5(14)-amphibine-B	$C_{39}H_{47}N_5O_5$	665
Scutianine-A (**136**)	5(14)-scutianine-A	$C_{39}H_{47}N_5O_5$	665
Amphibine-E (**133**)	5(14)-amphibine-B	$C_{38}H_{50}N_6O_5$	670
Hymenocardine (**139**)	5(14)-hymenocardine	$C_{37}H_{50}N_6O_6$	674
Nummularine-H (**38**)	5(13)-zizyphine-A	$C_{39}H_{47}N_5O_6$	681
Feretine (**137**)	5(14)-scutianine-A	$C_{41}H_{43}N_5O_5$	685
Jubanine-A (**39**)	5(13)-zizyphine-A	$C_{40}H_{49}N_5O_6$	695
Adouetine-Z (**138**)	5(14)-scutianine-A	$C_{42}H_{45}N_5O_5$	699
Lasiodine-A (**153**)	Linear	$C_{39}H_{49}N_5O_7$	699
Nummularine-O (**40**)	5(13)-zizyphine-A	$C_{42}H_{45}N_5O_6$	715
Jubanine-B (**41**)	5(13)-zizyphine-A	$C_{43}H_{47}N_5O_6$	729
Celenamide-C (**154**)	Linear	$C_{46}H_{48}BrN_5O_{13}$	959/957
Celenamide-B (**155**)	Linear	$C_{45}H_{46}BrN_5O_{14}$	961/959
Celenamide-A (**156**)	Linear	$C_{46}H_{48}BrN_5O_{14}$	975/973
Celenamide-D (**157**)	Linear	$C_{50}H_{52}N_4O_{20}$	1028

Table 8. *Alphabetical List of Cyclopeptide Alkaloids and Related Compounds Including Sources*

Compound	Sources (plant part)
Abyssinine-A (**142**)	*Zizyphus abyssinica* (sb), *Z. oenoplia* (sb)
Abyssinine-B (**141**)	*Zizyphus abyssinica* (sb), *Z. oenoplia* (sb)
Abyssinine-C (**140**)	*Zizyphus abyssinica* (sb)
Adouetine-X (**53**)	*Ceanothus americanus* (rb), *Waltheria americana* (wh)
Adouetine-Y (**100**)	*Ceanothus americanus* (rb), *Waltheria americana* (wh)
Adouetine-Y′ (**62**)	*Antidesma montana* (l, tb), *Ceanothus americanus* (rb), *Discaria febrifuga* (sb), *D. longispina* (r, rb), *Melochia corchorifolia* (ae, l), *Myrianthus arboreus* (l), *Waltheria americana* (wh)

Table 8. *(continued)*

Compound	Sources (plant part)
Adouetine-Z **(138)**	*Feretia apodanthera* (l), *Melochia pyramidata* (l), *Waltheria americana* (wh)
AM-1	see Adouetine-Y' **(62)**
AM-2 **(95)**	*Antidesma montana* (l, tb)
Americine **(65)**	*Ceanothus americanus* (rb)
Amphibine-A **(76)**	*Zizyphus amphibia* (sb), *Z. nummularia* (rb), *Z. spina-christi* (sb)
Amphibine-B **(132)**	*Zizyphus amphibia* (sb), *Z. mauritiana* (sb)
Amphibine-C **(130)**	*Zizyphus amphibia* (sb)
Amphibine-D **(131)**	*Zizyphus amphibia* (sb), *Z. juazeiro* (sb), *Z. mauritiana* (sb), *Z. rugosa* (sb), *Z. vulgaris* var. *spinosus* (sd)
Amphibine-E **(133)**	*Zizyphus amphibia* (sb), *Z. mauritiana* (sb), *Z. spina-christi* (sb)
Amphibine-F **(109)**	*Zizyphus amphibia* (sb), *Z. mauritiana* (sb), *Z. spina-christi* (sb)
Amphibine-G **(112)**	*Zizyphus amphibia* (sb)
Amphibine-H **(26)**	*Zizyphus amphibia* (sb), *Z. jujuba* (sb), *Z. nummularia* (rb), *Z. spina-christi* (sb), *Z. xylopyra* (sb)
Aralionine-A **(102)**	*Araliothamnus vaginatus* (l, sb)
Aralionine-B **(94)**	*Araliothamnus vaginatus* (l, sb)
Aralionine-C **(103)**	*Araliothamnus vaginatus* (sb)
5-Benzyl-8-*N*-(*N'*-methylpropyl)-9-isopropyl-phencyclopeptine	see Ceanothine-B **(56)**
5-Benzyl-8-*N*,*N*-dimethyl-isoleucyl-9-phenyl-phencyclopeptine	see Deoxo-aralionine-A **(98)**
5-*sec*-Butyl-8-N-(N'-methyl-phenyl-alanyl)-9-isopropyl-phencyclopeptine	see *N*-Desmethyl-myrianthine-B **(58)**
Canthiumine **(92)**	*Canthium euryoides* (?)
Ceanothamine-A	see Frangulanine **(54)**
Ceanothamine-B	see Adouetine-X **(53)**
Ceanothine-A **(59)**	*Ceanothus americanus* (rb)
Ceanothine-B **(56)**	*Ceanothus americanus* (rb), *C. sanguineus* (rb)
Ceanothine-C **(43)**	*Ceanothus americanus* (rb)
Ceanothine-D **(42)**	*Ceanothus americanus* (rb)
Ceanothine-E **(99)**	*Ceanothus americanus* (rb)
Celenamide-A **(156)**	*Cliona celata* (sponge)
Celenamide-B **(155)**	*Cliona celata* (sponge)
Celenamide-C **(154)**	*Cliona celata* (sponge)
Celenamide-D **(157)**	*Cliona celata* (sponge)
Compound 2 (*71*) **(3)**	*Zizyphus mucronata* (r)
Crenatine-A **(96)**	*Discaria crenata* (l, s)
Daechucyclopeptide-I **(10)**	*Zizyphus jujuba* var. *inermis* (fr, sb)
Daechuine-S1	see Frangufoline **(61)**
Daechuine-S2	see Frangulanine **(54)**

Table 8. *(continued)*

Compound	Sources (plant part)
Daechuine-S3 (**30**)	*Zizyphus jujuba* var. *inermis* (sb)
Daechuine-S4	see Franganine (**52**)
Daechuine-S5 (**47**)	*Zizyphus jujuba* var. *inermis* (sb)
Daechuine-S6 (**12**)	*Zizyphus jujuba* var. *inermis* (sb)
Daechuine-S7 (**2**)	*Zizyphus jujuba* var. *inermis* (sb)
Daechuine-S8-1 (**28**)	*Zizyphus jujuba* var. *inermis* (sb)
Daechuine-S9	see Mucronine-D (**37**)
Daechuine-S10 (**15**)	*Zizyphus jujuba* var. *inermis* (sb)
Daechuine-S26	see Daechucyclopeptide-I (**10**)
Daechuine-S27	see Nummularine-B (**23**)
Deoxo-aralionine-A (**98**)	*Ceanothus integerrimus* var. *californicus* (rb), *C. integerrimus* var. *integerrimus* (rb)
Deoxy-aralionine-C	see Deoxo-aralionine-A (**98**)
Desbenzoyl-aralionine-A (**81**)	*Araliothamnus vaginatus* (sb)
N-Desmethyl-abyssinine-B	see Abyssinine-C (**140**)
N-Desmethyl-adouetine-X	see Discarine-F (**49**)
N-Desmethyl-adouetine-Z	see Feretine (**137**)
N-Desmethyl-amphibine-H	see Nummularine-B (**23**)
N-Desmethyl-discarine-B	see Homoamericine (**69**)
N-Desmethyl-frangufoline	see Sanjoinine-B (**57**)
N-Desmethyl-frangulanine	see Hovenine (**50**)
N-Desmethyl-integerrenine	see Nummularine-D (**84**)
N-Desmethyl-integerrine (**101**)	*Ceanothus integerrimus* var. *integerrimus* (rb)
N-Desmethyl-jubanine-A	see Nummularine-H (**38**)
N-Desmethyl-jubanine-B	see Nummularine-O (**40**)
N-Desmethyl-lotusine-A	see Lotusine-D (**110**)
N-Desmethyl-mauritine-A	see Mauritine-F (**121**)
N-Desmethyl-mucronine-A	see Mucronine-B (**150**)
N-Desmethyl-mucronine-B	see Mucronine-H (**148**)
N-Desmethyl-mucronine-C	see Abyssinine-A (**141**)
N-Desmethyl-mucronine-D	see Nummularine-A (**35**)
N-Desmethyl-mucronine-E	see Mucronine-F (**145**)
N-Desmethyl-myrianthine-B (**58**)	*Ceanothus sanguineus* (rb)
N-Desmethyl-myrianthine-C (**44**)	*Plectronia odorata* (ae)
N-Desmethyl-scutianine-A	see Scutianine-F (**135**)
N-Desmethyl-texensine	see Homoamericine (**69**)
N-Desmethyl-zizyphine-A	see Zizyphine-B (**24**)
N-Desmethyl-zizyphine-D	see Zizyphine-E (**143**)
O-Desmethyl-mucronine-D (**36**)	*Zizyphus mucronata* (r)
O-Desmethyl-zizyphine-A	see Zizyphine-F (**25**)
Discarine-A (**75**)	*Discaria longispina* (r)
Discarine-B (**72**)	*Ceanothus integerrimus* var. *californicus* (rb), *C. integerrimus* var. *integerrimus* (rb), *C. sanguineus* (rb), *Discaria longispina* (r, rb)
Discarine-C (**88**)	*Discaria febrifuga* (sb)

Table 8. *(continued)*

Compound	Sources (plant part)
Discarine-D **(97)**	*Discaria febrifuga* (sb)
Discarine-E **(55)**	*Discaria febrifuga* (sb), *D. longispina* (rb)
Discarine-F **(49)**	*Discaria febrifuga* (rb)
Discarine-G **(118)**	*Discaria febrifuga* (rb)
Discarine-H **(113)**	*Discaria febrifuga* (rb)
Discarine-K **(120)**	*Discaria febrifuga* (r)
Discarine-L **(114)**	*Discaria febrifuga* (rb)
Discarine-X **(73)**	*Discaria longispina* (rb)
Feretine **(137)**	*Feretia apodanthera* (l)
Franganine **(52)**	*Discaria febrifuga* (r, sb), *Euonymus europaeus* (r, rb), *Melochia corchorifolia* (ae, l), *Rhamnus frangula* (sb), *Zizyphus spina-christi* (sb)
Frangufoline **(61)**	*Ceanothus sanguineus* (rb), *Discaria febrifuga* (sb), *D. longispina* (r), *Euonymus europaeus* (l), *Melochia corchorifolia* (l), *M. pyramidata* (l), *Rhamnus frangula* (sb), *Zizyphus jujuba* (sb), *Z. jujuba* var. *inermis* (sb), *Z. lotus* (ae), *Z. mauritiana* (sb), *Z. nummularia* (rb, sb), *Z. vulgaris* var. *spinosus* (sd)
Frangufoline amido-aldehyde	see Sanjoinine-G2 **(152)**
Frangulanine **(54)**	*Ceanothus americanus* (rb), *Discaria longispina* (r), *Euonymus europaeus* (l, r, rb, sb), *Hovenia dulcis* (rb), *H. tomentella* (rb), *Rhamnus frangula* (sb), *Zizyphus jujuba* var. *inermis* (sb), *Z. sativa* (sb)
Homoamericine **(69)**	*Ceanothus americanus* (rb), *Hovenia dulcis* (rb), *H. tomentella* (rb)
Hovenine **(50)**	*Hovenia dulcis* (rb), *H. tomentella* (rb)
Hymenocardine **(139)**	*Hymenocardia acida* (rb)
Hysodricanine-A **(126)**	*Zizyphus hutchinsonii* (sb), *Z. hysodrica* (sb)
5-β-Indolylmethyl-8-*N*-methyl-valyl-9-phenyl-phencyclopeptine	see *N*-Desmethyl-integerrine **(101)**
5-β-Indolylmethyl-8-*N*, *N*-dimethyl-valyl-9-isopropyl-phencyclopeptine	see *N*-Methyl-americine **(68)**
Integerrenine **(90)**	*Ceanothus integerrimus* (r), *C. integerrimus* var. *californicus* (rb), *Ceanothus integerrimus* var. *integerrimus* (rb), *Melochia pyramidata* (l), *Zizyphus nummularia* (rb)
Integerresine **(93)**	*Ceanothus integerrimus* (r)
Integerrine **(104)**	*Ceanothus integerrimus* (r), *Ceanothus integerrimus* var. *integerrimus* (rb)
5-Isobutyl-8-*N*-methyl-isoleucyl-9-phenyl-phencyclopeptine **(83)**	*Ceanothus integerrimus* var. *californicus* (rb), *C. integerrimus* var. *integerrimus* (rb)
Jubanine-A **(39)**	*Zizyphus jujuba* (sb), *Z. nummularia* (rb), *Z. spina-christi* (sb)

Table 8. *(continued)*

Compound	Sources (plant part)
Jubanine-B **(41)**	*Zizyphus jujuba* (sb), *Z. nummularia* (rb, sb)
Lasiodine-A **(153)**	*Lasiodiscus marmoratus* (l)
Lasiodine-B **(134)**	*Lasiodiscus marmoratus* (l)
Lotusanine-A **(63)**	*Zizyphus lotus* (ae)
Lotusanine-B **(160)**	*Zizyphus lotus* (ae)
Lotusine-A **(111)**	*Zizyphus lotus* (rb)
Lotusine-B **(129)**	*Zizyphus lotus* (rb)
Lotusine-C **(127)**	*Zizyphus lotus* (rb)
Lotusine-D **(110)**	*Zizyphus lotus* (rb)
Lotusine-E **(33)**	*Zizyphus lotus* (rb)
Lotusine-F **(8)**	*Zizyphus lotus* (rb)
Mauritine-A **(122)**	*Zizyphus jujuba* (sb), *Z. mauritiana* (sb), *Z. nummularia* (sb), *Z. spina-christi* (sb)
Mauritine-B **(128)**	*Zizyphus mauritiana* (sb)
Mauritine-C **(108)**	*Zizyphus mauritiana* (sb), *Z. nummularia* (rb), *Z. spina-christi* (sb)
Mauritine-D **(125)**	*Zizyphus mauritiana* (sb), *Z. nummularia* (sb), *Z. xylopyra* (sb)
Mauritine-E **(124)**	*Zizyphus mauritiana* (sb)
Mauritine-F **(121)**	*Zizyphus mauritiana* (sb), *Z. nummularia* (rb)
Mauritine-H **(123)**	*Zizyphus mauritiana* (sb)
Melofoline **(51)**	*Melochia corchorifolia* (ae)
Melonovine-A **(46)**	*Melochia tomentosa* (r)
Melonovine-B **(64)**	*Melochia tomentosa* (r)
4-Methoxy-abyssinine-A	see Mucronine-E **(149)**
4-Methoxy-abyssinine-C	see Mucronine-G **(146)**
N-Methyl-americine **(68)**	*Ceanothus integerrimus* var. *integerrimus* (rb), *C. sanguineus* (rb)
O-Methyl-sanjoinine-G1	see Sanjoinine-D **(119)**
Mucronine-A **(151)**	*Zizyphus abyssinica* (sb), *Z. mucronata* (sb)
Mucronine-B **(150)**	*Zizyphus abyssinica* (sb), *Z. mucronata* (sb)
Mucronine-C **(144)**	*Zizyphus abyssinica* (sb), *Z. mucronata* (sb)
Mucronine-D **(37)**	*Zizyphus jujuba* (sb), *Z. jujuba* var. *inermis* (sb), *Z. mucronata* (r, sb), *Z. nummularia* (rb), *Z. sativa* (sb)
Mucronine-E **(149)**	*Zizyphus mucronata* (sb)
Mucronine-F **(145)**	*Zizyphus mucronata* (sb)
Mucronine-G **(146)**	*Zizyphus mucronata* (sb)
Mucronine-H **(148)**	*Zizyphus mucronata* (sb)
Myrianthine-A **(89)**	*Myrianthus arboreus* (l)
Myrianthine-B	see Adouetine-Y′ **(62)**
Myrianthine-C **(48)**	*Myrianthus arboreus* (l), *Plectronia odorata* (ae)
Nummularine-A **(35)**	*Zizyphus jujuba* (sb), *Z. nummularia* (rb, sb)
Nummularine-B **(23)**	*Zizyphus jujuba* (sb), *Z. jujuba* var. *inermis* (sb), *Z. nummularia* (rb, sb), *Z. sativa* (sb), *Z. xylopyra* (sb)

Table 8. *(continued)*

Compound	Sources (plant part)
Nummularine-C **(11)**	*Zizyphus nummularia* (rb, sb)
Nummularine-D **(84)**	*Zizyphus nummularia* (rb, sb)
Nummularine-E **(86)**	*Zizyphus nummularia* (rb, sb), *Z. hysodrica* (sb)
Nummularine-F **(105)**	*Zizyphus nummularia* (rb, sb)
Nummularine-G **(87)**	*Zizyphus nummularia* (sb)
Nummularine-H **(38)**	*Zizyphus nummularia* (sb)
Nummularine-K **(74)**	*Zizyphus nummularia* (sb), *Z. xylopyra* (sb)
Nummularine-M **(91)**	*Zizyphus nummularia* (sb)
Nummularine-N **(22)**	*Zizyphus nummularia* (sb)
Nummularine-O **(40)**	*Zizyphus nummularia* (rb, sb)
Nummularine-P **(18)**	*Zizyphus nummularia* (sb)
Nummularine-R **(14)**	*Zizyphus nummularia* (sb)
Nummularine-S **(6)**	*Zizyphus nummularia* (sb)
Nummularine-T **(29)**	*Zizyphus nummularia* (sb)
Paliurine-B **(34)**	*Paliurus ramosissimus* (r, s)
Pandamine **(116)**	*Panda oleosa* (rb)
Pandaminine **(115)**	*Panda oleosa* (rb)
Pubescine-A **(45)**	*Discaria pubescens* (?)
Rugosanine-A **(21)**	*Zizyphus rugosa* (sb)
Rugosanine-B **(16)**	*Zizyphus rugosa* (sb)
Sanjoinenine **(158)**	*Zizyphus lotus* (ae), *Z. vulgaris* var. *spinosus* (sd)
Sanjoinine-A	see Frangufoline **(61)**
Sanjoinine-B **(57)**	*Zizyphus vulgaris* var. *spinosus* (sd)
Sanjoinine-D **(119)**	*Zizyphus vulgaris* var. *spinosus* (sd)
Sanjoinine-F **(67)**	*Zizyphus lotus* (ae), *Z. vulgaris* var. *spinosus* (sd)
Sanjoinine-G1 **(117)**	*Zizyphus vulgaris* var. *spinosus* (sd)
Sanjoinine-G2 **(152)**	*Zizyphus vulgaris* var. *spinosus* (sd)
Sativanine-A **(85)**	*Zizyphus sativa* (sb), *Z. spina-christi* (sb)
Sativanine-B **(82)**	*Zizyphus sativa* (sb)
Sativanine-C **(19)**	*Zizyphus sativa* (sb)
Sativanine-D **(20)**	*Zizyphus sativa* (sb)
Sativanine-E **(13)**	*Zizyphus sativa* (sb)
Sativanine-F **(31)**	*Zizyphus sativa* (sb)
Sativanine-G **(4)**	*Zizyphus sativa* (sb)
Sativanine-H **(17)**	*Zizyphus sativa* (sb)
Sativanine-K **(5)**	*Zizyphus sativa* (sb)
Scutianene-C **(159)**	*Scutia buxifolia* (r)
Scutianine-A **(136)**	*Scutia buxifolia* (sb)
Scutianine-B **(70)**	*Discaria febrifuga* (sb), *Melochia tomentosa* (r), *Scutia buxifolia* (r, sb)
Scutianine-C **(60)**	*Scutia buxifolia* (r, sb), *Zizyphus nummularia* (sb)
Scutianine-D **(77)**	*Scutia buxifolia* (r, sb)
Scutianine-E **(78)**	*Scutia buxifolia* (sb)
Scutianine-F **(135)**	*Scutia buxifolia* (sb)
Scutianine-G **(79)**	*Scutia buxifolia* (sb)

Table 8. *(continued)*

Compound	Sources (plant part)
Scutianine-H (**66**)	*Scutia buxifolia* (sb)
Scutianine-J (**80**)	*Scutia buxifolia* (sb)
Spinanine-A (**106**)	*Zizyphus spina-christi* (sb)
Subfraction I (*69*) (**9**)	*Sphaeranthus indicus* (fl)
Subfraction II (*69*) (**1**)	*Sphaeranthus indicus* (fl)
Texensine (**71**)	*Colubrina texensis* (ae)
Tscheschamine (**7**)	*Zizyphus sativa* (sb)
Zizyphine-A (**27**)	*Zizyphus oenoplia* (rb, sb)
Zizyphine-B (**24**)	*Zizyphus oenoplia* (rb, sb)
Zizyphine-C (**32**)	*Zizyphus oenoplia* (sb)
Zizyphine-D (**147**)	*Zizyphus oenoplia* (sb)
Zizyphine-E (**143**)	*Zizyphus oenoplia* (sb)
Zizyphine-F (**25**)	*Zizyphus oenoplia* (sb), *Z. spina-christi* (sb)
Zizyphine-G (**107**)	*Zizyphus oenoplia* (sb)
Zizyphinine	see Zizyphine-B (**24**)

ae = aerial parts, fl = flowers, fr = fruits, l = leaves, r = roots, rb = root bark, s = stems, sb = stem bark, sd = seeds, tb = terminal branches, wh = whole plant

Table 9. *Plant Index*

Asteraceae	*Sphaeranthus indicus* **1, 9**
Celastraceae	*Euonymus europaeus* **52, 54, 61**
Euphorbiaceae	*Antidesma montana* **62, 95**
	Hymenocardia acida **139**
Pandaceae	*Panda oleosa* **115, 116**
Rhamnaceae	*Araliothamnus vaginatus* **81, 94, 102, 103**
	Ceanothus
	C. americanus **42, 43, 53, 54, 56, 59, 65, 69, 99, 100**
	C. integerrimus **90, 93, 104**
	C. integerrimus var. *californicus* **72, 83, 90, 98**
	C. integerrimus var. *integerrimus* **68, 72, 83, 90, 98, 101, 104**
	C. sanguineus **56, 58, 61, 62, 68, 72**
	Colubrina texensis **71**
	Discaria
	D. crenata **96**
	D. febrifuga **49, 52, 55, 61, 62, 70, 88, 97, 113, 114, 118, 120**
	D. longispina **54, 55, 61, 62, 72, 73, 75**
	D. pubescens **45**
	Hovenia
	H. dulcis **50, 54**
	H. tomentella **50, 54**
	Lasiodiscus marmoratus **134, 153**
	Paliurus ramosissimus **34**

Table 9. (*continued*)

	Rhamnus frangula **52, 54, 61**
	Scutia buxifolia **60, 66, 70, 77, 78, 79, 80, 135, 136, 159**
	Zizyphus
	Z. abyssinica **140, 141, 142, 144, 150, 151**
	Z. amphibia **26, 76, 109, 112, 130, 131, 132, 133**
	Z. hutchinsonii **126**
	Z. juazeiro **131**
	Z. jujuba **23, 26, 35, 37, 39, 41, 61, 122, ****
	Z. jujuba var. *inermis* **2, 10, 12, 15, 23, 28, 30, 37, 47, 52, 54, 61**
	Z. lotus **8, 33, 61, 63, 67, 110, 111, 127, 129, 158, 160**
	Z. mauritiana **61, 108, 109, 121, 122, 123, 124, 125, 128, 131, 132, 133**
	Z. mucronata **3, 36, 37, 144, 145, 146, 148, 149, 150, 151**
	Z. nummularia **6, 11, 14, 18, 22, 23, 26, 29, 35, 37, 38, 40, 41, 60, 61, 74, 76, 84, 86, 87, 90, 91, 105, 121, 122, 125**
	Z. oenoplia **24, 25, 27, 32, 107, 141, 142, 143, 147**
	Z. rugosa **16, 21, 131**
	Z. sativa **4, 5, 7, 13, 17, 19, 20, 23, 31, 37, 54, 82, 85**
	Z. spina-christi **25, 26, 39, 52, 76, 85, 106, 108, 109, 122, 133**
	Z. vulgaris var. *spinosus* **57, 61, 67, 117, 119, 131, 152, 158**
	Z. xylopyra **23, 26, 74, 125**
Rubiaceae	*Canthium euryoides* **92**
	Feretia apodanthera **137, 138**
	Plectronia odorata **44, 48**
Sterculiaceae	*Melochia*
	M. corchorifolia **51, 52, 61, 62**
	M. pyramidata **61, 90, 138**
	M. tomentosa **46, 64, 70**
	Waltheria americana **53, 62, 100, 138**
Urticaceae	*Myrianthus arboreus* **48, 62, 89**

Addendum

It covers data published prior to 1995 that have been omitted, as well as new data appeared in 1996.

Physical and Spectral Data of Cyclopeptide Alkaloids and Related Compounds

5(13)-Zizyphine-A-Type Cyclopeptide Alkaloids

Zizyphine-K

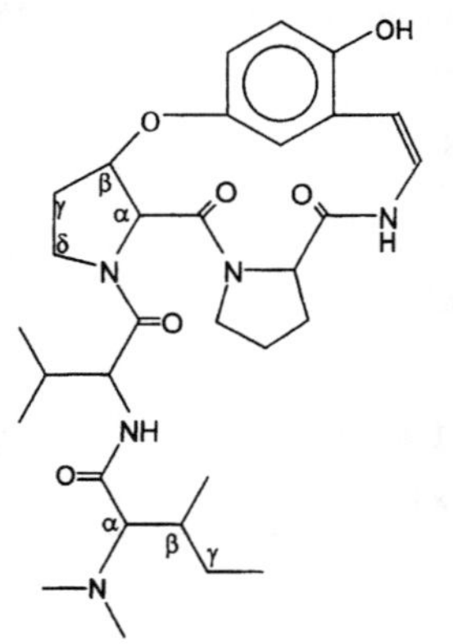

A = N,N-diMe-Ileu
B = β-OH-Pro
C = Pro
E = Val
$C_{31}H_{45}N_5O_6$, 583

Mp = 230 (*158*)
UV (MeOH) : 220, 260, 315 (*158*)
IR ($CHCl_3$) : 3350, 3000, 2970, 2780, 1675, 1635, 1590, 1450, 1225, 1050 (*158*)
MS : 583(M^+), 526(b), 470(c), 468(d), 371(f), 370(g), 344(h), 343(i), 342(j), 310(k), 274(u), 245(r), 219(s), 202(t), 193(p), 165(q), 151(x), 114(a=100%), 96(v), 86(p'), 70(q'), 68(w) (*158*)
^{1}H-NMR (60 and 90MHz, $CDCl_3$) : 0.75-1.30, 12H (4xC-Me), 2.15, 6H (NMe_2), 4.20 (δ-H-β-OH-Pro) (*158*)
Sources : *Zizyphus oenoplia* (Rhamnaceae)-stem bark (*158*)

Zizyphine-I

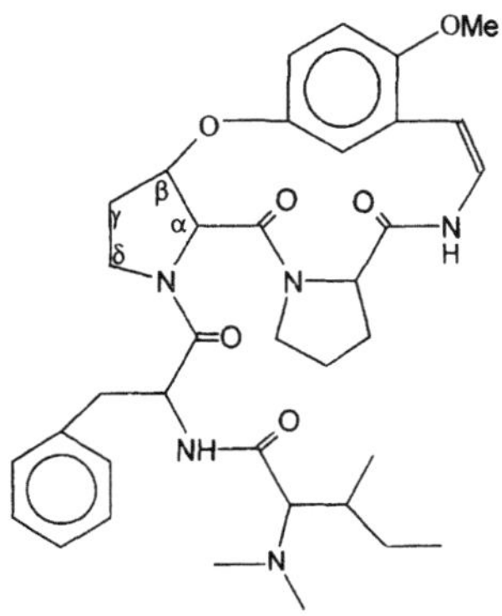

A = N,N-diMe-Ileu
B = β-OH-Pro
C = Pro
E = Phe
$C_{36}H_{47}N_5O_6$, 645

Mp = 135 (*159*)
UV (MeOH) : 255, 320 (*159*)
IR ($CHCl_3$) : 3335, 2950, 2875, 2745, 1670, 1630, 1610, 1480, 1225, 1030 (*159*)
MS : 645(M^+), 588(b), 532(c), 530(d), 385(f), 384(g), 358(h), 357(i), 356(j), 288(u), 259(r), 233(s), 216(t), 193(p), 165(q/x), 114(a=100%), 96(v), 86(p'), 70(q'), 68(w) (*159*)
^{1}H-NMR (60 and 90MHz, $CDCl_3$) : 0.90-1.30, 6H (2xC-Me), 2.30, 6H (NMe_2), 3.88, 3H (OMe), 4.39, 1H (α-H-β-OH-Pro), 6.70-7.20, 10H (olefinic + aromatic-H) (*159*)
Sources : *Zizyphus oenoplia* (Rhamnaceae)-stem bark (*159*)

4(14)-Frangulanine-Type Cyclopeptide Alkaloids

Anorldianine

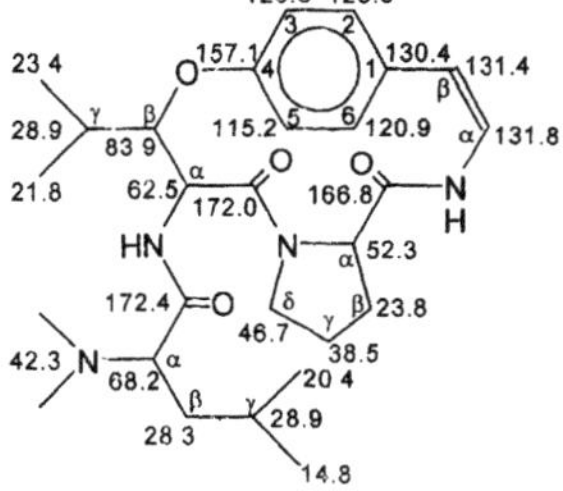

A = N,N-diMe-Leu
B = *threo*-β-OH-Leu[2]
C = Pro
$C_{27}H_{40}N_4O_4$, 484

Mp = 160 (*160*)
UV (EtOH) : 228 (4.80) (*160*)
IR (?) . 3400, 2925, 2850, 1720, 1625, 1270, 1040, 970, 700 (*160*)
MS : 485($M+H^+$), 484(M^+), 328(j), 190(f), 166(l), 135(i), 114(a=100%), 97(m), 57 (*160*)
^{1}H-NMR (360MHz, D_2O) : 2.00m, 1H (γ-H-β-OH-Leu), 2.12s, 6H (N-Me_2), 2.26dd, 12.0, 5.7 Hz, 1H (β-H-Pro), 2.69dd, 7.8, 6.6 Hz, 1H (δ-H-Pro), 3.32ddd, 10.1, 8.2, 2.1 Hz, 1H (β-H-N,N-diMe-Leu), 3.90dt, 10.1, 7.0 Hz, 1H (β-H-N,N-diMe-Leu), 4.24d, 7.4 Hz, 1H (α-H-Pro), 4.85d, 0.7 Hz, 2H (α-H- + β-H-β-OH-Leu), 6.13d, 10.7 Hz, 1H (NH-Sty), 6.36d, 7.6 Hz, 1H (β-H-Sty), 6.66dd, 10.4, 7.6 Hz, 1H (α-H-Sty), 7.00dd, 8.2, 1.9 Hz, 1H (5-H-Sty), 7.07dd, 8.5, 1.9 Hz, 1H (3-H-Sty), 7.11dd, 8.2, 2.5 Hz, 1H (6-H-Sty), 7.18dd, 8.5, 2.5 Hz, 1H (2-H-Sty) (*160*)
^{13}C-NMR (50.32MHz, $CDCl_3$) : see figure (*160*)
Sources : *Canthium anorldianum* (Rubiaceae)-stem bark (*160*)

Discarine-I

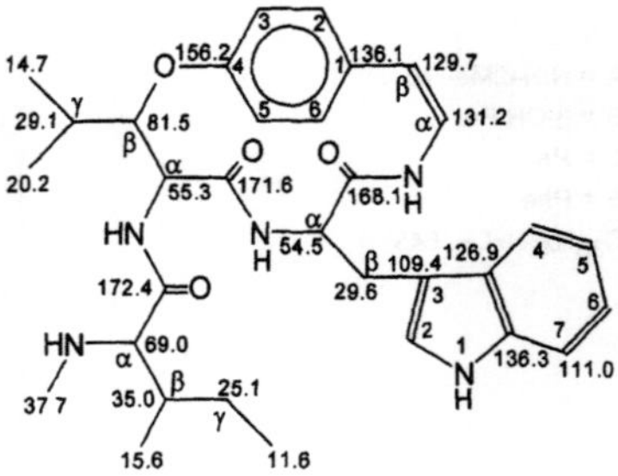

A = N-Me-Ileu
B = *erythro*-β-OH-Leu[2]
C = Trp
$C_{32}H_{41}N_5O_4$, 559

Mp = 140 (*161*)
$[a]_D$ (25) = -149 (c=0.1, MeOH) (*161*)
UV (MeOH) : 227.5, 270, 278, 289 (*161*)
IR (?) : 3400, 2800, 1690-1650, 1230 (*161*)
MS : 559 (M^+), 502(b), 460(g), 417(j), 304(v), 283(k), 255(l), 247(t), 190(f), 181(c), 170(r), 159(q), 153(d), 135(i), 130, 100(a=100%), 97(m), 71(n), 69, 57(p-Me) (*161*)
^{1}H-NMR (400MHz, $CDCl_3$ + 20% CD_3OD) : 0.80d, 7.0 Hz, 3H (CH-Me-N-Me-Ileu), 0.83t, 7.0 Hz, 3H (CH_2-Me-N-Me-Ileu), 0.92d, 7.0 Hz, 3H (Me-β-OH-Leu), 1.12d, 7.0 Hz, 3H (Me-β-OH-Leu), 1.88s, 3H (NH-Me), 2.00m, 1H (γ-H-β-OH-Leu), 2.62d, 5.0 Hz, 1H (α-H-N-Me-Ileu), 4.20t, 7.0 Hz, 1H (α-H-Trp); 4.42d, 8.0 Hz, 1H (α-H-β-OH-Leu), 4.78dd, 8.0, 2.0 Hz, 1H (β-H-β-OH-Leu), 6.14d, 7.0 Hz, 1H (β-H-Sty), 6.60d, 7.0 Hz, 1H (α-H-Sty), 7.27d, 8.0 Hz, 1H (7-H-Trp), 7.43d, 8.0 Hz, 1H (4-H-Trp) (*161*)
^{13}C-NMR (100MHz, $CDCl_3$ + 1% CD_3OD) : see figure (*161*)
Sources : *Discaria febrifuga* (Rhamnaceae)-root bark (*161*)

4(14)-Amphibine-F-Type Cyclopeptide Alkaloids

Mucronine-J

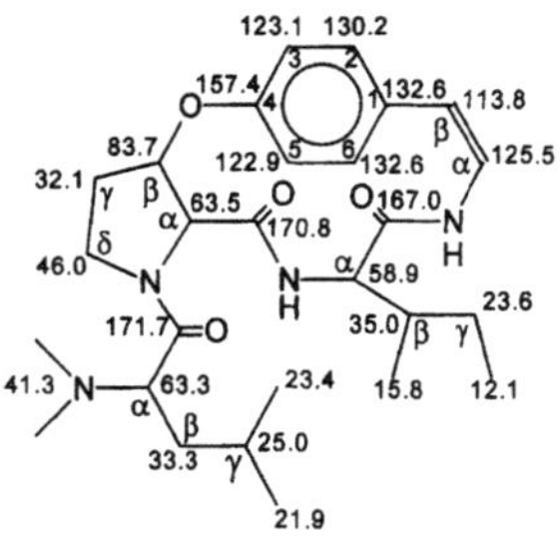

A = N,N-diMe-L-Leu[1] (*162*)
B = *trans*-β-OH-Pro[1] (*162*)
C = L-Ileu[1] (*162*)
$C_{27}H_{40}N_4O_4$, 484 (MS) (*162*)

[a]$_D$ (21) = -236 (c=1, $CHCl_3$) (*162*)
IR (neat) : 3398, 2963, 2931, 2880, 2791, 1658, 1632, 1511, 1460, 1229, 1171, 1106, 1031 (*162*)
MS : 427(b), 229(r), 209(p), 201(r-CO), 189(u')*, 161, 153(n), 135(i'), 114(a=100%), 86(p'/q'), 72(o') (*162*).
** Fragment u' corresponds to u of 4(14)-frangulanine- and -integerrine-type cyclopeptide alkaloids.*
MS (FAB+) : 485(M+H$^+$), 355(b'-CO+H$^+$), 341(j-H$^+$), 327(j-Me), 281, 267(y'+H$^+$), 221, 207 (*162*)
MS (HRFAB) : 485.3088(M+H$^+$) (*162*)
^{1}H-NMR (300.13MHz, $CDCl_3$) : 0.69d, 7.0 Hz, 3H (CH-Me-Ileu), 0.79d, 6.3 Hz, 3H (Me-N,N-diMe-Leu), 0.82t, 7.2 Hz, 3H (CH_2-Me-Ileu), 0.83d, 6.3 Hz, 3H (Me-N,N-diMe-Leu), 1.05m, 1H (γ-H-Ileu), 1.20m, 1H (γ-H-Ileu), 1.23m, 2H (β-H- + γ-H-N,N-diMe-Leu), 1.67dd, 9.4, 9.1 Hz, 1H (β-H-N,N-diMe-Leu), 2.13m, 1H (γ-H-β-OH-Pro), 2.18m, 1H (β-H-Ileu), 2.27s, 6H (N-Me_2), 2.55ddd, 12.2, 7.2, 5.2 Hz, 1H (γ-H-β-OH-Pro), 3.21dd, 9.4, 3.4 Hz, 1H (α-H-N,N-diMe-Leu), 3.32ddd, 13.1, 10.8, 5.2 Hz, 1H (δ-H-β-OH-Pro), 4.16dd, 8.8, 2.9 Hz, 1H (α-H-Ileu), 4.28dd, 10.8, 8.2 Hz, 1H (δ-H-β-OH-Pro), 4.31d, 5.3 Hz, 1H (α-H-β-OH-Pro), 5.59ddd, 9.8, 7.2, 5.3 Hz, 1H (β-H-β-OH-Pro), 6.27d, 7.7 Hz, 1H (β-H-Sty), 6.52d, 10.8 Hz, 1H (NH-Sty), 6.70d, 8.8 Hz, 1H (NH-Ileu), 6.74dd, 10.8, 7.7 Hz, 1H (α-H-Sty), 7.06m, 1H (6-H-Sty), 7.10m, 2H (2-H- + 3-H-Sty), 7.28dd, 9.0, 2.6 Hz, 1H (5-H-Sty) (*162*)
^{13}C-NMR (75.47MHz, $CDCl_3$) : see figure (*162*)
NOESY, COSY, HMBC and HMQC-NMR spectra (*162*)
Sources : *Zizyphus mucronata* (Rhamnaceae)-root bark (*162*)

(–)-Nummularine-F (Synthetic Product) (*163*)

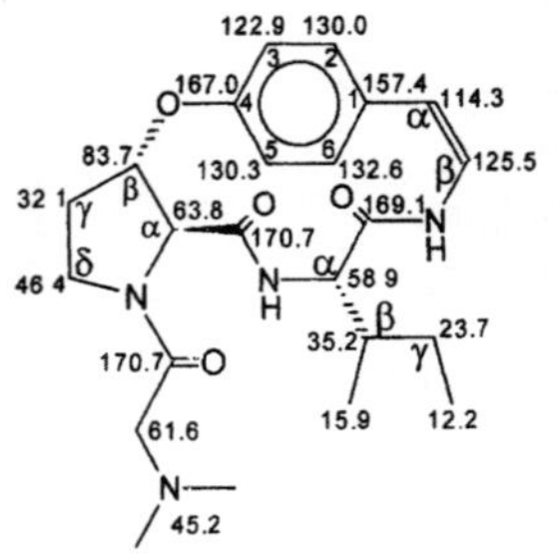

A = N,N-diMe-Gly
B = β-OH-Pro
C = Ileu
$C_{23}H_{33}N_4O_4$, 428 (MS) (*163*)

Mp = 152-154 (*163*)
[a]$_D$ (20) = -197 (c=0.45, MeOH) (*163*)
IR ($CHCl_3$) : 3410, 2980, 2950, 2890, 2870, 2840, 2790, 1690, 1625, 1500, 1480, 1455, 1360, 1310, 1255, 1170, 1115, 1095, 1080, 1020, 860 (*163*)
MS (HRMS) : 429.255($M+H^+$) (*163*)
^{1}H-NMR (500MHz, $CDCl_3$) : 0.75d, 6.9 Hz, 3H (CH-Me-Ileu), 0.85t, 7.3 Hz, 3H (CH_2-Me-Ileu), 1.08-1.14m, 1H (γ-H-Ileu), 1.22-1.24m, 1H (γ-H-Ileu), 2.11-2.16m, 10.6, 8.3, 7.1, 5.4 Hz, 1H (*pro-R*-γ-H-β-OH-Pro), 2.17-2.25m, 1H (β-H-Ileu), 2.28s, 6H (N-Me_2), 2.55-2.60m, 12.9, 10.6, 9.8 Hz, 1H (*pro-S*-γ-H-β-OH-Pro), 3.04d and 3.15d, 14.2 Hz, 2H (CH_2-N,N-diMe-Gly), 3.42-3.48m, 12.9, 11.5, 5.4 Hz, 1H (*pro-S*-δ-H-β-OH-Pro), 4.08dd, 11.4, 8.3 Hz, 1H (*pro-R*-δ-H-β-OH-Pro), 4.18dd, 8.7, 3.1 Hz, 1H (α-H-Ileu), 4.29d, 5.4 Hz, 1H (α-H-β-OH-Pro), 5.56-5.61m, 9.8, 7.1, 5.4 Hz, 1H (β-H-β-OH-Pro), 6.30d, 7.8 Hz, 1H (β-H-Sty), 6.52d, 10.6 Hz, 1H (NH-Sty), 6.62d, 8.7 Hz, 1H (NH-Ileu), 6.76dd, 10.6, 7.8 Hz, 1H (α-H-Sty), 7.10d, 8.8 Hz, 1H (Ar-H-Sty), 7.14s, 2H (2xAr-H-Sty), 7.30d, 8.8 Hz, 1H (Ar-H-Sty) (*163*)
^{13}C-NMR (125MHz, $CDCl_3$) : see figure (*163*)

4(14)-Pandamine-Type Cyclopeptide Alkaloids

Sanjoinine-G1

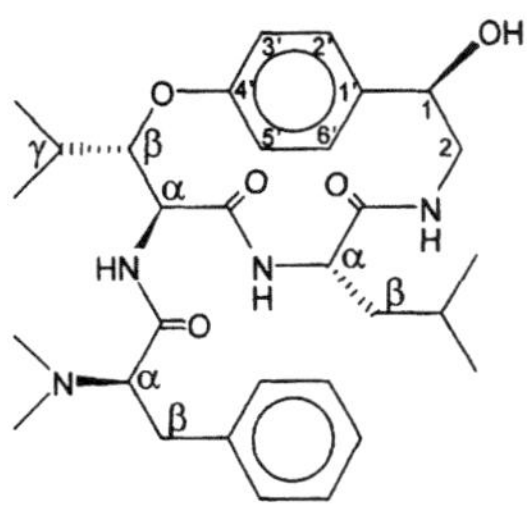

A = N,N-diMe-Phe
B = *erythro*-β-OH-Leu2
C = Leu
$C_{31}H_{44}N_4O_5$, 552.3273 (MS) (*164*)

Mp = 236 (*164*)
[a]D (20) = -68.6 (c=0.175, $CHCl_3$) (*164*)
UV (MeCN) : 232 (3.97), 278 (3.30) (*164*)
CD (MeCN) : +2.1(275) (*164*)
IR (KBr) : 3300, 3270, 2770, 1670, 1230 (*164*)
MS : 552(M^+), 537(M^+-CH_3), 509(M^+-C_3H_7), 491(M^+-C_3H_7-H_2O), 461(b), 443(b"), 210(k), 208(-), 207(f'-H^+), 195(c), 190(f), 189(f-H^+), 182(l), 167(d), 153(i'), 148(a=100%), 135(i), 97(m/r), 86(p/q) (*164*)
^{1}H-NMR (360MHz, $CDCl_3$) : 0.76d, 5.4 Hz, 6H (2xMe-Leu), 0.95d, 6.5 Hz, 3H (Me-β-OH-Leu), 1.18d, 6.5 Hz, 3H (Me-β-OH-Leu), 1.27dd, 6.7, 5.8 Hz, 2H (2xβ-H-Leu), 1.31m, 1H (γ-H-Leu), 1.92m, 1H (γ-H-β-OH-Leu), 2.23s, 6H (N-Me_2), 2.90dd, 14.0, 6.2 Hz, 1H (β-H-N,N-diMe-Phe), 3.06d, 14.4 Hz, 1H (2α-H-Phe-Et), 3.12dd, 14.0, 6.2 Hz, 1H (β-H-N,N-diMe-Phe), 3.22t, 6.2 Hz, 1H (α-H-N,N-diMe-Phe), 3.95dt, 9.2, 6.7 Hz, 1H (α-H-Leu), 4.27ddd, 14.4, 11.0, 3.8 Hz, 1H (2β-H-Phe-Et), 4.32dd, 9.8, 8.6 Hz, 1H (α-H-β-OH-Leu), 4.79dd, 8.6, 1.8 Hz, 1H (β-H-β-OH-Leu), 5.17d, 3.8 Hz, 1H (1α-H-Phe-Et), 5.66d, 11.0 Hz, 1H (NH-Phe-Et), 6.05d, 9.2 Hz, 1H (NH-Leu), 6.43s, 1H (OH), 6.81dd, 8.0, 2.5 Hz, 1H (3'-H-Phe-Et), 6.94dd, 8.2, 2.5 Hz, 1H (5'-H-Phe-Et), 6.97dd, 8.0, 2.0 Hz, 1H (2'-H-Phe-Et), 7.19-7.22m, 5H (Ar-H-N,N-diMe-Phe), 7.32dd, 8.2, 2.0 Hz, 1H (6'-H-Phe-Et), 7.40d, 9.8 Hz, 1H (NH-β-OH-Leu) (*164*)
Derivatives : Sanjoinine-G1-monoacetate (*164*)
Sanjoinine-G1-benzoate (*164*)
Sanjoinine-G1-*p*-bromobenzoate (*164*)
Sources : *Zizyphus vulgaris* var. *spinosus* (Rhamnaceae)-seeds (*164*)

5(14)-Amphibine-B-Type Cyclopeptide Alkaloids

Mauritine-J (= *N*-Desmethyl-amphibine-E)

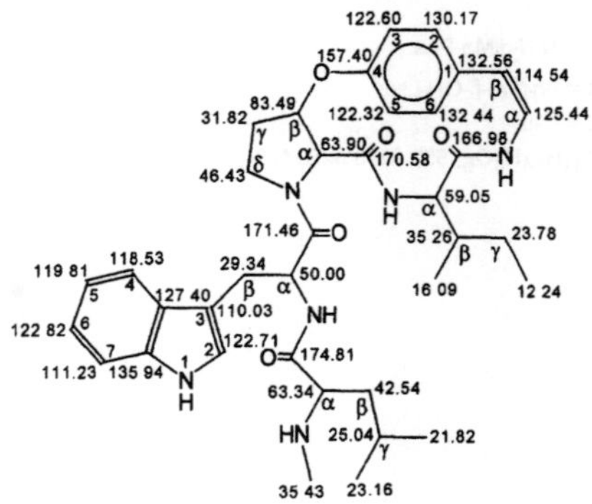

A = N-Me-Leu
B = *trans*-β-OH-Pro[1] (*165*)
C = Ileu
E = Trp
$C_{37}H_{48}N_6O_5$, 656

Mp = amorphous (*165*)
[a]D (?) = -175.9 (c=1, MeOH) (*165*)
UV (MeOH) : 220, 270, 281, 290 (*165*)
IR (KBr) : 3390, 1691, 1235, 1040(*165*)
MS (CI) : 658(M+2H$^+$), 657(M+H$^+$=100%) (*165*)
^{1}H-NMR (300MHz, $CDCl_3$) : 0.82d, 7.0 Hz, 3H (CH-Me-Ileu), 0.87d, 6.5 Hz, 3H (Me-N-Me-Leu), 0.88t, 6.5 Hz, 3H (CH_2-Me-Ileu), 0.89d, 6.5 Hz, 3H (Me-N-Me-Leu), 1.15m, 1H (γ-H-Ileu), 1.25m, 1H (β-H-N-Me-Leu), 1.30m, 1H (γ-H-Ileu), 1.42m, 1H (β-H-N-Me-Leu), 1.59m, 1H (γ-H-N-Me-Leu), 1.90br s, 1H (NH-N-Me-Leu), 2.00m, 1H (γ-H-β-OH-Pro), 2.22m, 1H (β-H-Ileu), 2.32m, 1H (γ-H-β-OH-Pro), 2.38s, 3H (NH-Me), 2.50m, 1H (δ-H-β-OH-Pro), 2.90dd, 14.5, 8.0 Hz, 1H (β-H-Trp), 2.98dd, 9.2, 5.0 Hz, 1H (α-H-N-Me-Leu), 3.13dd, 14.5, 6.2 Hz, 1H (β-H-Trp), 3.79dd, 11.5, 8.3 Hz, 1H (δ-H-β-OH-Pro), 4.20dd, 8.5, 3.0 Hz, 1H (α-H-Ileu), 4.25d, 5.5 Hz, 1H (α-H-β-OH-Pro), 5.00ddd, 8.5, 8.0, 6.2 Hz, 1H (α-H-Trp), 5.43m, 1H (β-H-β-OH-Pro), 6.30d, 7.7 Hz, 1H (β-H-Sty), 6.51d, 10.5 Hz, 1H (NH-Sty), 6.58d, 8.5 Hz, 1H (NH-Ileu), 6.73dd, 10.5, 7.7 Hz, 1H (α-H-Sty), 6.76d, 2.5 Hz, 1H (2-H-Trp), 7.05m, 1H (6-H-Sty), 7.09m, 1H (2-H-Sty), 7.10m, 1H (6-H-Trp), 7.12m, 1H (5-H-Trp), 7.18m, 1H (5-H-Sty), 7.24m, 1H (3-H-Sty), 7.31d, 8.0 Hz, 1H (7-H-Trp), 7.68d, 8.0 Hz, 1H (4-H-Trp), 7.73d, 8.5 Hz, 1H (NH-Trp), 8.12br s, 1H (1-H-Trp) (*165*)
^{13}C-NMR (75MHz, $CDCl_3$) : see figure (*165*)
COSY, NOESY, HMBC NMR spectra (*165*)
Sources : *Zizyphus mauritiana* (Rhamnaceae)-root bark (*165*)

133. Amphibine-E

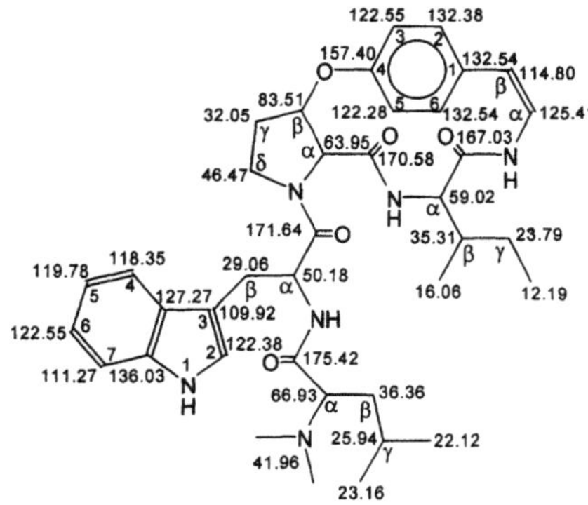

A = N,N-diMe-Leu
B = *trans*-β-OH-Pro[1] (*36*)
C = Ileu
E = Trp
$C_{38}H_{50}N_6O_5$, 670

[a]D (?) = -187.2 (c=1, MeOH) (*165*)
MS (CI) : 672(M+2H$^+$), 671(M+H$^+$=100%) (*165*)
^{1}H-NMR (300MHz, $CDCl_3$) : 0.80d, 7.0 Hz, 3H (CH-Me-Ileu), 0.85d, 6.5 Hz, 6H (2xMe-N,N-diMe-Leu), 0.89t, 6.5 Hz, 3H (CH_2-Me-Ileu), 1.08m, 1H (γ-H-Ileu), 1.25m, 1H (β-H-N,N-diMe-Leu), 1.29m, 1H (γ-H-Ileu), 1.48m, 1H (β-H-N,N-diMe-Leu), 1.59m, 1H (γ-H-N,N-diMe-Leu), 2.02m, 1H (γ-H-β-OH-Pro), 2.15s, 6H (N-Me_2), 2.19m, 1H (β-H-Ileu), 2.30m, 1H (γ-H-β-OH-Pro), 2.55m, 1H (δ-H-β-OH-Pro), 2.88dd, 9.2, 5.0 Hz, 1H (α-H-N,N-diMe-Leu), 2.90dd, 14.5, 8.0 Hz, 1H (β-H-Trp), 3.10dd, 14.5, 6.2 Hz, 1H (β-H-Trp), 4.18dd, 8.5, 3.0 Hz, 1H (α-H-Ileu), 4.25d, 5.5 Hz, 1H (α-H-β-OH-Pro), 5.05ddd, 8.5, 8.0, 6.2 Hz, 1H (α-H-Trp), 5.42m, 1H (β-H-β-OH-Pro), 6.29d, 7.7 Hz, 1H (β-H-Sty), 6.52d, 10.5 Hz, 1H (NH-Sty), 6.62d, 8.5 Hz, 1H (NH-Ileu), 6.72dd, 10.5, 7.7 Hz, 1H (α-H-Sty), 6.75d, 2.5 Hz, 1H (2-H-Trp), 7.05m, 1H (6-H-Sty), 7.08m, 1H (2-H-Sty), 7.10m, 1H (5-H- + 6-H-Trp), 7.20m, 1H (5-H-Sty), 7.22m, 1H (3-H-Sty), 7.28d, 8.0 Hz, 1H (7-H-Trp), 7.55d, 8.5 Hz, 1H (NH-Trp), 7.64d, 8.0 Hz, 1H (4-H-Trp), 8.41br s, 1H (1-H-Trp)(*165*)
^{13}C-NMR (75MHz, $CDCl_3$) : see figure (*165*)
Sources : *Zizyphus mauritiana* (Rhamnaceae)-root bark (*165*)

Synthesis

The reported synthesis of frangulanine (*166*), the major representative of this type, constitutes the first total synthesis of a 4(14)-membered cyclopeptide alkaloid. This synthesis (>26 steps in an overall yield of *ca* 0.30%), starting from (2*S*,3*S*)-2-hydroxymethyl-3-isopropyloxirane, comprises the preparation of a oxazolidinone derivative, subsequently conversed to the corresponding aryloxy compound, the coupling of the (*Z*)-leucine, the ring closure, the installation of the insaturation of the styrylamine unit and, finally, the insertion of the (*S*)-*N*,*N*-dimethylisoleucine moiety.

The total synthesis of (−)-nummularine-F (*163*), a 4(14)-amphibine-F-type cyclopeptide alkaloid, from D-serine was accomplished in 25 steps and an overall yield of 0.48%. This approach includes a stereoselective synthesis of a *cis*-2-(hydroxymethyl)-3-pyrrolidinol derivative, the inversion of the *cis*-3-hydroxyl function to the desired β-aryl ether, the cyclization to a pair of rigid para ansa 14-membered cyclopeptides and the installation of the double bond of the enamide unit.

The total synthesis of sanjoinine-G1, a 4(14)-pandamine-type cyclopeptide alkaloid, was achieved in 17 overall steps (1.36% overall yield) starting from D-serine (*167*). A novel synthetic protocol was developed, including highly diastereoselective synthesis of the (*S*,*S*)-β-phenoxyleucine unit.

A novel strategy towards the total synthesis of cyclopeptide alkaloids is based upon the hitherto unknown ring-forming process in which formation of aryl-alkyl ether bond is the crucial step (*168*).

Isolation of Known Compounds from New Sources

Abyssinine-A (**142**): *Zizyphus mucronata* (Rhamnaceae)-root bark (*162*)

Amphibine-B (**132**): *Zizyphus oenoplia* (Rhamnaceae)-stem bark (*169*)

Amphibine-E (**133**): *Zizyphus mauritiana* (Rhamnaceae)-root bark (*165*)

Discarine-B (**72**): *Discaria febrifuga* (Rhamnaceae)-root bark (*161*)

Frangufoline (**61**): *Zizyphus oenoplia* (Rhamnaceae)-stem bark (*169*)

Mauritine-D (**125**): *Zizyphus oenoplia* (Rhamnaceae)-stem bark (*169*)

Mucronine-D (**37**): *Zizyphus mucronata* (Rhamnaceae)-root bark (*162*)

Our study had already been sent for publication when a review article entitled "Macrocyclic Peptide Alkaloids from Plants" appeared covering cyclopeptide alkaloids (*170*). However from the total of 82 compounds presented as newly found (daechucyclopeptide-I = daechuine-S26) and accompanied by their chemical formulas and their sources, the following 32, namely adouetine-Y′ (**62**), amphibine-B (**132**), -E (**133**), -H (**26**), discarine-A (**75**), -B (**72**), franganine (**52**), frangufoline (**61**), frangulanine (**54**), integerrenine (**90**), jubanine-A (**39**), -B (**41**), mauritine-C (**108**), -D (**125**), mucronine-D (**37**), nummularine-B (**23**), -C (**11**), -E (**86**), -F (**105**), -G (**87**), -K (**74**), -M (**91**), -N (**22**), sativanine-A (**85**), -B (**82**), -C (**19**), -D (**20**), -E (**13**), -F (**31**), -G (**4**), zizyphine-F (**25**), -G (**107**), have already been reported in previous reviews (*1*, *2*, *4*, *6*). In addition 4 other new compounds found in our study, paliurine-B (**34**), melofoline (**51**), scutianine-J (**80**) and AM-2 (**95**), have been omitted.

References

1. Schmidt, U., A. Lieberknecht, and E. Haslinger: Peptide Alkaloids in "The Alkaloids". A. Brossi (Ed.), Academic Press, New York, Vol. 26, p. 299 (1985).
2. Tschesche, R., and E.U. Kaussmann: The Cyclopeptide Alkaloids in "The Alkaloids". R.H.F. Manske (Ed.), Academic Press, New York, Vol. 15, Chapter 4, p. 165 (1975).
3. Tschesche, R.: Cyclopeptide Alkaloids Containing a Styrylamine Ring Unit. Heterocycles, **4**, 107 (1976).
4. Shah, A.H., and V.B. Pandey: Progress in the Chemistry of Cyclopeptide Alkaloids Containing a Styrylamine Unit. J. Chem. Soc. Pak., **7**, 363 (1985).
5. Bhat, K.L., and M.M. Joullié: Cyclopeptide Alkaloids. J. Chem. Educ., **64**, 21 (1987).
6. Joullié, M.M., and R.F. Nutt: Cyclopeptide Alkaloids in "Alkaloids : Chemical and Biological Perspectives". S. W. Pelletier (Ed.), Wiley, New York, Vol. 3, p. 113 (1984).
7. Warnhoff, E.W.: Peptide Alkaloids. Fortschr. Chem. Org. Naturst., **28**, 162 (1970).
8. Broadbent, T.A., and E.G. Paul: Carbon-13 Nuclear Magnetic Resonance in Alkaloid Chemistry. Heterocycles, **20**, 863 (1983).
9. Païs, M., J. Mainil, and R. Goutarel: Les Adouétines X, Y et Z, Alcaloïdes du *Waltheria americana* L. (Sterculiacées). Ann. Pharm. Fr., **21**, 139 (1963).
10. Menard, E.L., J.M. Müller, A.F. Thomas, S.S. Bhatnagar, and N.J. Dastoor: Über die Inhaltstoffe von *Zizyphus oenoplia* MILL. 1. Mitteilung: Isolierung der Inhaltstoffe. Helv. Chim. Acta, **46**, 1801 (1963).
11. Païs, M., F.-X. Jarreau, X. Lusinchi, and R. Goutarel: Alcaloïdes Peptidiques, III (I). Pandamine et Pandaminine, Alcaloïdes du *Panda oleosa* Pierre (Pandacées). Ann. Chim., **1**, 83 (1966).
12. Gonzalez Sierra, M., O.A. Mascaretti, F.J. Diaz, and E.A. Rúveda: The Stereochemistry of the β-Hydroxyleucine Unit of Frangulanine. J. Chem. Soc., Chem. Commun., 915 (1972).
13. Marchand, J., M. Païs, and F.-X. Jarreau: Étude par spectrométrie de RMN de quelques acides α-aminés β-hydroxylés et de leur éthers. Bull. Soc. Chim. Fr., **10**, 3742 (1971).
14. Marchand, J., F. Rocchiccioli, M. Païs, and F.-X. Jarreau: Alcaloïdes peptidiques. IX(1). - Détermination de la configuration relative du fragment β-hydroxyleucine de la lasiodine B par la synthèse stéréospécifique des thréo et érythro-β-(p-tolyloxy)-leucines. Bull. Soc. Chim. Fr., **12**, 4699 (1972).
15. Morel, A.F., R.V.F. Bravo, F.A.M. Reis, and E.A. Rúveda: Peptide Alkaloids of *Scutia buxifolia*, Phytochemistry, **18**, 473 (1979).
16. Païs, M., F.-X. Jarreau, M. Gonzalez Sierra, O.A. Mascaretti, E.A. Rúveda, C.-J. Chang, E.W. Hagaman, and E. Wenkert: Carbon-13 NMR Analysis of Cyclic Peptide Alkaloids. Phytochemistry, **18**, 1869 (1979).
17. Medina, E., and G. Spiteller: Über Inhaltsstoffe von *Melochia pyramidata* L. Liebigs Ann. Chem., 538 (1981).
18. Tschesche, R., and E. Ammermann: Scutianin-C, -D und -E, drei weitere Cyclopeptidalkaloide aus *Scutia buxifolia* Reiss. Chem. Ber., **107**, 2274 (1974).
19. Tschesche, R., and D. Hillebrand: Scutianin-G, ein weiteres Cyclopeptidalkaloid aus *Scutia buxifolia*. Phytochemistry, **16**, 1817 (1977).
20. Tschesche, R., D. Hillebrand, and I.R.C. Bick: Pubescine A, a Cyclopeptide Alkaloid from *Discaria pubescens*. Phytochemistry, **19**, 1000 (1980).

21. TAKAI, M., K.-I. KAWAI, Y. OGIHARA, Y. IITAKA, and S. SHIBATA: X-ray Analysis of Tri N-methylfrangulanine Methiodide. J. Chem. Soc., Chem. Commun., 653 (1974).

22. TAKAI, M., Y. OGIHARA, Y. IITAKA, and S. SHIBATA: Peptides in Higher Plants. I. The Conformation of Frangulanine. Chem. Pharm. Bull., **23**, 2256 (1975).

23. KIRFEL, A., G. WILL, R. TSCHESCHE, and H. WILHELM: Die Kristallstruktur des Mauritins-A. Z. Naturforsch., Teil B, **31**, 279 (1976).

24. PAÏS, M., X. MONSEUR, X. LUSINCHI, and R. GOUTAREL: Alcaloïdes peptidiques. II. – Structure de la pandamine, alcaloïde du *Panda oleosa* Pierre (Pandacées). Bull. Soc. Chim. Fr., 817 (1964).

25. TSCHESCHE, R., R. WELTERS, and H.-W. FEHLHABER: Scutianin, ein cyclisches Peptid-Alkaloid aus *Scutia buxifolia* Reiss. Chem. Ber., **100**, 323 (1967).

26. TSCHESCHE, R., J. RHEINGANS, H.-W. FEHLHABER, and G. LEGLER: Integerressin und Integerrenin, zwei Peptid-Alkaloide aus *Ceanothus integerrimus* Hook. et Arn. Chem. Ber., **100**, 3924 (1967).

27. TSCHESCHE, R., H. LAST, and H.-W. FEHLHABER: Frangulanin, ein Peptid-Alkaloid aus *Rhamnus frangula* L. Chem. Ber., **100**, 3937 (1967).

28. FEHLHABER, H.-W.: Massenspektrometrische Strukturermittlung von Peptid-Alkaloiden. Z. Anal. Chem., **235**, 91 (1968).

29. PAÏS, M., J. MARCHAND, G. RATLE, and F.-X. JARREAU: Alcaloïdes peptidiques. VI (1). – L' hymenocardine, alcaloïde de l' *Hymenocardia acida* Tul. (Hymenocardiacées). Bull. Soc. Chim. Fr., 2979 (1968).

30. TSCHESCHE, R., L. BEHRENDT, and H.-W. FEHLHABER: Aralionin, ein Peptid-Alkaloid aus *Araliothamnus vaginatus* Perrier. Chem. Ber., **102**, 50 (1969).

31. BOULVIN, G., R. OTTINGER, M. PAÏS, and G. CHIURDOGLU: Alcaloïdes Peptidiques. IX (1). La Canthiumine, Alcaloïde du *Canthium euryoides* Bull. (Rubiacées). Bull. Soc. Chim. Belges, **78**, 583 (1969).

32. TSCHESCHE, R., L. FROHBERG, and H.-W. FEHLHABER: Aralionin-B, ein Nebenalkaloid aus *Araliothamnus vaginata* Perrier. Chem. Ber., **103**, 2501 (1970).

33. MASCARETTI, O.A., V.M. MERKUZA, G.E. FERRARO, and E.A. RÚVEDA: Peptide Alkaloids of *Discaria longispina*. Phytochemistry, **11**, 1133 (1972).

34. TSCHESCHE, R., E.U. KAUSSMANN, and H.-W. FEHLHABER: Alkaloide aus Rhamnaceen, XI. Amphibin-A, ein cyclisches Peptidalkaloid aus *Zizyphus amphibia* A. Cheval. Tetrahedron Lett., **9**, 865 (1972).

35. TSCHESCHE, R., H. WILHELM, and H.-W. FEHLHABER: Alkaloide aus Rhamnaceen, XIV. Mauritin-A und Mauritin-B, zwei Peptidalkaloide aus *Zizyphus mauritiana* Lam. Tetrahedron Lett., **26**, 2609 (1972).

36. TSCHESCHE, R., E.U. KAUSSMANN, and H.-W. FEHLHABER: Alkaloide aus Rhamnaceen, XIII. Amphibin-B, -C, -D und -E, vier Peptidalkaloide aus *Zizyphus amphibia* A. Cheval. Chem. Ber., **105**, 3094 (1972).

37. TSCHESCHE, R., S.T. DAVID, J. UHLENDORF, und H.-W. FEHLHABER: Alkaloide aus Rhamnaceen, XV. Mucronin-D, ein weiteres cyclisches Peptid-Alkaloid aus *Zizyphus mucronata* Willd. Chem. Ber., **105**, 3106 (1972).

38. TSCHESCHE, R., E.U. KAUSSMANN, and G. ECKHARDT: Alkaloide aus Rhamnaceen, XVI. Über die Struktur des Zizyphins-A. Tetrahedron Lett., **28**, 2577 (1973).

39. WANI, M.C., H.L. TAYLOR, and M.E. WALL: Plant Antitumour Agents. XII. Texensine, a new Peptide Alkaloid from *Colubrina texensis*. Tetrahedron Lett., **47**, 4675 (1973).

40. TSCHESCHE. R., I. KHOKHAR, C. SPILLES, G. ECKHARDT, and B.K. CASSELS: Alkaloide aus Rhamnaceen, XXVI. Zizyphin-F und -G, neue Cyclopeptidalkaloide aus *Zizyphus oenoplia* Mill. Tetrahedron Lett., **34**, 2941 (1974).

41. TSCHESCHE, R., C. SPILLES, and G. ECKHARDT: Alkaloide aus Rhamnaceen, XVIII. Amphibin-F, -G und -H, weitere Peptidalkaloide aus *Zizyphus amphibia* A. Cheval. Chem. Ber., **107**, 686 (1974).
42. TSCHESCHE, R., H. WILHELM, E.U. KAUSSMANN, and G. ECKHARDT: Alkaloide aus Rhamnaceen, XVII. Mauritin-C, -D, -E und -F; neue Peptidalkaloide aus *Zizyphus mauritiana* Lam. Liebigs Ann. Chem., 1694 (1974).
43. TSCHESCHE, R., M. ELGAMAL, G.A. MIANA, and G. ECKHARDT: Alkaloids from Rhamnaceae-XXVI. Nummularine-D, -E and -F, new Cyclopeptide Alkaloids from *Zizyphus nummularia*. Tetrahedron, **31**, 2944 (1975).
44. KAPADIA, G.J., Y.N. SHUKLA, J.F. MORTON, and H.A. LLOYD: New Cyclopeptide Alkaloids from *Melochia tomentosa*. Phytochemistry, **16**, 1431 (1977).
45. LAGARIAS, J.C., D. GOFF, F.K. KLEIN, and H. RAPOPORT: Cyclopeptide Alkaloids. Phencyclopeptines from the Polymorphic Species *Ceanothus integerrimus*. J. Nat. Prod., **42**, 220 (1979).
46. LAGARIAS, J.C., D. GOFF, and H. RAPOPORT: Cyclopeptide Alkaloids. Phencyclopeptines from *Ceanothus sanguineus*. J. Nat. Prod., **42**, 663 (1979).
47. MOREL, A.F., R. HERZOG, J. BIERMANN, and W. VŒLTER: Ein neues Peptidalkaloid aus *Discaria febrifuga* Mart. Z. Naturforsch., Teil B, **39**, 1825 (1984).
48. MOREL, A.F., R. HERZOG, and W. VŒLTER: Discarin-E, ein neues Peptidalkaloid aus *Discaria febrifuga* Mart. Chimia, **39**, 98 (1985).
49. SHAH, A.H., V.B. PANDEY, G. ECKHARDT, and R. TSCHESCHE: Sativanine-E, a new 13-membered Cyclopeptide Alkaloid containing a Short Side-Chain, from *Zizyphus sativa*. J. Nat. Prod., **48**, 555 (1985).
50. SHAH, A.H., G. ECKHARDT, and R. TSCHESCHE: Mass Spectrometric Fragmentation of Sativanine-A and -B. J. Chem. Soc. Pak., **7**, 79 (1985).
51. SHAH, A.H., V.B. PANDEY, G. ECKHARDT, and R. TSCHESCHE: A 13-membered Cyclopeptide Alkaloid from *Zizyphus sativa*. Phytochemistry, **24**, 2765 (1985).
52. SHAH, A.H., V.B. PANDEY, G. ECKHARDT, and R. TSCHESCHE: An N-Formyl Cyclopeptide Alkaloid from the bark of *Zizyphus sativa*. *Phytochemistry*, **24**, 2768 (1985).
53. BHAKUNI, R.S., Y.N. SHUKLA, and R.S. THAKUR: Cyclopeptide Alkaloids from *Melochia corchorifolia*. Phytochemistry, **26**, 324 (1987).
54. SHAH, A.H., M.A. AL-YAHYA, S. DEVI, and V.B. PANDEY: Sativanine-K: An Additional N-Formyl Cyclopeptide Alkaloid from *Zizyphus sativa*. Phytochemistry, **26**, 1230 (1987).
55. VŒLTER, W., A.F. MOREL, A.-U. RAHMAN, and M.M. QURESHI: Studies on the Peptide Alkaloids of *Discaria febrifuga*. Z. Naturforsch., Teil B, **42**, 467 (1987).
56. PANDEY, V.B., Y.C. TRIPATHI, S. DEVI, J.P. SINGH, and A.H. SHAH: A Cyclopeptide Alkaloid from the bark of *Zizyphus rugosa*. Phytochemistry, **27**, 1915 (1988).
57. SHAH, A.H., R.M.A. KHAN, S.K. MAURYA, and V.P. SINGH: Nummularine-S: A Cyclopeptide Alkaloid from stem bark of *Zizyphus nummularia*. Phytochemistry, **28**, 305 (1989).
58. GHEDIRA, K., R. CHEMLI, B. RICHARD, J.-M. NUZILLARD, M. ZECHES, and L. LE MEN-OLIVIER: Two Cyclopeptide Alkaloids from *Zizyphus lotus*. Phytochemistry, **32**, 1591 (1993).
59. ABU-ZARGA, M., S. SABRI, A. AL-ABOUDI, M. SALEH AJAZ, N. SULTANA, and ATTA-UR-RAHMAN: New Cyclopeptide Alkaloids from *Zizyphus lotus*. J. Nat. Prod., **58**, 504 (1995).
60. STONARD, R.J., and R.J. ANDERSEN: Celenamides A and B, Linear Peptide Alkaloids from the Sponge *Cliona celata*. J. Org. Chem., **45**, 3687 (1980).

61. STONARD, R.J., and R.J. ANDERSEN: Linear Peptide Alkaloids from the Sponge *Cliona celata* (Grant). Celenamides C and D. Can. J. Chem., **58**, 2121 (1980).
62. MARCHAND, J., M. PAÏS, X. MONSEUR, and F.-X. JARREAU: Alcaloïdes peptidiques-VII. Les Lasiodines A et B, Alcaloides du *Lasiodiscus marmoratus* C.H. Wright (Rhamnacées). Tetrahedron, **25**, 937 (1969).
63. HAN, B.H., M.H. PARK, and Y.N. HAN: Cyclic Peptide and Peptide Alkaloids from seeds of *Zizyphus vulgaris*. Phytochemistry, **29**, 3315 (1990).
64. GONZALEZ SIERRA, M., O.A. MASCARETTI, V.M. MERKUZA, E.L. TOSTI, and E.A. RÚVEDA: Peptide Alkaloids of *Scutia buxifolia*. Phytochemistry, **13**, 2865 (1974).
65. SHAH, A.H., G.A. MIANA, and V.B. PANDEY: Cyclopeptide Alkaloids. Cyclization of the Open Side Chain in Nummularine-B into Imidazolidinone Ring. J. Chem. Soc. Pak., **7**, 341 (1985).
66. TSCHESCHE, R., M. ELGAMAL, and G. ECKHARDT: Alkaloide aus Rhamnaceen, XXVIII. Nummularin-G, -H und -K, weitere Peptidalkaloide aus *Zizyphus nummularia*. Chem. Ber., **110**, 2649 (1977).
67. TSCHESCHE, R., A.H. SHAH, and G. ECKHARDT: Sativanine-A and Sativanine-B, Two New Cyclopeptide Alkaloids from the Bark of *Zizyphus sativa*. Phytochemistry, **18**, 702 (1979).
68. SINGH, B., and V.B. PANDEY: An N-Formyl Cyclopeptide Alkaloid from *Zizyphus nummularia* bark. Phytochemistry, **38**, 271 (1995).
69. CHUGTAI, M.I.D., I. KHOKHAR, and A. AHMAD: Isolation, Purification and Structural Determination of Alkaloids from the Flowers of *Sphaeranthus indicus*. Sci. Int. (Lahore), **4**, 151 (1992).
70. HAN, B.H., M.H. PARK, and J.H. PARK: Chemical and Pharmacological Studies on Sedative Cyclopeptide Alkaloids in Some Rhamnaceae Plants. Pure Appl. Chem., **61**, 443 (1989).
71. BARBONI, L., P. GARIBOLDI, E. TORREGIANI, and L. VEROTTA: Cyclopeptide Alkaloids from *Zizyphus mucronata*. Phytochemistry, **35**, 1579 (1994).
72. SHAH, A.H., V.B. PANDEY, J.P. SINGH, K.N. SINGH, and G. ECKHARDT: Sativanine- G, a Cyclopeptide Alkaloid from *Zizyphus sativa*. Phytochemistry, **23**, 2120 (1984).
73. PANDEY, V.B., and S. DEVI: Biologically Active Cyclopeptide Alkaloids from Rhamnaceae Plants. Planta Med., **56**, 649 (1990).
74. SHAH, A.H., V.B. PANDEY, G. ECKHARDT, and G.A. MIANA: Tscheschamine – A new Cyclopeptide Alkaloid from the Bark of *Zizyphus sativa* Gaertn. Heterocycles, **27**, 2777 (1988).
75. GHEDIRA, K., R. CHEMLI, C. CARON, J.-M. NUZILLARD, M. ZECHES, and L. LE MEN-OLIVIER: Four Cyclopeptide Alkaloids from *Zizyphus lotus*. Phytochemistry, **38**, 767 (1995).
76. TSCHESCHE, R., G.A. MIANA, and G. ECKHARDT: Alkaloide aus Rhamnaceen, XXV. Nummularin-A, -B und -C, drei neue 13gliedrige Peptidalkaloide aus *Zizyphus nummularia*. Chem. Ber., **107**, 3180 (1974).
77. DEVI, S., V.B. PANDEY, J.P. SINGH, and A.H. SHAH: Peptide Alkaloids from *Zizyphus* Species. Phytochemistry, **26**, 3374 (1987).
78. TRIPATHI, Y.C., S.K. MAURYA, V.P. SINGH, and V.B. PANDEY: Cyclopeptide Alkaloids from *Zizyphus rugosa* bark. Phytochemistry, **28**, 1563 (1989).
79. SHAH, A.H., G.A. MIANA, S. DEVI, and V.B. PANDEY: Sativanine-H: A new Alkaloid from the Bark of *Zizyphus sativa*. Planta Med., **52**, 500 (1986).
80. DWIVEDI, S.P.D., V.B. PANDEY, A.H. SHAH, and G. ECKHARDT: Cyclopeptide Alkaloids from *Zizyphus nummularia*. J. Nat. Prod., **50**, 235 (1987).
81. SHAH, A.H., V.B. PANDEY, G. ECKHARDT, and R. TSCHESCHE: Sativanine-C: A Cyclopeptide Alkaloid from the Bark of *Zizyphus sativa*. Phytochemistry, **23**, 931 (1984).

82. PANDEY, V.B., J.P. SINGH, K.K. SETH, A.H. SHAH, and G. ECKHARDT: Cyclopeptide Alkaloids from *Zizyphus nummularia*. Phytochemistry, **23**, 2118 (1984).

83. SHAH, A.H., G.A. MIANA, V.B. PANDEY, R. WAGNER, and R. TSCHESCHE: Alkaloide aus Rhamnaceen, 36: ^{1}H-NMR-Spektroskopische Untersuchungen an Nummularin-B. J. Chem. Soc. Pak., **7**, 37 (1985).

84. TSCHESCHE, R., I. KHOKHAR, H. WILHELM, and G. ECKHARDT: Jubanin-A und Jubanin-B, neue Cyclopeptidalkaloide aus *Zizyphus jujuba*. Phytochemistry, **15**, 541 (1976).

85. PANDEY, V.B., S. DEVI, J.P. SINGH, and A.H. SHAH: Cyclopeptide Alkaloids from *Zizyphus xylopyra*. J. Nat. Prod., **49**, 939 (1986).

86. PAILER, M., E. HASLINGER, and E. ZBIRAL: Die Konstitution des Peptidalkaloides Zizyphinin (aus *Zizyphus oenoplia* Mill.). Monatsh. Chem., **100**, 1608 (1969).

87. CASSELS, B.K., G. ECKHARDT, E.U. KAUSSMANN, and R. TSCHESCHE: Cyclopeptide Alkaloids of *Zizyphus oenoplia*. Tetrahedron, **30**, 2461 (1974).

88. ABDEL-GALIL, F.M., and M.A. EL-JISSRY: Cyclopeptide Alkaloids from *Zizyphus spina- christi*. Phytochemistry, **30**, 1348 (1991).

89. TSCHESCHE, R., C. SPILLES, and G. ECKHARDT: Alkaloide aus Rhamnaceen, XXII. Amphibin-I, ein neues Alkaloid aus *Zizyphus amphibia* A. Cheval. Chem. Ber., **107**, 1329 (1974).

90. ZBIRAL, E., E.L. MENARD, and J.M. MÜLLER: Über die Inhaltstoffe von *Zizyphus oenoplia* MILL. 2. Mitteilung: Zur Konstitutionsermittlung des Zizyphins. Helv. Chim. Acta, **48**, 404 (1965).

91. HASLINGER, E., and W. ROBIEN: NMR Spectroscopic Studies on Peptide Alkaloids ^{1}H and ^{13}C Spectra of Zizyphin A and Frangulanin. Monatsh. Chem., **113**, 95 (1982).

92. HINDENLANG, D.M., M. SHAMMA, G.A. MIANA, A.H. SHAH, and B.K. CASSELS: The ^{13}C-NMR Spectra of Cyclopeptide Alkaloids. Liebigs Ann. Chem., **41**, 447 (1980).

93. YU, C., Y.-Y. TSENG, and S.-S. LEE: Calculating Three-dimensional Molecular Structure of Paliurine B from Atom-Atom Distance and Restrained Energy Minimization. Biochim. Biophys. Acta, **1156**, 334 (1993).

94. MIANA, G.A., and A.H. SHAH: Isolation of Jubanine-A, -B and Mauritine-C from the Root Bark of *Zizyphus nummularia*. Fitoterapia, **56**, 363 (1985).

95. PANDEY, V.B., S.P.D. DWIVEDI, A.H. SHAH, and G. ECKHARDT: Nummularine-O, a Cyclopeptide Alkaloid from *Zizyphus nummularia*. Phytochemistry, **25**, 2690 (1986).

96. SERVIS, R.E., A.I. KOSAK, R. TSCHESCHE, E. FROHBERG, and H.-W. FEHLHABER: Peptide Alkaloids from *Ceanothus americanus* L. (Rhamnaceae). J. Am.Chem. Soc., **91**, 5619 (1969).

97. WARNHOFF, E.W., S.K. PRADHAN, and J.C.N. MA: *Ceanothus* Alkaloids. I. Isolation, Separation and Characterization. Can. J. Chem., **43**, 2594 (1965).

98. GOURNELIS, D.: Iridoïdes et Alcaloïdes de *Plectronia odorata* Benth. et Hook. (Rubiacées). Thèse de Doctorat, Université René Descartes, Paris, pp. 62–67 and 127–128 (1988).

99. GOURNELIS, D., A.-L. SKALTSOUNIS, F. TILLEQUIN, M. KOCH, J. PUSSET, and S. LABARRE: Plantes de Nouvelle-Calédonie, CXXI. Iridoïdes et Alcaloïdes de *Plectronia odorata*. J. Nat. Prod., **52**, 306 (1989).

100. MARCHAND, J., X. MONSEUR, and M. PAÏS: Alcaloïdes Peptidiques, VII (5). – Les Myrianthines A, B et C, Alcaloïdes du *Myrianthus arboreus* P. Beauv. (Urticacées). Ann. Pharm. Fr., **26**, 771 (1968).

101. TAKAI, M., Y. OGIHARA, and S. SHIBATA: New Peptide Alkaloids from *Hovenia dulcis* and *H. tomentella*. Phytohemistry, **12**, 2985 (1973).

102. TSCHESCHE, R., and H.LAST: Alkaloide aus Rhamnaceen, V. Franganin und Frangufolin, zwei weitere Peptid-Alkaloide aus *Rhamnus frangula* L. Tetrahedron Lett., **25**, 2993 (1968).
103. BHAKUNI, R.S., Y.N. SHUKLA, and R.S. THAKUR: Melochicorine, a Pseudooxindole Alkaloid from *Melochia corchorifolia*. Phytochemistry, **30**, 3159 (1991).
104. BISHAY, D.W., Z. KOWALEWSKI, and J.D. PHILLIPSON: Peptide and Tetrahydroisoquinoline Alkaloids from *Euonymus europaeus*. Phytochemistry, **12**, 693 (1973).
105. DIGEL, M., A.F. MOREL, H. LAYER, J. BIERMANN, and W. VŒLTER: Peptidalkaloide aus *Discaria febrifuga* Mart. Hoppe-Seyler's Z. Physiol. Chem., **364**, 1641 (1983).
106. SHAH, A.H., A.M. AGEEL, M. TARIQ, J.S. MOSSA, and M.A. AL-YAHYA: Chemical Constituents of the Stem Bark of *Zizyphus spina-christi*. Fitoterapia, **57**, 452 (1986).
107. TSCHESCHE, R., und I. REUTEL: Alkaloide aus Sterculiaceen, I. Über Peptidalkaloide aus *Melochia corchorifolia*. Tetrahedron Lett., **35**, 3817 (1968).
108. PAÏS, M., J. MARCHAND, F.-X. JARREAU, and R. GOUTAREL: Alcaloïdes Peptidiques. V. – Structures des Adouétines X, Y, Y′ et Z, Alcaloïdes du *Waltheria americana* L. (Sterculiacées). Bull. Soc. Chim. Fr., **3**, 1145 (1968).
109. CHANG, C.-J., E.W. HAGAMAN, E. WENKERT, M. GONZALEZ SIERRA, O.A. MASCARETTI, V.M. MERKUZA, and E.A. RÚVEDA: PMR Spectral Analysis of Some Peptide Alkaloids. Phytochemistry, **13**, 1273 (1974).
110. HASLINGER, E.: Zur Konformation von Frangulanin. Untersuchung mit Hilfe der ^{13}C- und ^{1}H-NMR-Spektroskopie. Tetrahedron, **34**, 685 (1978).
111. HASLINGER, E.: Partiell relaxierte ^{1}H-NMR-Spektren von Frangulanin. Monatsh. Chem., **109**, 523 (1978).
112. HASLINGER, E., and W. ROBIEN: Protonen-Spin-Gitter-Relaxation und interne Beweglichkeit von Molekülen. Monatsh. Chem., **110**, 1011 (1979).
113. MACHADO, E.C., A.A. FILHO, A.F. MOREL, and F. DELLE MONACHE: Four Cyclopeptide Alkaloids from *Discaria longispina*. J. Nat. Prod., **58**, 548 (1995).
114. WARNHOFF, E.W., J.C.N. MA, and P. REYNOLDS-WARNHOFF: Ceanothine-B, a Naturally Occurring Oxazacyclononadiene. J. Am. Chem. Soc., **87**, 4198 (1965).
115. SERVIS, R.E., and A.I. KOSAK: A Revised Structure of Ceanothine-B. J. Am. Chem. Soc., **90**, 4179 (1968).
116. KLEIN, F.K., and H. RAPOPORT: The Structure of Ceanothine-B. J. Am. Chem. Soc., **90**, 3576 (1968).
117. TSCHESCHE, R., D. HILLEBRAND, H. WILHELM, E. AMMERMANN, and G. ECKHARDT: Hysodricanin-A, Mauritin-H, Scutianin-F und Aralionin-C, vier weitere Cyclopeptidalkaloide aus *Zizyphus*, *Scutia* und *Araliothamnus*. Phytochemistry, **16**, 1025 (1977).
118. MENEZES, A.S., M.A. MOSTARDEIRO, N. ZANATTA, and A.F. MOREL: Scutianine-J, a Cyclopeptide Alkaloid Isolated from *Scutia buxifolia*. Phytochemistry, **38**, 783 (1995).
119. MERKUZA, V.M., M. GONZALEZ SIERRA, O.A. MASCARETTI, E.A. RÚVEDA, C.-J. CHANG, and E. WENKERT: Peptide Alkaloids of *Discaria longispina* and *Scutia buxifolia*. Phytochemistry, **13**, 1279 (1974).
120. ARBAIN, D., and W.C. TAYLOR: Cyclopeptide Alkaloids from *Antidesma montana*, Phytochemistry, **33**, 1263 (1993).
121. KLEIN, F.K., and H. RAPOPORT: *Ceanothus* Alkaloids. Americine. J. Am. Chem. Soc., **90**, 2398 (1968).
122. TSCHESCHE, R., E. AMMERMANN, and H.-W. FEHLHABER: Alkaloide aus Rhamnaceen, X. Scutianin-B, ein weiteres Peptidalkaloid aus *Scutia buxifolia* Reiss. Tetrahedron Lett., **46**, 4405 (1971).

123. TSCHESCHE, R., I. KHOKHAR, C. SPILLES, and M. VON RADLOFF: Peptide Alkaloids from *Zizyphus spina-christi*. Phytochemistry, **13**, 1633 (1974).

124. TSCHESCHE, R., A.H. SHAH, V.B. PANDEY, J.P. SINGH, M. VON RADLOFF, and G. ECKHARDT: Alkaloids of Rhamnaceae. Pharmazie, **36**, 511 (1981).

125. SILVA, M., D.S. BHAKUNI, P.G. SAMMES, M. PAÏS, and F.-X. JARREAU: A new Peptide Alkaloid from *Discaria crenata*. Phytochemistry, **13**, 861 (1974).

126. TSCHESCHE, R., E. FROHBERG, and H.-W. FEHLHABER: Alkaloide aus Rhamnaceen, IV. Integerrin, ein weiteres Peptid-Alkaloid aus *Ceanothus integerrimus* Hook et Arn. Tetrahedron Lett., **11**, 1311 (1968).

127. HERZOG, R., A.F. MOREL, J. BIERMANN, and W. VŒLTER: Discarin-H, ein neues Peptidalkaloid aus *Discaria febrifuga*. Chem.-Ztg., **108**, 406 (1984).

128. MOREL, A.F., E.C. MACHADO, and L.A. WESSJOHANN: Cyclopeptide Alkaloids of *Discaria febrifuga* (Rhamnaceae). Phytochemistry, **39**, 431 (1995).

129. HERZOG, R., A.F. MOREL, J. BIERMANN, and W. VŒLTER: Ein neues Peptidalkaloid aus *Discaria febrifuga* Mart. Hoppe-Seyler's Z. Physiol. Chem., **365**, 1351 (1984).

130. BAILLEUL, F., and P. DELAVEAU: La Féretine, Alcaloïde Peptidique du *Feretia apodanthera* Del. (Rubiacées). C. R. Acad. Sci. Paris, Série C, **279**, 949 (1974).

131. PAÏS, M., J. MARCHAND, X. MONSEUR, F.-X. JARREAU, and R. GOUTAREL: Alcaloïdes Peptidiques. Structure de l' hymenocardine, alcaloïde de l' *Hymenocardia acida* Tul. (Euphorbiacées). C. R. Acad. Sci. Paris, Série C, **264**, 1409 (1967).

132. TSCHESCHE, R., S.T. DAVID, R. ZERBES, M. VON RADLOFF, E.U. KAUSSMANN, and G. ECKHARDT: Alkaloide aus Rhamnaceen, XIX. Mucronin-E, -F, -G und -H sowie Abyssenin-A, -B und -C, weitere 15gliedrige Cyclopeptidalkaloide. Liebigs Ann. Chem., 1915 (1974).

133. FEHLHABER, H.-W., J. UHLENDORF, S.T. DAVID, and R. TSCHESCHE: Alkaloide aus Rhamnaceen, XII. Mucronin-A, -B und -C, Peptid-Alkaloide eines neuen Strukturtyps aus *Zizyphus mucronata*. Liebigs Ann. Chem., 759, 195 (1972).

134. CHUGHTAI, M.I.D., I. KHOKHAR, and F. TAHIRA: Cyclopeptide-Alkaloids from the Leaves of *Zizyphus jujuba*. Pak. J. Sci., **30**, 136 (1978).

135. CHUGHTAI, M.I.D., I. KHOKHAR, A. AHMAD, U. GHANI, and M. ANWAR: Studies in Medicinal Plants of Pakistan-Part I: Alkaloids from the Leaves of *Cocculus villosus*. Pak. J. Sci. Res., **31**, 79 (1979).

136. CHUGHTAI, M.I.D., I. KHOKHAR, A. AHMAD, I. AHMAD, and A. REHMAN: Studies in Medicinal Plants of Pakistan-II. Alkaloids from the Leaves of *Cocculus villosus*. Pak. J. Sci. Res., **31**, 237 (1979).

137. SCHMIDT, U.: Synthesis of Cyclopeptides from Plants, Fungi and Sea Animals. Pure Appl. Chem., **58**, 295 (1986).

138. LIPSHUTZ, B.H., B. HUFF, K. MCCARTHY, S.M.J. MUKARRAM, and W. VACCARO: A Non-Peptidal Approach to the Cyclopeptide Alkaloids via Oxazolophane Intermediates. Abstr. Pap. Am. Chem. Soc., **193**, 84 (1987).

139. FLANAGAN, D.M., K.L. BHAT, and M.M. JOULLIÉ: Synthesis of Dipeptides Related to Cyclopeptide Alkaloids. J. Prakt. Chem., **329**, 915 (1987).

140. FLANAGAN, D.M., and M.M. JOULLIÉ: Observations on the Stereochemical Outcome of the Ugi Four-Component Condensation. Synth. Commun., **19**, 1 (1989).

141. HEFFNER, R.J., and M.M. JOULLIÉ: Studies Directed Toward the Total Synthesis of 14-Membered Cyclopeptide Alkaloids: Synthesis of a Cyclic Precursor to Nummularine-F. Tetrahedron Lett., **30**, 7021 (1989).

142. BOWERS, M.M., P. CARROLL, and M.M. JOULLIÉ: Model Studies Directed Toward the Total Synthesis of 14-Membered Cyclopeptide Alkaloids: Synthesis of Prolyl Peptides *via* a Four-Component Condensation. J. Chem. Soc., Perkin Trans. I, **5**, 857 (1989).

143. HEFFNER, R.J., and M.M. JOULLIÉ: Studies Directed Toward a Total Synthesis of 14-Membered Cyclopeptide Alkaloids: Synthesis of a Cyclic Precursor to Nummularine-F. Abstr. Pap. Am. Chem. Soc., **200**, 175 (1990).

144. FLANAGAN, D.M., and M.M. JOULLIÉ: Studies Directed Toward the Total Synthesis of 14-Membered Cyclopeptide Alkaloids : Synthesis of a Linear Precursor to Nummularine-F. Synth. Commun., **20**, 459 (1990).

145. LIPSHUTZ, B.H., B.E. HUFF, K.E. MCCARTHY, T.A. MILLER, S.M.J. MUKARRAM, T.J. SIAHAAN, W.D. VACCARO, H. WEBB and A.M. FALICK: Oxazolophanes as Masked Cyclopeptide Alkaloid Equivalents: Cyclic Peptide Chemistry Without Peptide Couplings. J. Am. Chem. Soc., **112**, 7032 (1990).

146. WILLIAMS, L., Z. ZHANG, X. DING, and M.M. JOULLIÉ: A Practical Stereoselective Synthesis of (2*S*, 3*S*)-3-Hydroxyleucine. Tetrahedron Lett., **36**, 7031 (1995).

147. SCHMIDT, U., and U. SCHANBACHER: Total Synthesis of Mucronin B. Angew. Chem. Int. Ed. Engl., **22**, 152 (1983).

148. SCHMIDT, U., A. LIEBERKNECHT, H. BÖKENS, and H. GRIESSER: Total Synthesis of Zizyphine A. Synthesis of Peptide Alkaloids. 8. Amino Acids and Peptides. 40. J. Org. Chem., **48**, 2680 (1983).

149. SCHMIDT, U., H. BÖKENS, A. LIEBERKNECHT, and H. GRIESSER: Synthesis of Peptide Alkaloids-III. Amino Acids and Peptides-XXXII. Synthesis of Dihydro-Zizyphine A and B. Tetrahedron Lett., **22**, 4949 (1981).

150. NUTT, R.F., K.-M. CHEN, and M.M. JOULLIÉ: Synthesis of Dihydromauritine A, a Reduced Cyclopeptide Alkaloid. J. Org. Chem., **49**, 1013 (1984).

151. SCHMIDT, U., A. LIEBERKNECHT, H. GRIESSER, and J. HÄUSLER: Synthesis of Dihydrozizyphin G. Angew. Chem. Int. Ed. Engl., **20**, 281 (1981).

152. SCHMIDT, U., A. LIEBERKNECHT, H. GRIESSER, and J. HÄUSLER: Totalsynthese von 9,10-Dihydrozizyphin G. Liebigs Ann. Chem., 2153 (1982).

153. SCHMIDT, U., and J. WILD: Total Synthesis of Hexaacetylcelenamide A. Angew. Chem. Int. Ed. Engl., **23**, 991 (1984).

154. BLANPIN, O., M. PAÏS, and M.A. QUEVAUVILLER: Etude Pharmacodynamique de l' Adouétine Z, Alcaloïde du *Waltheria americana* L. (Sterculiacées). Ann. Pharm. Fr., **21**, 147 (1963).

155. HAN, B.H., and M.H. PARK: Alkaloids are the Sedative Principles of the Seeds of *Zizyphus vulgaris* var. *spinosus*. Arch. Pharmacal. Res., **10**, 203 (1987).

156. HAN, B.H., and M.H. PARK: Sedative Activity and the Active Components of *Zizyphus* Fructus. Arch. Pharmacal. Res., **10**, 208 (1987).

157. BAIG, M.A., D.V. BANTHORPE, A.A. COLEMN, M.D. TAMPION, J. TAMPION, and J.J. WHITE: Accumulation of Tetrapeptide Precursors of Macrocyclic Alkaloids by Callus of *Ceanothus americanus*. Phytochemistry, **34**, 171 (1993).

158. KHOKHAR, I., and A. AHMAD: Spectrometric Studies on a New 13-Membered Cyclopeptide Alkaloid from *Zizyphus oenoplia*. Mill. J. Nat. Sci. Math., **36**, 171 (1994).

159. KHOKHAR, I., and A. AHMAD: Spectrometric Studies on a New 13-Membered Cyclopeptide Alkaloid from *Zizyphus oenoplia.* Pak. J. Sci., **45**, 54 (1993).

160. DONGO, E., J.F. AYAFOR, B.L. SONDENGAM, and J.D. CONNOLLY: A New Peptide Alkaloid from *Canthium anorldianum.* J. Nat. Prod., **52**, 840 (1989).

161. HENNIG, P., A. MOREL, and W. VŒLTER: Discarin-I, ein neues Peptidalkalid aus *Discaria febrifuga* Martius. Z. Naturforsch., Teil B, **41**, 1180 (1986).

162. AUVIN, C., F. LEZENVEN, A. BLOND, I. AUGEVEN-BOUR., J.-L. POUSSET, B. BODO, and J. CAMARA: Mucronine J, a 14-Membered Cyclopeptide Alkaloid from *Zizyphus mucronata.* J. Nat. Prod., **59**, 676 (1996).

163. HEFFNER, R.J., J. JIANG, and M. JOULLIÉ: Total Synthesis of (–)-Nummularine-F. J. Am. Chem. Soc., **114**, 10181 (1992).

164. PARK, M.H., D.-Y. SUH, and B.H. HAN: Absolute Configuration of a Cyclopeptide Alkaloid, Sanjoinine-G1, from *Zizyphus vulgaris* var. *spinosus*. Phytochemistry, **43**, 701 (1996).

165. JOSSANG, A., A. ZAHIR, and D. DIAKITE: Mauritine J, a Cyclopeptide Alkaloid from *Zizyphus mauritiana*. Phytochemistry, **42**, 565 (1996).

166. SCHMIDT, U., M. ZÄH, and A. LIEBERKNECHT: The Total Synthesis of Frangulanine. J. Chem. Soc., Chem. Commun., 1002 (1991).

167. HAN, B.H., Y.C. KIM, M.K. PARK, J.H. PARK, H.J. GO, H.O. YANG, D.-Y. SUH, and Y.-H. KANG: Total Synthesis of Sanjoinine-G1. Heterocycles, **41**, 1909 (1995).

168. SHU, J.H., T. LAÏB, J. CHASTANET, and R. BEUGELMANS: A Novel Strategy Towards the Total Synthesis of Cyclopeptide Alkaloids. Angew. Chem. Int. Ed. Engl., **35**, 2517 (1996).

169. MAURYA, S.K., D.P. PANDEY, J.P. SINGH, and V.B. PANDEY: Pharmazie, **50**, 372 (1995).

170. ITOKAWA, H., K. TAKEYA, Y. HITOTSUYANAGI, and H. MORITA: Macrocyclic Peptide Alkaloids from Plants in "The Alkaloids: Chemistry and Pharmacology". G.A. Cordell (Ed.), Academic Press, San Diego, London, Boston, New York, Sydney, Tokyo, Toronto, Vol. 49, chapter 4, p. 301 (1997).

(Received December 12, 1996, in revised form July 10, 1997)

Naturally Occurring 6-Substituted 5,6-Dihydro-α-Pyrones

L. A. COLLETT, M. T. DAVIES-COLEMAN, and D. E. A. RIVETT
Department of Chemistry, Rhodes University,
Grahamstown, South Africa

Contents

1. Introduction

6-Substituted derivatives of 5,6-dihydro-α-pyrones (dihydropyran-2-ones or more specifically 2H-dihydropyran-2-ones) occur widely in nature, particularly in plants and bacteria. They possess an α,β-unsaturated-δ-lactone ring (**1**) with an alkyl, alkenyl or aryl substituent at C-6 and occasionally a varied substitution pattern around the ring. Many of these compounds are biologically active, exhibiting phytotoxi-

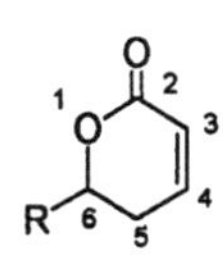

1

city, cytotoxicity against tumour cells and antifungal or antimicrobial activity.

This review follows the pattern of its predecessor (*1*) and covers the literature included in Chemical Abstracts to December, 1996 with occasional references to later journals. As before, steroids containing a 5,6-dihydro-α-pyrone ring, such as the withanolides, are excluded. In view of the recent comprehensive review (*2*) of fungal pyrones and 5,6-dihydro-α-pyrones isolated up to December 1991 the latter compounds, of fungal origin, will only be reviewed from this time onwards. The general emphasis in our review will again be on distribution, structure elucidation, absolute stereochemistry and bioactivity of naturally occurring 5,6-dihydro-α-pyrones. There has been considerable synthetic activity in this field and we have tried to cover the most important aspects. Although details are not provided the titles of the relevant articles often indicate the methodology employed.

2. 6-Alkyl-5,6-dihydro-α-pyrones

While the plant families Annonaceae, Lamiaceae, and Lauraceae continue to be excellent sources of 6-substituted-5,6 dihydro-α-pyrones they are not exclusive sources of these compounds. Gerberin (**2**) has been isolated in 3.7% yield from commercially grown *Gerbera jamesonii hybrida* (Compositae) (*3*). Standard spectroscopic and chemical techniques were used to determine the structure of (**2**) while hydrogenation of the tetraacetyl derivative of (**2**) afforded the known tetracetyl parasorboside (**3**). The (6*S*)-stereochemistry followed from the positive sign of the n → π* Cotton effect in the circular dichroism (CD) spectrum of its acetate derivative (**4**) (*4*). It must be noted here that there is an error pertaining to the CD of 5,6-dihydro-α-pyrones in our earlier review (*1*). The structures on page 25, (**82**) and (**83**), give rise to negative and positive Cotton effects respectively and not the other way around. A related polyhydroxylated compound (**5**) has been isolated as its triacetate (**6**) from a liquid culture of the soil fungus *Taleromyces flavus* (*5*). The

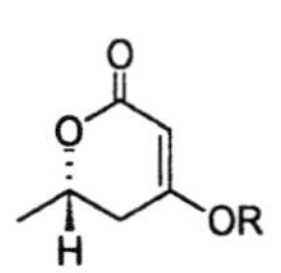

2 R = β - D-glucopyranosyl
4 R = Ac

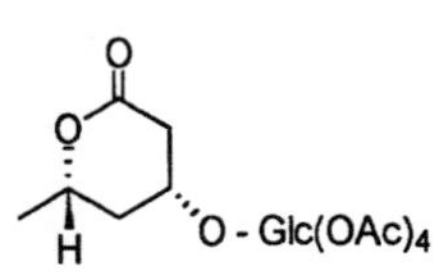

3

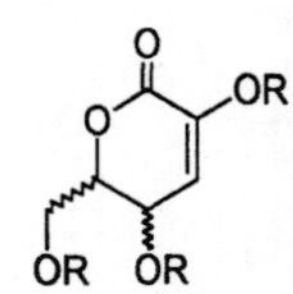

5 R = H
6 R = Ac

stereochemistry of the two chiral centres in this compound was not established. Although the fungus displays antifungal activity against Verticillium wilt of eggplant (*Solanum melongena* L.), a disease caused by the fungus *Verticillium dahliae*, it appears that (**5**) is not responsible for the antifungal properties of *T. flavus*.

Several syntheses of parasorbic acid (**7**) have been reported (*6–15*). Osmundalactone (**8**) has also been isolated from the fungus *Paxillus atrotomentosus* (*16*). A synthesis of chiral aspyrone (**9**) has confirmed its absolute stereochemistry (*17*) and studies on its biosynthesis have continued (*18*).

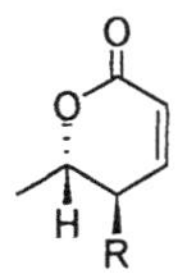

7 R = H
8 R = OH

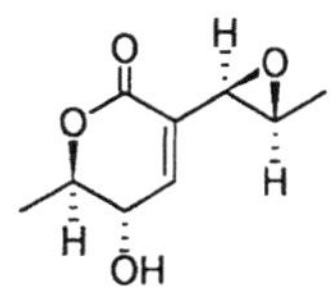

9

The absolute stereochemistry of alternaric acid (**10**), first isolated in 1949 from the fungus *Alternaria solani* (*19*), has been established by total synthesis (*20, 21*). Another five 6-methyl-5,6-dihydro-α-pyrones (**11–15**), structurally related to alternaric acid, have recently been obtained from this fungus (*22, 23*). Spectroscopic data indicated a close relationship between compounds (**10–15**) and a (6*R*)-stereochemistry for the lactone ring. In addition, the stereochemistry of (**13**) was assigned by a combination of spectroscopic methods, chemical degradations and interconversions. Oxidative cleavage of the ketone adjacent to the pyrone ring gave the diacid (**16**). Esterification of (**16**) followed by application of the modified Mosher method (*24*) afforded the stereochemistry of the secondary alcohol moiety. To determine the relative stereochemistry at C-8″, the two ester groups were first reduced to primary alcohols with $LiAlH_4$. Examination of the H-8″ and H-9″ coupling constant in the isopropylidene derivative of the resulting 1,3-diol indicated a *trans*-diaxial orientation for these two protons. The configurations at C-4″ and C-10" were determined by comparison of chemical degradation products with synthetic analogs of known stereochemistry. Catalytic hydrogenation of (**13**) yielded (**12**) thus establishing the absolute stereochemistry of the latter compound. Treatment of the fungus with specific cytochrome P-450 inhibitors resulted in the generation of a plausible precursor named proalternaric acid (**17**) whose structure was proved by synthesis from (**13**). The origins of the oxygen and hydrogen atoms in the alternaric acids were investigated by following the incorporation of sodium [1-^{13}C,^{18}O]- and [1-^{13}C, $^{2}H_3$]-acetate into these metabolites. A biosynthetic pathway for alternaric acid was then proposed.

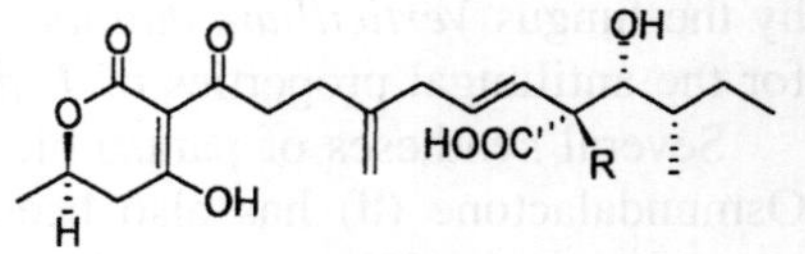

10 R = OH
11 R = H

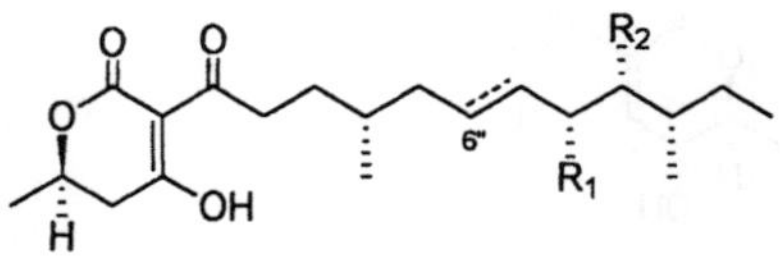

12 R_1 = COOH, R_2 = OH
13 $\Delta^{6''}$ (E), R_1 = COOH, R_2 = OH
14 $\Delta^{6''}$ (E), R_1 = COOH, R_2 = H
17 $\Delta^{6''}$ (E), R_1 = CH_2OH, R_2 = OH

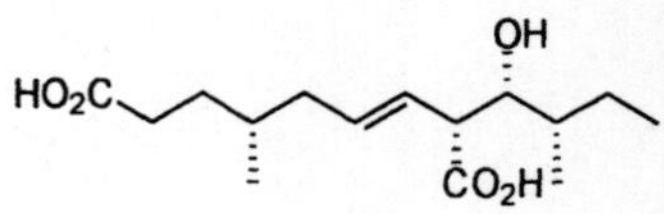

16

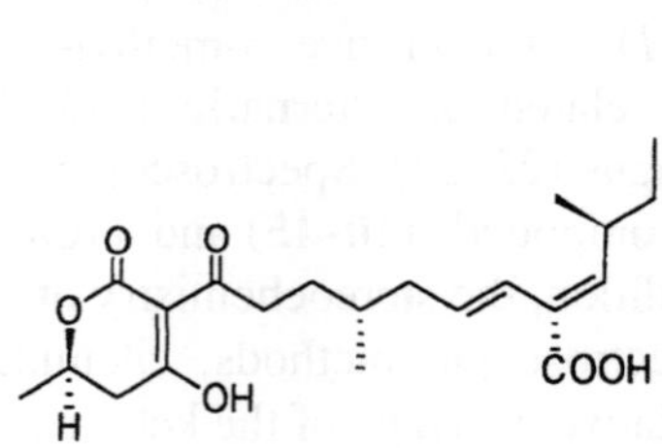

15

There have been very few examples of 5,6-dihydro-α-pyrones isolated from the marine environment. In 1992, HAMADA *et al.* obtained lobatrienolide (**18**) from an Okinawan soft coral *Sinularia flexibilis* (*25*). The structure of lobatrienolide was established by spectral techniques and confirmed by photo-oxidation of lobatriene (**19**), isolated from the same soft coral collection, to (**18**). The structure of (**19**) was proved as follows (*Scheme 1*). Lithium-ethylamine reduction of (**19**) produced the diol (**20**) which on ozonolysis, MCPBA oxidation to the acetate (**21**) and hydrolysis gave the alcohol (**22**). The *S*-stereochemistry of the alcoholic group in (**22**) was determined by the modified Mosher method (*24*), which together with NMR correlations in lobatriene, afforded the complete stereochemistry of (**19**) and hence also that of lobatrienolide

(**18**) (*26*). The absolute stereochemistry of the cyclohexane ring in (**18**) is the same as that in the well known sesquiterpene (+)-β-elemene (**23**).

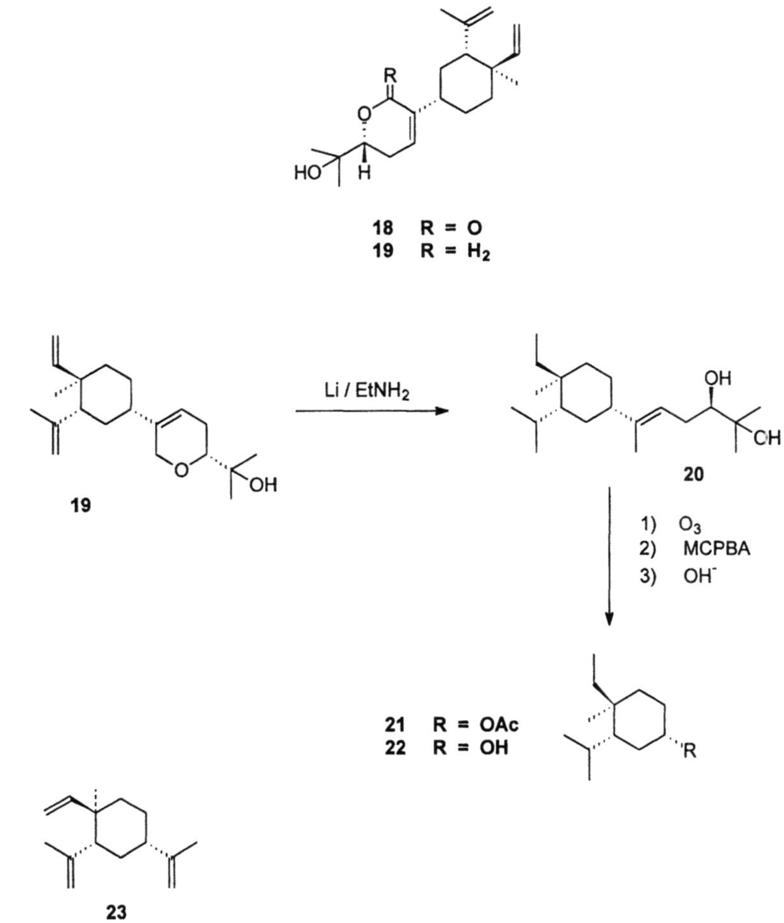

Scheme 1

In the same year Tsuda *et al.* (*27*) reported the isolation of cytotoxic sesterterpenes from the Okinawan marine sponge *Luffariella sp.* among which was a compound with a 5,6-dihydro-α-pyrone moiety, luffariolide E. This compound showed cytotoxicity against murine leukaemia L1210 cells (IC_{50} 1.1–7.8 μg/ml) in vitro. The original structure, which rested on chemical and spectroscopic evidence, has been corrected by synthesis to (**24**) (*28*).

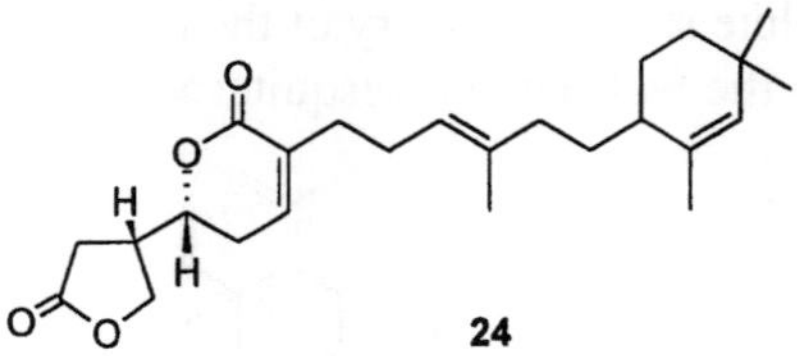

24

Both the natural (−) and unnatural (+) enantiomers of the antifungal podoblastins A, B and C (**25–27**) and of synthetic podoblastin S (**28**) have been synthesized from the corresponding R(−)- and S(+)-glycerol derivatives (*29*). The (6S)-isomers were considerably less active against rice blast than the (6R)-isomers.

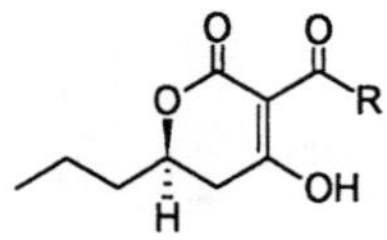

25 R = $(CH_2)_{10}CH_3$
26 R = $(CH_2)_9$ - CH = CH_2
27 R = $(CH_2)_{12}CH_3$
28 R = $(CH_2)_8CH_3$

Further syntheses of (−)-tarchonanthus lactone (**29**) have appeared (*30*, *31*). Several syntheses of racemic and enantiomerically pure massoialactone (**30**) have been reported (*9*, *11*, 32–37). Pestalotin (**31**) and hydroxypestalotin (LL-P880β) (**32**) have been isolated from *Pestalotiopsis oenotherae* (*38*). Syntheses of natural and racemic pestalotin as well as the (6*R*)-isomer have been published (*39–41*). All

29

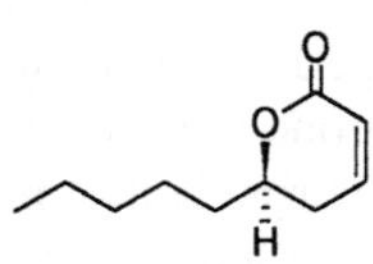

30

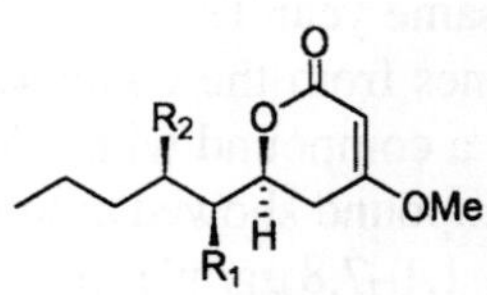

31 R_1 = OH R_2 = H
32 R_1 = R_2 = OH

the possible stereoisomers of the fungal metabolite LL-P880β have been synthesized from either D-glucose, D-idose or (+)-tartaric acid (*42–45*).

6-Heptyl-5,6-dihydro-α-pyrones with various levels of oxygenation in the heptyl side chain occur widely. Two such compounds gamahonolide A (**33**) and B (**34**) were isolated by KOSHINO and co-workers (*46*) from the grass *Phleum pratense* infected with the choke disease caused by the phytopathogenic fungus *Epichloe typhinia.* Interestingly, plants suffering from the disease are resistant to the leaf spot fungus *Cladosporium herbarum*. Compounds (**33**) and (**34**), which are more abundant in plants suffering from the disease, are thought to be responsible for the systemic and mutualistic behaviour attributed to *E. typhinia*, since they are active against *C. herbarum*. The structures of (**33**) and (**34**) were determined by NMR spectroscopic methods including extensive proton decoupling experiments. ^{13}C NMR data suggested the presence of the ethyl succinate moiety in (**34**), the ethyl ester of which is probably an isolation artifact formed during ethanolic extraction. A comparison of the ORD spectra of (**33**) and (**34**) with that of massoialactone (**30**) indicated a (6*R*)-stereochemistry for the former compounds. The (6′*R*)-stereochemistry in (**33**) was established by TROST'S method (*47*, *48*) from an examination of the ^{1}H NMR chemical shift difference of the C-7′ methyl group in the (R)- and the (S)-O-methylmandelyl esters.

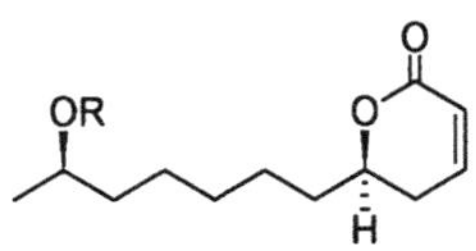

33 R = H
34 R = $COCH_2CH_2CO_2Et$

Syndenolide (**35**) together with the known 5,6-dihydro-α-pyrone, deacetylboronolide (**36**), was isolated from *Syncolostemon densiflorus* (Lamiaceae) (*49*). The presence of a triol in (**35**) was demonstrated by D_2O induced collapse of three hydroxyl proton signals in its ^{1}H NMR spectrum and the presence of four acetate methyl signals in the ^{1}H NMR spectrum of the peracetylated derivative (**37**). A (6*R*)-stereochemistry for (**35**) was deduced from the positive sign of the $n \rightarrow \pi^*$ Cotton effect in its CD spectrum and the stereochemistry of the 1′, 2′ and 3′ chiral centres was proposed to be the same as in (**36**) from biosynthetic arguments. This stereochemical assignment was supported by a comparison of the relevant coupling constants of (**37**) with those of boronolide (**38**). The stereochemistry at C-5′ followed from a detailed NMR analysis of the diacetonide (**39**) (*50*). The ^{13}C NMR spectrum indicated the presence of a six membered acetonide ring in the chair conformation with one methyl equatorial (δ30.0) and one axial (δ19.8) (*51*), as well as a puckered 5-membered acetonide ring with equivalent methyl groups. NOE difference experiments supported this structure (**39**).

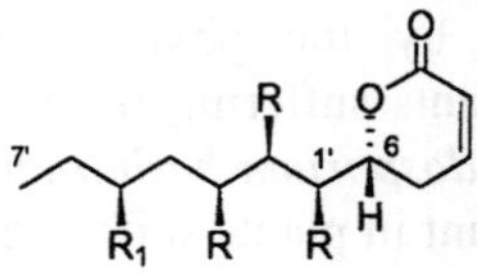

35 R = OH, R_1 = OAc
36 R = OH, R_1 = H
37 R = R_1 = OAc
38 R = OAc, R_1 = H

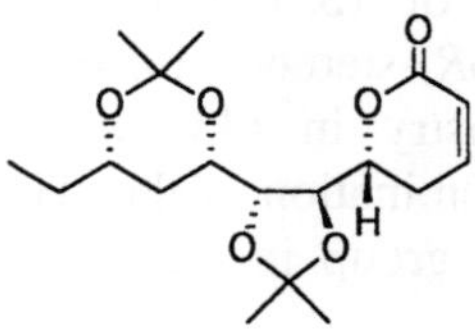

39

Three syntheses of boronolide (**38**) have appeared. The first synthesis from an acrolein dimer, afforded the racemate (*52*). Naturally occurring (+)-boronolide has been obtained from D-glucose (*53*) and employing the SHARPLESS catalytic asymmetric dihydroxylation reaction (*54*).

The presence of acetate substituents on the side chain at C-6 is very common. Compounds (**40**) and (**41**) were isolated by DREWES *et al.* from the bark of *Cryptocarya latifolia* trees in their investigations of the chemistry of plants used for magical and medicinal purposes by the Zulu people (*55*). The CD spectra of these two compounds show a positive Cotton effect and hence they possess the (6*R*)-configuration. Saponification followed by acetonide formation of (**40**) afforded two acetonides (**42**) and (**43**) (*50*). A (2′*R*,6′*S*)-stereochemistry followed from application of the MTPA determination rule (*24*) to the (*R*)- and (*S*)-MTPA esters of these acetonides. A (4′*S*)-stereochemistry was assigned from the *syn*-diol relationship of the two acetal oxygen atoms in (**42**) and (**43**) as determined from their ^{13}C NMR spectra (*51*). Thus the acetonide rings in both compounds possessed chair conformations with the alkyl substituents equatorial and the one methyl group axial (δ20) and the other equatorial (δ30).

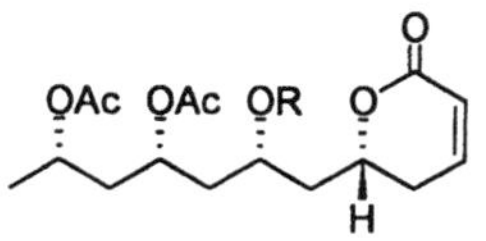

40 R = Ac
46 R = H

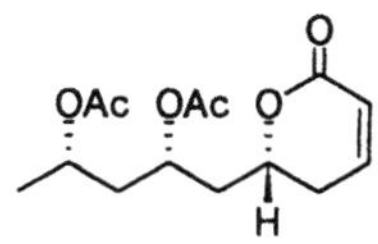

41

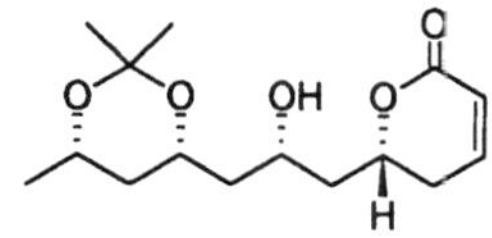

42

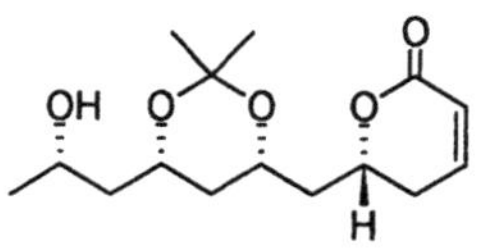

43

The *syn* stereochemistry of the acetoxy groups in the diacetate (**41**) was determined via its acetonide derivative in the same way. However, since the stereochemistry of the C-2′ methine hydrogen cannot be correlated with that at C-6 because of the rotating C-1′ methylene group, the absolute stereochemistry of (**41**) remains unknown but is probably (2′*S*, 4′*S*) as in (**40**). Drewes *et al.* have suggested that the triacetate (**40**) is the precursor of cryptocaryolone diacetate (**44**), which together with cryptocaryolone (**45**), also occurs in *Cryptocarya latifolia*, and is formed by Michael addition of the 2′-acetoxy group in (**40**). In fact Horn has recently converted (**46**), a minor constituent of the plant, to cryptocaryolone diacetate with sodium hydride in dichloromethane (*56*). This transformation is similar to an earlier synthesis of the related styryl lactone, goniopyrone (**116**) (*57*) (vide infra), and suggests structure (**44**) for cryptocaryolone diacetate. Fang *et al.* (*58*) have indicated from molecular modeling studies that the 6-membered rings in compounds such as (**116**) and hence also in (**44**), must be *cis*-fused and are in a semichair conformation.

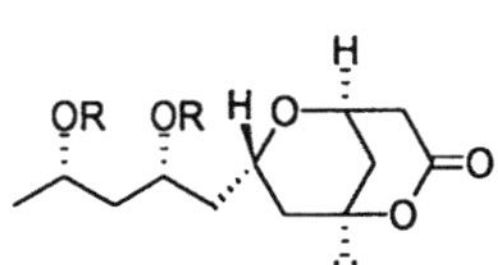

44 R = Ac
45 R = H

The trihydroxyheneicosyl-5,6-dihydro-α-pyrone (**47**) from *Eupatorium pilosum* has been synthesised from (−)-malic acid (*59*). Syntheses have been reported for two rare examples of 5,6-dihydro-α-pyrones, *viz*, a 6,6-disubstituted compound (**48**) present in *Chrysanthemum flosculosum* (*60*) and the alkaloid dumetorine (**49**) (*61*).

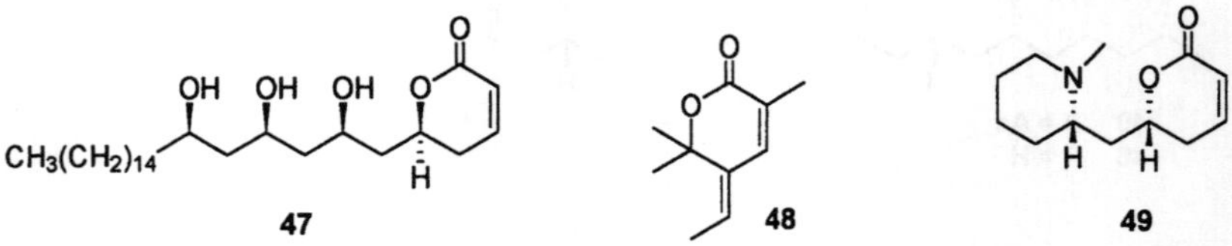

47 **48** **49**

3. 6-Alkenyl-5,6-dihydro-α-pyrones

The antibiotic phomalactone (**50**) has been obtained from a *Drechslera* species (*62*) and the fungus *Hirsutella thompsonii* var. *synnematosa* (*63*) and found to be active against a wide range of microorganisms. Phomalactone acetate (**51**) has been synthesized from 2-furylcarbinols (*64*, *65*). The absolute stereochemistry of (+)-asperlin (**52**) has been established by several syntheses (*65–70*). Phomopsolide B (**53**), an antifeedant against elm bark beetle, isolated from the fungus *Phomopsis oblonga* (*71*), has been prepared from D-glucal triacetate, confirming its absolute configuration (*72*).

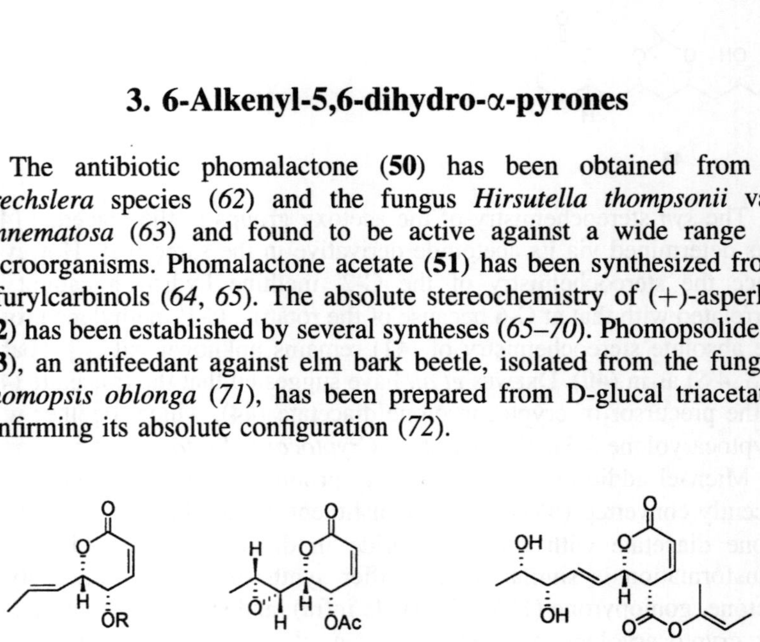

50 R = H
51 R = Ac
52
53

6-Heptenyl-5,6-dihydro-α-pyrones are as common as their saturated counterparts. Several syntheses (*73–77*) of argentilactone (**54**), which also occurs in the Brazilian medicinal plant, *Chorisia crispiflora* (*78*), have been reported. Umuravumbolide (**55**) and deacetylumuravumbolide (**56**) were initially isolated from *Tetradenia riparia* by VAN PUYVELDE *et al.* (*79*). A *trans* configuration was assigned to the double bond based on tenuous IR evidence, but a subsequent high field NMR examination of (*56*), isolated from a South African *Tetradenia sp.*, showed that the double bond was *cis* (*80*). A (3′*S*)-configuration for the single acyclic chiral centre in (**56**) was deduced from the application of the modified Mosher method (*24*).

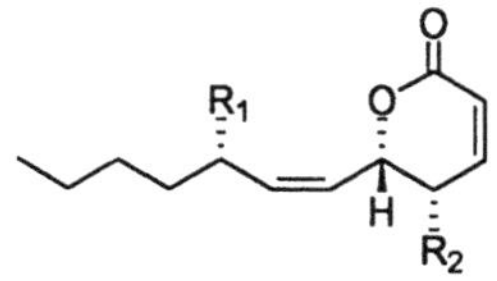

54 $R_1 = R_2 = H$
55 $R_1 = OAc, R_2 = H$
56 $R_1 = OH, R_2 = H$
57 $R_1 = R_2 = OAc$
58 $R_1 = OAc, R_2 = OH$
59 $R_1 = OH, R_2 = OAc$

Pectinolide A (**57**) and its monodeacetyl analogs pectinolide B (**58**) and pectinolide C (**59**) have been isolated from *Hyptis pectinata* (*81*). These compounds exhibit antimicrobial activity and strong cytotoxicity against a variety of tumour cells ($ED_{50} < 4$ μg/ml). The structure of (**57**) was established from spectral and chemical evidence with a $J_{1',2'}$ coupling constant of 10.5 Hz implying (*E*)-stereochemistry for the double bond. A pseudo-equatorial orientation of the C-6 side chain could be inferred from the $J_{5,6}$ coupling constant of 2.9 Hz validating (*4*) the assignment of a (6*S*)-configuration from the CD data. Ozonolysis of (**57**) yielded 2-acetyloxyhexanoic acid and a (3′*S*)-configuration was determined from the CD spectrum of the α-hydroxy acid which revealed a weak negative CD maximum at $\Delta\varepsilon_{244} = 0.01$ and a positive Cotton effect at $\Delta\varepsilon_{209} = +1.58$. Acetylation of (**58**) and (**59**) afforded (**57**) and hence defined their respective stereochemistry.

Structurally related compounds hypurticin (**60**) and synparvolide B (**61**) have been isolated from yet another *Hyptis* species, *H. urticoides*, and *Syncolostemon parviflorus*, respectively (*82, 83*). The structures of (**60**) and (**61**) were determined spectroscopically with the stereochemistry of (**61**) being derived from its biosynthetic relationship to synparvolide A (**70**).

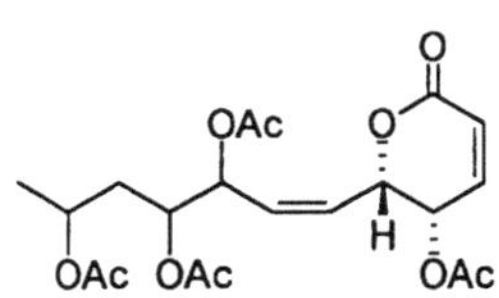

60

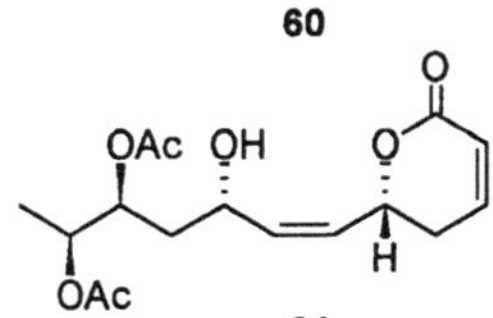

61

In their investigations of Brazilian medicinal plants, Matsuda *et al.* have reported the bioactivity guided isolation of two cytotoxic compounds, (**62**) and (**63**), with conjugated diene side chains, from *Chorisia crispiflora* (*78*). These plants are used as folk medicines for rheumatism and menorrhagia. The $\Delta^{1'}$, $\Delta^{3'}$-Z,E-stereochemistry of the dienone (**62**) was determined from the coupling constants ($J_{1',2'} = 11.0$, $J_{3',4'} = 15.0$ Hz). Split Cotton effects ($\Delta\varepsilon_{264} - 4.7$ and $\Delta\varepsilon_{209} + 9.0$), due to the interaction of the $\pi \rightarrow \pi^*$ transitions of the enone and the dienone, were observed in the CD spectrum of (**62**) implying a (6*R*)-configuration. The structure of (**63**) was determined by spectral techniques and confirmed by oxidation with CrO_3 to give (**62**). The (5′*R*)-configuration in (**63**) was determined using the modified Mosher method.

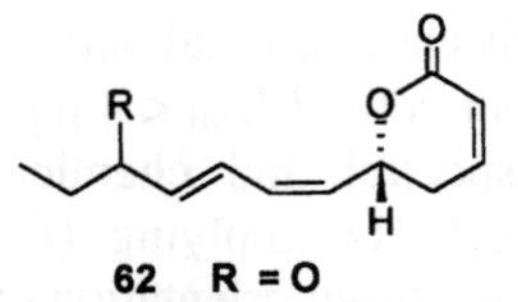

62 R = O
63 R = βOH

5-Deacetoxy-5′-epiolguine (**64**) and three minor related compounds (**65**), (**66**) and (**67**) are bioactive and have been isolated from *Hyptis oblongifolia* (*84*). Spectral and chiroptical data indicated that they were closely related and provided evidence for their structures. The relative stereochemistry of (**64**) was determined by X-ray analysis which together with the positive CD peak at 257 nm established its absolute (6*R*)-configuration. Acid catalysed hydrolysis and methanolysis of (**64**) gave (**65**) and (**66**) respectively, while acetylation of (**65**) afforded (**67**), thus proving the structures of compounds (**65–67**). From NMR and melting point evidence spicigera lactone obtained from *H. spicigera* (*85*) must be 5-deacetoxy-5′-epiolguine (**64**).

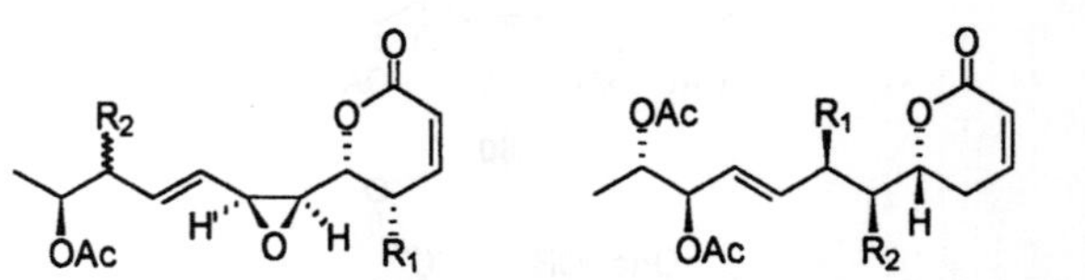

64 R_1 = H, R_2 = αOAc
68 R_1 = OAc, R_2 = βOAc
69 R_1 = OAc, R_2 = αOAc

65 R_1 = R_2 = OH
66 R_1 = OMe, R_2 = OH
67 R_1 = R_2 = OAc

Olguine (**68**) has been known for many years (*1*) and 5′-epiolguine (**69**) has now been obtained from *Hyptis capitata* (*86*) and from *Rabdosia ternifolia* (*87*). Other compounds with acetate groups on the 6-heptyl side chain are synparvolides A (**70**) and C (**71**) isolated from the leaves of *Syncolostemon parviflorus* (*83*), a plant traditionally used by the Zulus as an emetic to treat appetite loss in adults and children. The chemical structures of (**70**) and (**71**) were determined by standard spectroscopic techniques. A comparison of the ^{13}C NMR data of (**70**) and 5-deacetoxy-5′-epi-olguine (**64**) suggested the presence of a β-oxirane ring in the former. CD analysis indicated a (6*R*)-configuration for (**70**) and (**71**). The absolute stereochemistry of the secondary alcohol in (**70**) was established as (3′*S*) using the modified Mosher method and the stereochemistry of both C-5′ and C-6′ was assigned from biosynthetic arguments. Acetylation of (**71**) gave a triacetate (**72**) with well-resolved signals in the ^{1}H NMR spectrum, thus making this compound suitable for NOE difference experiments. These NMR experiments enabled the relative stereochemistry of the cyclic ether moiety of (**71**) to be determined. The 1′,3′-*syn*diol structure was deduced from the chemical shifts of the methyl groups in the ^{13}C NMR spectrum of the acetonide (**73**) (*51*). Monodesacetyl-synparvolide A (**74**) is a probable precursor of (**71**). A (6′*S*)-stereochemistry was assigned to (**71**) by application of the modified Mosher method to the free 6′-hydroxy group in the acetonide derivative of saponified (**71**).

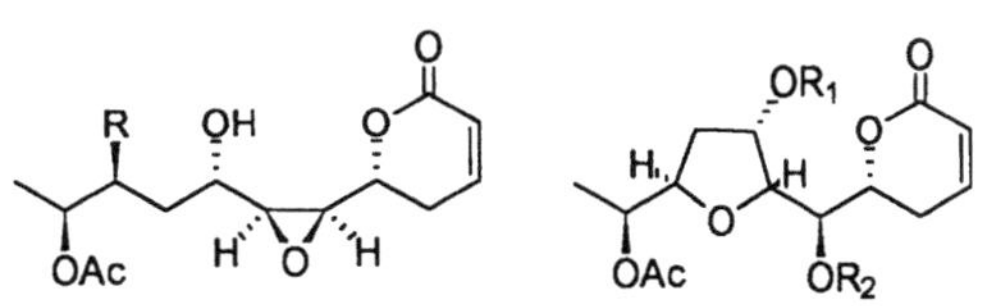

70 R = OAc
74 R = OH

71 $R_1 = R_2 =$ H
72 $R_1 = R_2 =$ Ac
73 $R_1 + R_2 =$ 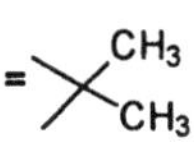

Collett has isolated five closely related compounds, synargentolides A-E (**75–79**), from *Syncolostemon argenteus* and assigned their structures using CD and NMR techniques (*88*). Synargentolide D (**77**) was thermally unstable and paucity of material prevented stereochemical investigations. Synargentolide B (**76**) is epimeric at C-5′ with (**65**).

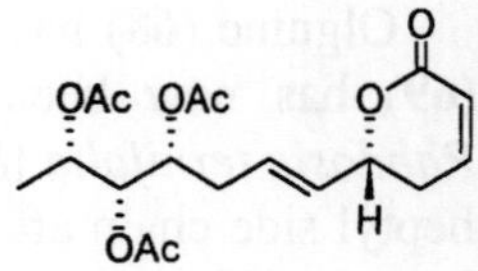

75

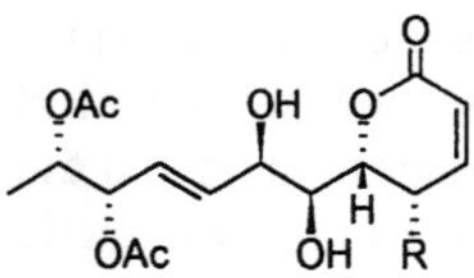

76 R = H
77 R = OAc

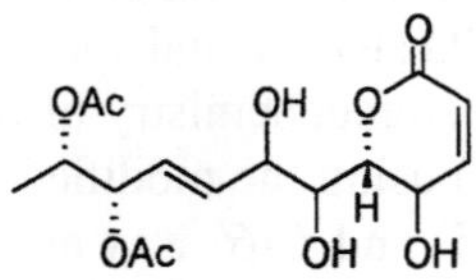

78

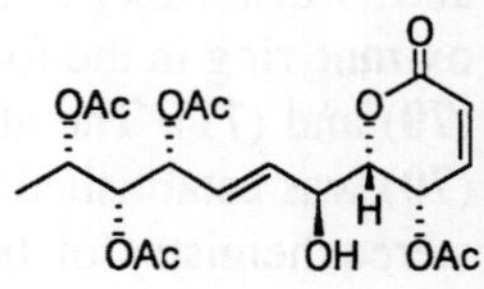

79

Compound (**80**), a structural isomer of umuravarumbolide (**55**), is a minor constituent of *Cryptocarya latifolia* (*89*). The immunosuppressive agent, (−)-PA-48153 (**81**), has been isolated from *Streptomyces prunicolor* sp. PA 48153 (*90*). The same compound, obtained from *Streptomyces* sp. NK10958 and shown to be a plant growth regulator, has been named pironetin, and the structure based on spectral data, confirmed by X-ray analysis (*91*, *92*) and synthesis (*93*).

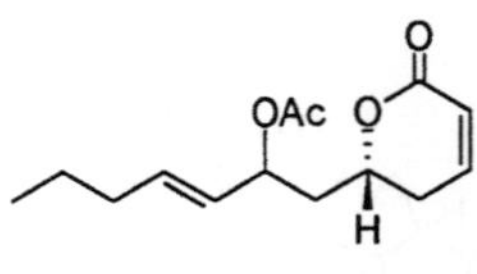

80

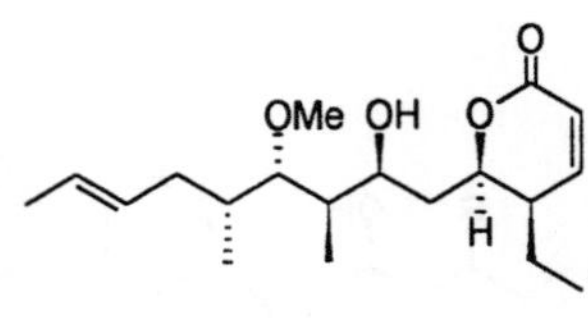

81

Oncorhyncolide (**82**) is the first 5,6-dihydro-α-pyrone to be isolated from marine bacteria. It was obtained from a bacterial isolate taken from surface sea water near a chinook salmon (*Oncorhyncus tshawytscha*) net pen farm by Needham and co-workers (*94*). Thermally unstable (**82**) was acetylated to afford its stable diacetate whose structure was determined using two dimensional NMR techniques. NOE experiments were used to determine the stereochemistry of the double bonds, but the configurations of the two chiral centres were surprisingly not assigned.

The structure (**83**) of toxin I from the fungus *Alternaria citri* (*1*) has been verified by synthesis of its 3-methyl enol ether (**84**) from glucose (*95*).

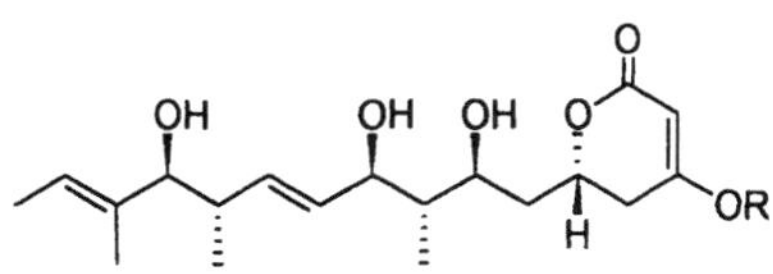

83 R = H
84 R = CH_3

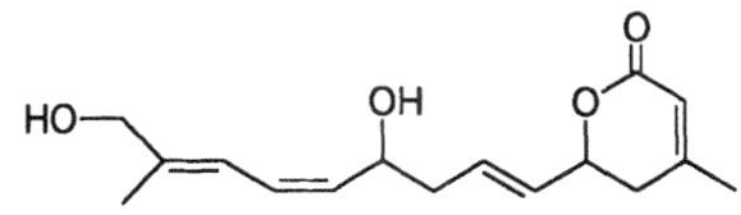

82

Cytostatin (**85**) is a novel inhibitor of cell adhesion to components of the extracellular matrix produced by *Streptomyces sp.* MJ654-NF4. Its structure, without stereochemistry, was determined from high field NMR and high resolution FAB mass spectrometry data (*96*). Sultriecin (**86**), an antifungal and antitumor antibiotic from *Streptomyces roseiscleroticus,* is closely related to (**85**) and also contains a conjugated triene in the side chain (*97*). The antibiotic reductoleptomycin A (**87**), isolated from *Streptomyces sp.* MJ132-NF5 (*98*), is closely related to leptomycin A (**88**) (*2*). Leptolstatin (**89**), the 5-demethyl analog of (**87**) obtained from *Streptomyces sp.* SAM1595, inhibits the progression of the mammalian cell cycle (*99*).

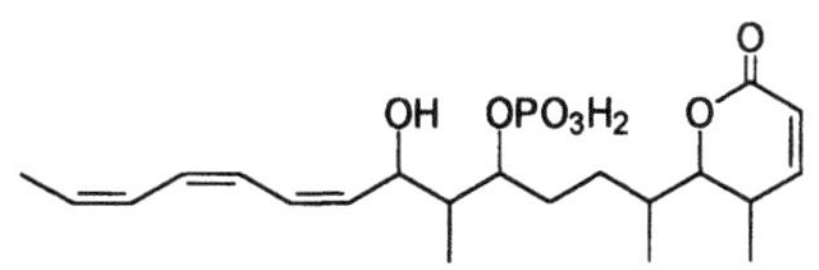

85

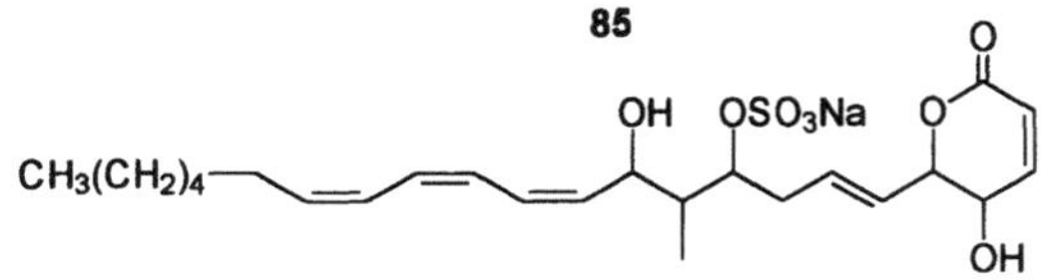

86

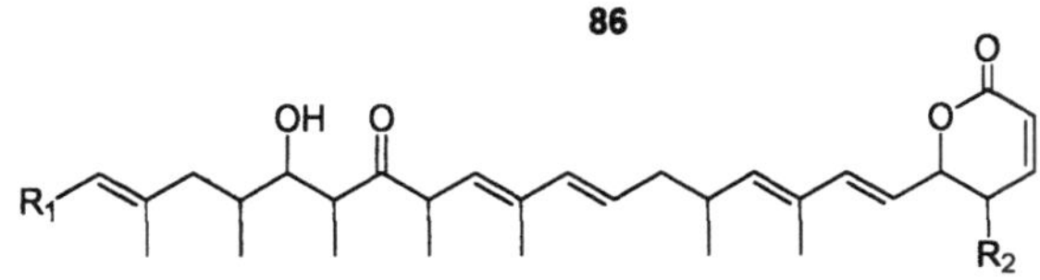

87 $R_1 = CH_2OH$, $R_2 = CH_3$
88 $R_1 = COOH$, $R_2 = CH_3$
89 $R_1 = CH_2OH$, $R_2 = H$

Leustroducsins A, B and C (**90–92**), produced by *Streptomyces platensis* SANK 60191, induce the production of colony stimulating factors by bone marrow stromal cells (*100*) and are closely related to the phoslactomycins, metabolites of other *Streptomyces* species (2).

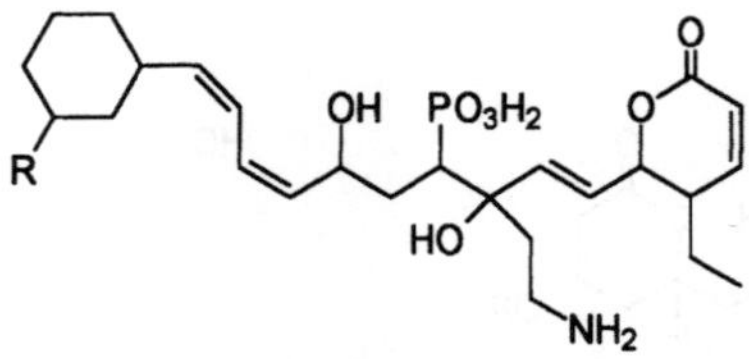

90 R = OCO(CH$_2$)$_3$CH(CH$_3$)$_2$
91 R = OCO(CH$_2$)$_4$CH(CH$_3$)CH$_2$CH$_3$
92 R = OCO(CH$_2$)$_5$CH(CH$_3$)$_2$

An antibiotic with a very narrow spectrum of activity, ratjadone (**93**), has been isolated from the culture broth of a myxobacterium *Sorangium cellulosum*. The growth of some important phytopathogenic fungi, especially species of Oomycetes, was inhibited at very low concentrations of (**93**) (*101*). The structure of (**93**) was determined by spectroscopic methods (*102*).

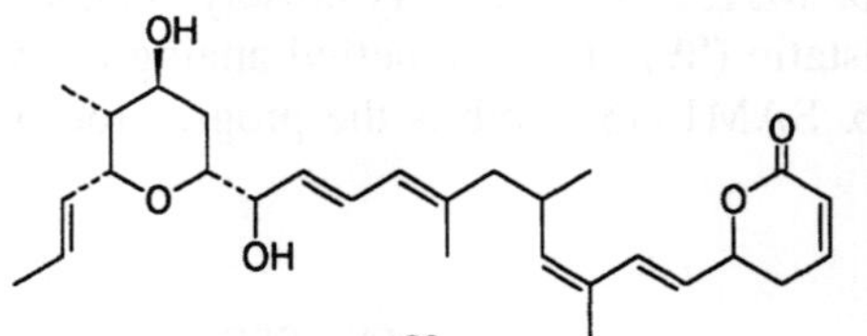

93

Compound (**94**) is the main component of freshly dissected Dufour's gland in the worker ant, *Tetramorium acleatum*. This compound is thought to be responsible for the skin irritating (urticating) properties of the ant and it has been proposed that the lipophilicity of the side chain aids penetration through the skin, or the insect exoskeleton. The structures of (**94**) and its cyclisation product (**95**), which slowly forms from (**94**) in solution, were based on spectral evidence. The relative configurations of C-3, C-6 and C-2′ in (**95**) were derived from a careful analysis of its ^{1}H NMR spectrum and the analogous stereochemistry of (**94**) followed accordingly (*103*).

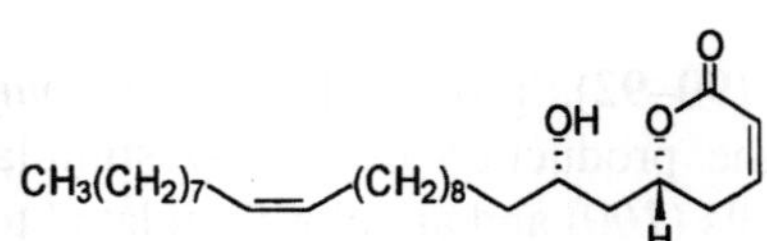

94

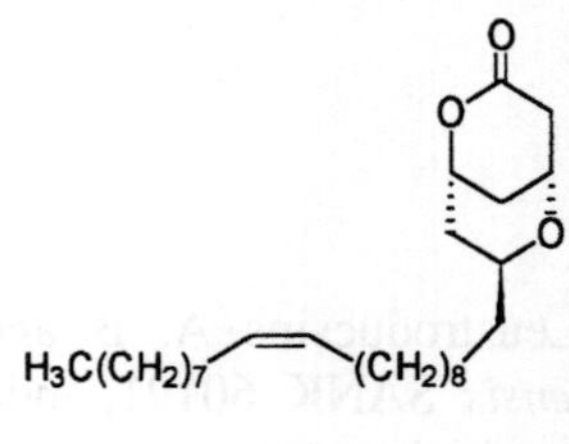

95

4. 6-Aryl-5,6-dihydro-α-pyrones

Two pairs of epimeric secolabdanes, hebecliniolide (**96**) and (**97**) and 3β-hydroxyhebecliniolide (**98**) and (**99**), have been isolated from *Hebeclinium macrophylum* (Compositae). The epimeric mixture of (**98**) and (**99**) was separated as their acetates (**100**) and (**101**) and their structures determined by comparison of their NMR spectra with the spectrum of the mixture of (**96**) and (**97**). NOE difference experiments indicated the relative stereochemistry of the acetates while their absolute configurations were deduced from the negative Cotton effects in the CD spectra of the mixture of (**96**) and (**97**) and chemically transformed products (*104*). Subsequently *Tamauilipa azurea* (also in Compositae) yielded (**96**), (**97**) and another hebecliniolide derivative, 5α-hydroxyhebecliniolide (**102**). A CD study defined the (12*S*)- (usually 6*S*) configuration of (**102**) (*105*).

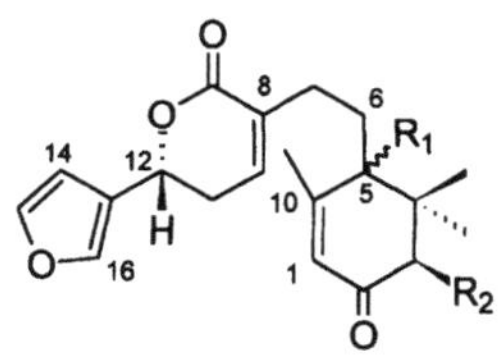

96 R_1 = αH, R_2 = H
97 R_1 = βH, R_2 = H
98 R_1 = αH, R_2 = OH
99 R_1 = βH, R_2 = OH
100 R_1 = αH, R_2 = OAc
101 R_1 = βH, R_2 = OAc
102 R_1 = αOH, R_2 = H

There has been considerable progress in the isolation and synthesis of the bioactive compounds from *Goniothalamus* species. Goniothalamin (**103**) has been isolated from *Goniothalamus sesquipedalis* (*106*) and together with the previously unknown 5-acetylgoniothalamin (**104**) from *G. uvaroides* (*107*). The corresponding desacetyl compound (**105**) occurs in *G. dolichocarpus*; its structure has been established by X-ray analysis and partial synthesis (*108*). Goniodiol (**106**) has been obtained from *G. giganteus* and the structure confirmed by X-ray analysis of the diacetate (*109*). The relative stereochemistry of goniodiol 1′-monoacetate (**107**) from *G. amuyon* has also been determined by crystallographic analysis (*110*). The related compounds goniodiol 2′-monoacetate (**108**) and goniotriol (**109**) are additional α-pyrone metabolites from this *Goniothalamus* species (*111*). The relative stereochemistry of (**109**) from *G. giganteus* was established by X-ray analysis (*112*); 2′-

acetylgoniotriol (**110**) is also present in this plant. Goniothalamin (**103**) and goniothalamin epoxide (**111**) have been isolated from *G. dolichocarpus* (*113*). MCPA epoxidation of (**103**) provided (**111**) and isogoniothalamin epoxide (**112**). The relative configuration of isogoniothalamin epoxide was determined by X-ray analysis and since the absolute configuration of (**103**) is known the absolute structure of isogoniothalamin epoxide is (**112**), and the diastereomeric goniothalamin epoxide must be (**111**). Acid hydrolysis of isogoniothalamin epoxide gave goniodiol (**106**), further supporting structure (**112**) for the former compound. 5-Acetoxygoniothalamin epoxide (**113**) occurs in *G. sesquipedalis*. This compound was misinterpreted as being derived from isogoniothalamin epoxide (*114*).

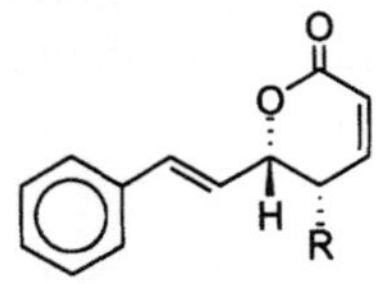

103 R = H
104 R = OAc
105 R = OH

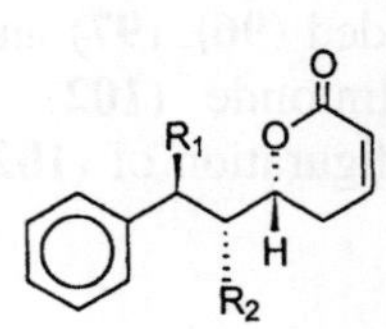

106 $R_1 = R_2 = OH$
107 $R_1 = OH$, $R_2 = OAc$
108 $R_1 = OAc$, $R_2 = OH$

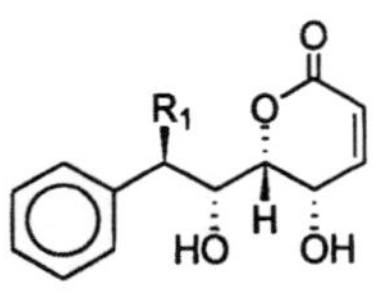

109 R_1 = OH
110 R_1 = OAc

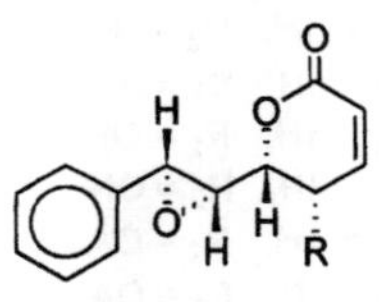

111 R = H
113 R = OAc

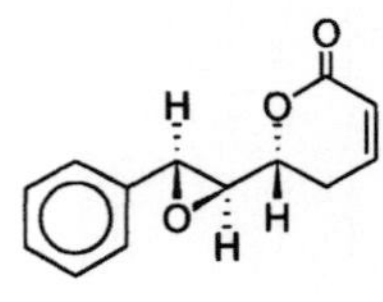

112

Several related dihydro-α-pyrones occur in *Goniothalamus* species (see reference *115* for an extensive bibliography). Altholactone (**114**) is such a bicyclic compound while compounds (**115–118**) with a [3.3.1]-structure are the products of Michael-type cyclizations involving the 2′-OH and the α,β-unsaturated lactone of the 6-substituted 5,6-dihydro-α-pyrone precursor. The structures and relative configurations of goniopyrone (**115**) and 5-deoxygoniopyrone (**116**) were deduced by McLAUGHLIN *et al.* from NMR spectral evidence and X-ray analysis (*58, 109*) and the absolute configurations established by SHING's synthesis from D-glycero-D-gulo-heptono-γ-lactone (*57*). Iso-5-deoxygoniopyrone (**117**), the 1′-epimer of (**116**), has been obtained from *G. dolichocarpus* (*113*). The structure (**118**) proposed for leiocarpin isolated from *G. leiocarpus* with the unusual α-configuration at H-2′ rests on sound ^{1}H NMR evidence (*116*) but an X-ray analysis would nevertheless be desirable. Recent syntheses of goniothalamin (*64, 75, 76, 77*), goniodiol (*117, 118*), goniotriol (*119*), altholactone (*120, 121*), goniopyrone and 5-deoxygoniopyrone (*122*), and (±)-5-deoxygoniopyrone (*123*) have been published.

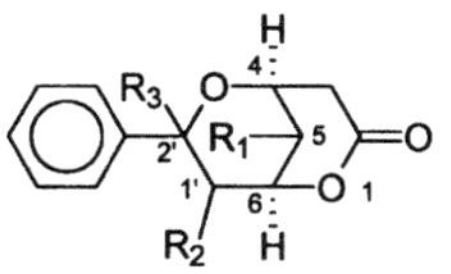

115 R_1 = R_2 = αOH, R_3 = βH
116 R_1 = H, R_2 = αOH, R_3 = βH
117 R_1 = H, R_2 = βOH, R_3 = βH
118 R_1 = H, R_2 = αOH, R_3 = αH

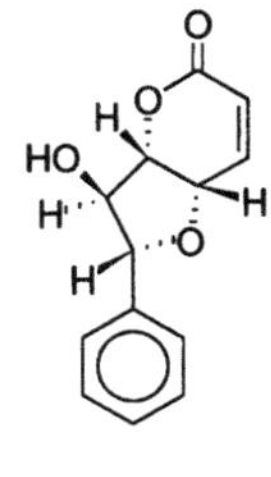

114

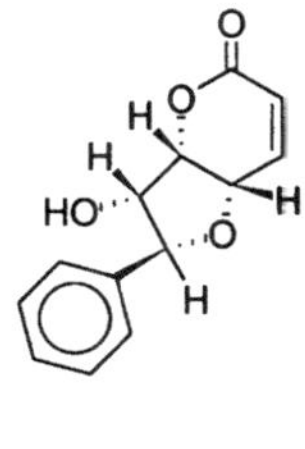

119

Isoaltholactone (**119**) has been obtained from *G. malayanus*, *G. montanus* and *G. tapis* and its relative stereochemistry assigned from NMR data and X-ray analysis. A synthesis from L-arabinose provided the absolute stereochemistry of (**119**). The biosynthetic pathway in Scheme 2 was proposed to account for the formation of altholactone (**114**) and isoaltholactone (**119**) from the same postulated precursor, 5-hydroxygoniothalamin (**105**) (*124*).

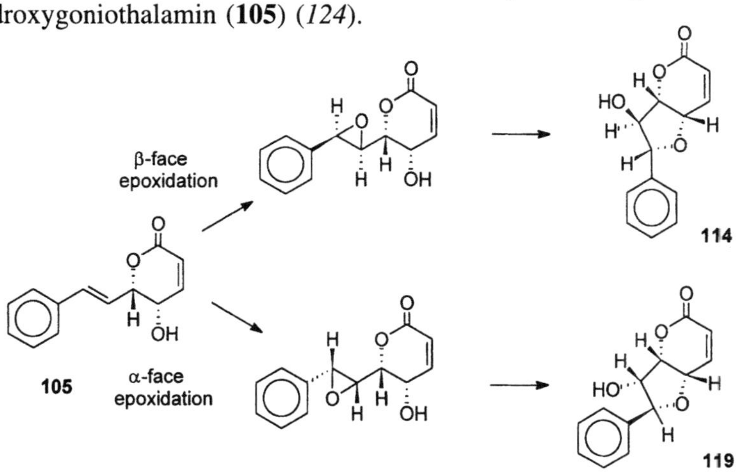

Scheme 2

The total syntheses of (±)-dihydrokawain-5-ol (**120**) and cryptocaryalactone (**121**) have been described (*125, 126*). *Cryptocarya* species contain a further number of related compounds. Cryptocaryalactone (**121**), desacetylcryptocaryalactone (**122**) and the derived [3,3,1]-bicyclo compound (**123**) have been obtained from *C. wyliei*. The structure of (**123**) rests on its formation from desacetylcryptocaryalactone by reaction with sodium hydride in methylene chloride at room temperature (*127*) and an X-ray analysis (*56*). Cryptofolione (**124**) occurs in *C. latifolia* (*128*) and *C. myrtifolia* (*127*) while *C. liebertiana* contains (**124**), (**125**) and 5′-dehydro-cryptofolione (**126**) (*129*). The structure of (**124**) was determined by ^{1}H NMR spectroscopy. The (4′*R*,6′*S*)-stereochemistry followed from a ^{1}H NMR analysis of the corresponding acetonide derivative. A closely related compound, kurzilactone (**127**), with marked cytotoxicity against KB cancer cells, has been isolated from *C. kurzii* (*130*) but in the absence of CD data the absolute stereochemistry at C-6 is unproven.

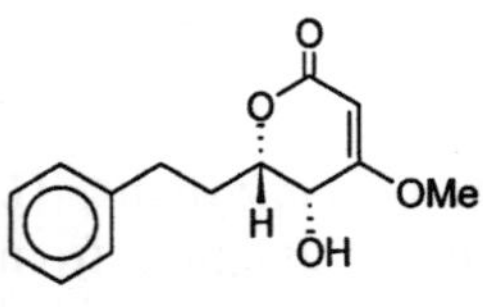

120

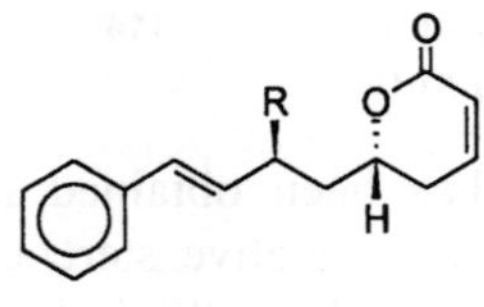

121 R = OAc
122 R = OH

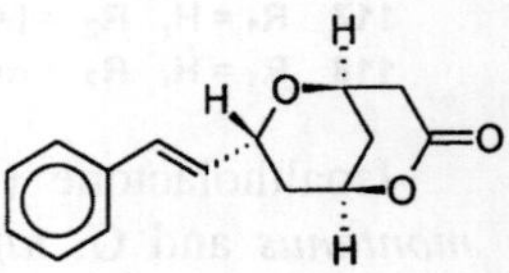

123

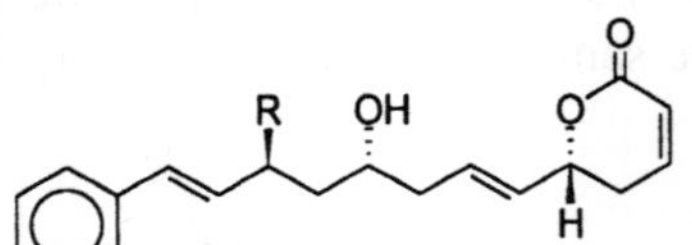

124 R = OH
125 R = OAc

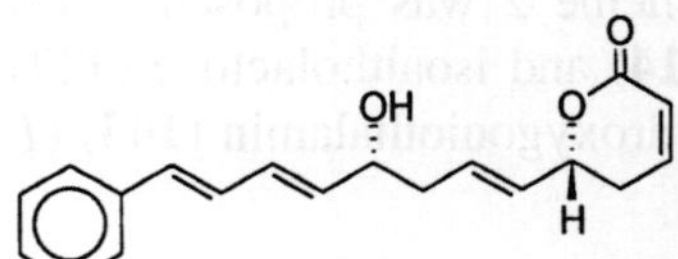

126

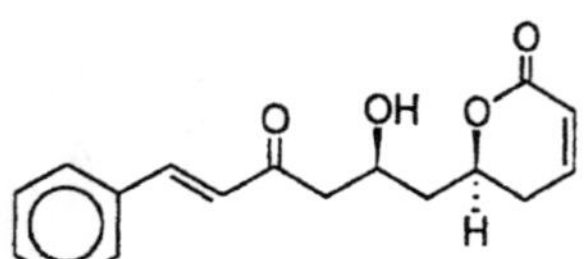

127

A geranylgeraniol derivative with a heterocyclic furanyl substituent, conyzaleucolide (**128**), has been isolated by Zdero *et al.* from *Conyza*

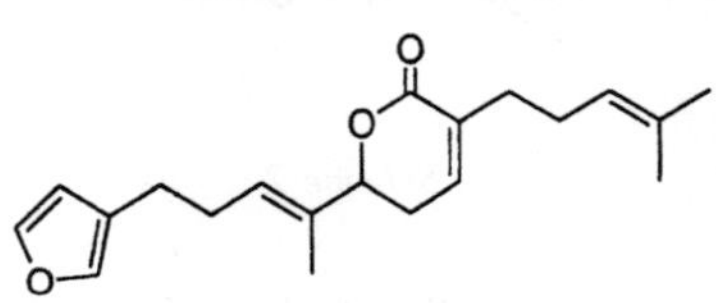

128

hypoleuca (Compositae). Its structure was elucidated by high field NMR techniques (*131*).

5. Physical Methods of Structure Determination

In this review we have drawn attention to the wide application of physical methods, especially high field NMR spectroscopy, in the determination of the structures and relative stereochemistry of 5,6-dihydro-α-pyrones. The modified Mosher method (*24*) has proved to be a useful tool to explore the absolute stereochemistry of hydroxylated chiral centres in the side chains routinely found in this group of compounds. The contiguous arrangement of these hydroxylated chiral centers in many of the side chains has also seen increased use made of ^{13}C

Table 1. *The n → π* CD and ORD Data for 6-Substituted-5,6-dihydro-α-pyrones*

Compound	No.	CD Δε	n →π* λ nm	ORD Φ	n →π* λ nm	Stereochemistry at C-6 implied from Snatske's rules
Gerberin acetate	**4**	+7.78	256			(*S*)
8′-Deoxyalternaric acid	**11**	−3.9	266			(*S*)
Acid	**12**	−3.2	259			(*S*)
Acid	**13**	−3.9	260			(*S*)
Acid	**14**	−3.2	260			(*S*)
Acid	**15**	−3.3	260			(*S*)
Proalternaric acid	**17**	−3.5	260			(*S*)
Gamahonolide A	**33**	−1.96	249	−6176	268	(*R*)
Gamahonolide B	**34**	−0.93	249	−2380	267	(*R*)
Syndenolide	**35**	+2.35	257			(*R*)
Triacetate	**40**	+2.5	256			(*R*)
Diacetate	**41**	+2.8	256			(*R*)
Desacetylumuravumbolide	**56**	+0.9	255			(*R*)
Pectinolide A	**57**	+2.4	265			(*S*)
Synparvolide B	**61**	+2.97	258			(*R*)
5-Deacetoxy-5′-epiolguine	**64**	+2.8	256			(*R*)
C_{12} Lactone	**67**	+2.4	257			(*R*)
Synparvolide A	**70**	+2.37	255			(*R*)
Synparvolide C	**71**	+2.34	256			(*R*)
Synargentolide A	**75**	+3.5	265			(*R*)
Synargentolide B	**76**	+3.2	258			(*R*)
Synargentolide C	**77**	+2.0	266			(*R*)
Synargentolide E	**79**	+1.1	265			(*S*)
5α-Hydroxyhebecliniolide	**102**	−2.5	324			(*S*)

chemical shift information from acetonide derivatives (*51*) to establish the relative stereochemistry of vicinal diols. Circular dichroism continues to provide a simple means of defining absolute configuration at C-6. Such data are presented in Table 1 to supplement those supplied previously (*1*). Unfortunately, despite their usefulness, CD data are still not routinely published.

References

1. Davies-Coleman, M.T., and D.E.A. Rivett: Naturally Occurring 6-Substituted 5,6-dihydro-α-pyrones. Fortschr. Chem. organ. Naturstoffe, **55**, 1 (1989).

2. Dickinson, J.M.: Microbial Pyran-2-ones and Dihydropyran-2-ones. Nat. Prod. Reports, **10,** 71 (1993).

3. Nagumo, S.T., T. Toyonaga, T. Inoue, and M. Nagai: New Glucosides of a 4-Hydroxy-5-methylcoumarin and a Dihydro-α-pyrone from *Gerbera jamesonii hybrida*. Chem. Pharm. Bull. (Japan), **37**, 2621 (1989).

4. Snatzke, G.: Circular Dichroism and Optical Rotatory Dispersion – Principles and Application to the Investigation of the Stereochemistry of Natural Products. Angew. Chem. Internat. Edit., **7**, 14 (1968).

5. Ayer, W.A., and J.S. Racok: The Metabolites of *Talaromyces flavus*: Part 2. Biological Activity and Biosynthetic Studies. Can. J. Chem., **68**, 2095 (1990).

6. Sato, T.: Synthesis of Parasorbic Acid, the Component of *Sorbus aukuparia*. Heterocycles, **24**, 2133 (1986).

7. Eicher, T., R. Graf, and R. Pick: Improved Synthesis of Racemic Parasorbic Acid. Arch. Pharmaz., **319**, 91 (1986).

8. Procter, G., A.T. Russell, P.J. Murphey, T.S. Tan, and A.N. Mather: Epoxy-silanes in Organic Synthesis. Tetrahedron, **44**, 3953 (1988).

9. Stevenson, R., and J.V. Weber: Synthesis of (±)-Parasorbic acid and (±)-Massoia-lactone from Meldrum's acid. J. Nat. Prod., **51**, 1215 (1988).

10. Gopalan, A.S., and H.K. Jacobs: Synthesis of *S*(+)-Parasorbic Acid and *S*(+)-2-Tridecanol Acetate. Tetrahedron Letters, **31**, 5575 (1990).

11. Hoeyer, T., A. Kjaer, and J. Lykkestedt: A Convenient Synthesis of Homochiral δ-Alkylated αβ-Unsaturated δ-Lactones. Coll. Czech. Chem. Comm., **56**, 1042 (1991).

12. Shao, L., T. Seki, H. Kawano, and M. Saburi: Asymmetric Hydrogenation of Methyl 3,5-Dioxohexanoate Catalysed by Ruthenium-binap Complex : A Short Step Asymmetric Synthesis of Dihydro-6-methyl-2H-pyran-2-one. Tetrahedron Letters, **32**, 7699 (1991).

13. Bernard, R., and D. Ghiringhelli: Synthesis of Enantiomerically Pure (*S*)-5,6-dihydro- and (*S*)-tetrahydro-6-methyl-2H-pyran-2-one. Gazz. chim. ital., **122**, 395 (1992).

14. Robin, S., and F. Huet: Preparation of Lactones with Several Ring Sizes via the Same Intermediate. Tetrahedron Letters, **34,** 2945 (1993).

15. Tiedemann, R., F. Narjes, and E. Schumann: 3-Methoxy-1-phenylthio-1-propene as d^1/d^3 Synthon : Application to an Asymmetric Synthesis of (*S*)-(+)-Parasorbic acid. Synlett, 594 (1994).

16. Buchanan, M.S., T. Hashimoto, S. Takaoka, and Y. Asakawa: (+)-Osmundalactone, γ-Lactones and Spiromentins from the Fungus *Paxillus atrotomentosus.* Phytochem., **40**, 1251 (1995).

17. SUGIYAMA, T., T. MURAYAMA, K. YAMASHITA, and T. ORITANI: Synthesis of Chiral Aspyrone, a Multi-functional Dihydropyranone Antibiotic. Biosci. Biotechnol. Biochem., **59**, 1921 (1995).
18. HILL, A.M., A. JACOBS, and J. STAUNTON: Investigation of the Stereochemistry of the Tri- and Tetra-ketide Hydroxyacyl Intermediates in the Biosynthesis of the Polyketide Aspyrone in *Aspergillus melleus* using Deuterium Labelling and Deuterium NMR Spectroscopy. Chem. Commun. 859 (1995).
19. BRIAN, P.W., P.J. CURTIS, H.G. HEMMING, C.H. UNWIN, and J.M. WRIGHT: Alternaric acid, a Biologically Active Metabolic Product of the Fungus *Alternaria solani.* Nature, **164**, 534 (1949).
20. TABUCHI, H., and A. ICHIHARA: Stereochemistry of Alternaric Acid; Synthesis of the C(9)-C(14) Fragment. Tetrahedron Letters, **33**, 4933 (1992).
21. TABUCHI, H., T. HAMAMOTO, S. MIKI, T. TEJIMA, and A. ICHIHARA: Total Synthesis of Alternaric Acid. Tetrahedron Letters, **34**, 2327 (1993).
22. TABUCHI, H., and A. ICHIHARA: Structures and Stereochemistries of New Compounds Related to Alternaric Acid. J. Chem. Soc. (London) Perkin Trans. 1, 125 (1994).
23. TABUCHI, H., O. HIDEAKI, and I. AKITAMI: Biosynthetic study of Alternaric Acid : Isolation of Plausible Biosynthetic Intermediates and Origins of the Hydrogen and Oxygen Atoms. J. Chem. Soc. (London) Perkin Trans. 1, 2833 (1994).
24. OHTANI, I., T. KUSUMI, Y. KASHMAN, and H. KAKISAWA: Highfield FT NMR Application of Mosher's Method. The Absolute Configuration of Marine Terpenoids. J. Amer. Chem. Soc., **113**, 4092 (1991).
25. HAMADA, T., T. KUSUMI, M.O. ISHITSUKA, and H. KAKISAWA: Structures and Absolute Configurations of New Lobane Diterpenoids from the Okinawan Soft Coral *Sinularia flexibilis.* Chem. Letters, 33 (1992).
26. KUSANI, T., T. HAMADA, M.O. ISHITSUKA, I. OHTANI, and H. KAKISAWA: Elucidation of the Relative and Absolute Stereochemistry of Lobatriene, a Marine Diterpene, by a Modified Mosher's Method. J. Organ. Chem. (USA), **57**, 1033 (1992).
27. TSUDA, M., H. SHIGEMORI, M. ISHIBASHI, T. SASAKI, and J. KOBAYASHI: Luffariolides A-E, New Cytotoxic Sesterpenes from the Okinawan Marine Sponge *Luffariella sp.* J. Organ. Chem. (USA), **57**, 3503 (1992).
28. HAREAU-VITTINI, G., and P.J. KOCIENSKI: A Synthesis of (3*S*,4*R*)-Luffariolide. Synlett 893 (1995).
29. ICHIMOTO, I., K. MACHIYA, M. KIRIHATA, and H. UEDA: Stereoselective Synthesis of Podoblastine and their Antiblast Activity. J. Pesticide Sci., **13**, 605 (1988).
30. MORI, Y., H. KAGEYAWA, and M. SUZUKI: Synthesis of (−)-Tarchonanthus Lactone, a Syn-1,3-polyol-derived α,β-Unsaturated δ-Lactone. Chem. Pharm. Bull. (Japan), **38**, 2574 (1990).
31. SOLLADIE, G., and L. GRESSOT-KEMPF: Chiral Sulfoxides in Asymmetric Synthesis: Enantioslective Synthesis of (−)(5S,7R)-Tarchonanthus Lactone. Tetrahedron Asymmetry, **1**, 2371 (1996).
32. ASOAKA, M., S. HAYASHIBE, S. SONODA, and H. TAKEI: Synthesis and Utilization of Optically Active 2-Substituted 4-(Trimethylsilyl)cyclopentanones: Synthesis of (−)-Massoialactone and (+)-β-Cuparenone. Tetrahedron Letters, **31**, 4761 (1990).
33. BENNETT, F., D.W. KNIGHT, and G. FENTON: Total Synthesis of Natural (+)-(4*R*,6*R*)-4-Hydroxy-6-pentylvalerolactone and of (−)-(6*R*)-Massoialactone. J. Chem. Soc. (London) Perkin **1**, 1543 (1991).
34. ROMEYICE, Y., M. KELLER, H. KLUGE, S. GRABLEY, and P. HAMMAN: Enantioselective Synthesis of δ-Lactones from Streptenol A, a Chiral Building Block from Streptomyces. Tetrahedron, **47**, 3335 (1991).

35. BONINI, C., P. PUCCI, R. RACIOPPI, and L. VIGGIANI: Enzyme Catalysed Lactonization of 3,5-Dihydroxy Esters: Enantioselective Synthesis of Naturally Occurring 3-Hydroxy-5-decanolide, (−)-Massoialactone and 3-Hydroxy-5-eicosanolide. Tetrahedron Asymmetry, **3**, 29 (1992).
36. TAKANO, S., M. SIETOH, and K. OGASAWARA: An Enantiospecific Route to (6*R*)-Massoialactone and (4*R*,6*R*)-(+)-4-Hydroxy-6-pentylvalerolactone. Tetrahedron Asymmetry, **3**, 533 (1992).
37. YU, L., and Z. WANG: Enantioselective Synthesis of 6*R*(−)-Massoialactone (from Mannitol). Chinese Chemical Letters **4**, 1 (1993).
38. VENKATASUBBAIAH, P., C.G. VAN DYKE, and W.S. CHILTON: Phytotoxins Produced by *Pestalotiopsis oenotherae*. Phytochem., **30**, 1471 (1991).
39. HONDA, T., A. OKUYAMA, T. HAYAKAWA, H. KONDOH, and M. TSUBUKI: A Stereoselective Synthesis of (±)-Pestalotin. Chem. Pharm. Bull. (Japan), **39**, 1866 (1991).
40. ZHANG, J., and D.P. CURRAN: Stereoselective Synthesis of 1,2-Diols by the Cycloadditive Strategy : Total Synthesis of (±)-Exobrevicomin and (±) and (−)-Pestalotin. J. Chem. Soc. (London) Perkin Trans. 1, 2627 (1991).
41. HAGIWARA, H., K. KIMURA, and H. UDA: High Diastereoselection in the Aldol Reaction of the Bistrimethylsilyl Enol Ether of Methyl Acetoacetate with 2-Benzyloxyhexanal : Synthesis of (−)-Pestalotin. J. Chem. Soc. (London) Perkin Trans. 1, 693 (1992).
42. KIRIHATA, M., K. OHTA, I. ICHIMOTO, and H. UEDA: Total Synthesis of (6*S*, 1′*S*, 2′*R*)-6-(1,2-Dihydroxypentyl)-4-methoxy-5,6-dihydropyran-2-one (LL-P880β) and its C_6-Epimer, a Fungal Metabolite from *Penicillium sp.* Agric. Biol. Chem., **54**, 5401 (1990).
43. KIRIHATA, M., Y. KAMIHISA, I. ICHIMOTO, and H. UEDA: Stereoselective Synthesis of (6*R*, 1′*R*, 2′*S*)- and (6*S*, 1′*R*, 2′*S*)-LL-P880β, Stereoisomers of the Fungal Metabolite from *Penicillium* strains. Chem. Express 7, 837 (1992).
44. KIRIHATA, M., M. OHE, I. ICHIMOTO, and H. UEDA: Stereoselective Synthesis of Unnatural Stereoisomers of LL-P880β and LL-P880β, Pestalotin Analogs from *Penicillium sp.* Biosci. Biotechnol. Biochem., **56**, 1825 (1992).
45. MASAKI, Y., T. IMAEDA, and M. KAWAI: Highly Stereoselective Synthesis and Structural Confirmation of a Fungal Metabolite LL-P880β. Chem. Pharm. Bull. (Japan), **42**, 179 (1994).
46. KOSHINO, H., T. YOSHIHARA, M. OKUNO, S. SAKAMURA, A. TAJIMI, and T. SHIMANUKI: Gamahonolides A, B, and Gamahorin, Novel Antifungal Compounds from Stromata of *Epichloe typhina* on *Phleum pratense*. Biosci. Biotechnol. Biochem., **56**, 1096 (1992).
47. DALE, J.A., and H.S. MOSHER: Nuclear Magnetic Resonance Enantiomer Reagents. Configurational Correlations via Nuclear Magnetic Resonance Chemical Shifts of Diastereomeric Mandelate, O-Methylmandelate, and α-Methoxy-α-trifluoromethylphenylacetate (MTPA) Esters. J. Amer. Chem. Soc., **95**, 512 (1973).
48. TROST, B.M., J.L. BELLETIRE, S. GODLESKI, G.S. PONTICELLO, S.L. VARGA, and J.P. SPRINGER: On the use of the O-methylmandelate Ester for the Establishment of the Absolute Configuration of Secondary Alcohols. J. Organ. Chem. (USA), **51**, 2370 (1986).
49. DAVIES-COLEMAN, M.T., and D.E.A. RIVETT: An α-Pyrone from *Syncolostemon densiflorus*. Phytochem., **35**, 1590 (1994).
50. COLLETT, L.A., M.T. DAVIES-COLEMAN, D.E.A. RIVETT, S.E. DREWES, and M.M. HORN: Absolute Configuration of α-Pyrones from *Cryptocarya latifolia* and *Syncolostemon* densiflorus. Phytochem., **4**, 935 (1997).

51. RYCHNOVSKY, S.D., and D.J. SKALITZKY: Stereochemistry of Alternating Polyol Chains: NMR Analysis of 1,3-Diol Acetonides. Tetrahedron, **31**, 945 (1990).
52. JEFFORD, C.W., and M.C. MOULIN: The Synthesis of Boronolide. Helv. Chim Acta, **74**, 336 (1991).
53. NAGANO, H., and H. YASUI: Synthesis of (+)-Boronolide from D-glucose. Chemistry Letters, 1045 (1992).
54. HONDA, T., S. HORIUCHI, H. MIZUTANI, and K. KANAI: Enantiocontrolled Synthesis of (+)-Boronolide. J. Organ. Chem. (USA), **61**, 4944 (1996).
55. DREWES, S.E., B.M. SEHLAPELO, M.M. HORN, R. SCOTT-SHAW, and P. SANDOR: 5,6-Dihydro-α-pyrones and Two Bicyclic Tetrahydro-α-pyrone Derivatives from *Cryptocarya latifolia.* Phytochem., **38**, 1427 (1995).
56. HORN, M.M.: M.Sc. thesis, University of Natal, Pietermaritzburg, South Africa, 1996.
57. SHING, T.K.M., H.C. TSUI, and Z.H. ZHOU: Enantiospecific Syntheses of (+)-Goniofufurone, (+)-7-epi-Goniofufurone, (+)-Goniobutenolide A, (−)-Goniobutenolide B, (+)-Goniopyrone, (+)-Altholactone, (+)-Goniotriol, and (+)-7-Acetylgoniotriol. J. Organ. Chem. (USA), **60**, 3121 (1995).
58. FANG, X.P., J.E. ANDERSON, C.J. CHANG, P.E. FANWICK, and J.L. MCLAUGHLIN: Novel Bioactive Styryl-lactones : Goniopyrone, and δ-Acetylgoniotriol from *Goniothalamus giganteus* (Annonaceae). X-Ray Molecular Structure of Goniofufurone and of Goniopyrone. J. Chem. Soc. (London) Perkin Trans. 1, 1655 (1990).
59. NAKATA, T., T. SUENAGA, K. NAKASHIMA, and T. OISHI: Total Synthesis of Natural Products having 1,3-*syn*-Polyol. δ-Lactone of (2*Z*,5*S*,7*S*,9*R*,11*R*)-Tetra-hydroxyhexacos-2-enoic Acid and 4,6,8,10,12,14,16,18,20-all-*syn*-Nonamethoxy-1-pentacosene. Tetrahedron Letters, **30**, 6529 (1989).
60. ISHIBASHI, H., H. NAKATANI, T.S. SO, T. FUJITA, M. IKEDA: Alkylative Lactonization of γ,δ-Unsaturated Esters with α-Chlorosulfides. A Concise Synthesis of the Monoterpene Lactone from *Chrysanthemum flosculosum.* Heterocycles, **31**, 215 (1990).
61. AMARASEKARA, A.S., and A. HASSNER: Stereospecific Synthesis and Stereochemical Structure Confirmation of Dumetorine. Tetrahedron Letters, **28**, 3151 (1987).
62. KRIVOBOK, S., F. THOMASSON, F. SEIGLE-MURANDI, R. STEIMAN, and C. BOTTEX-GAUTHIER: 6-Allyl-5,6-dihydro-5-hydroxypyran-2-one, Lactone Produced by a New *Drechslera* species : Specified ^{1}H and ^{13}C NMR assignments, Mutagenic and Immunomodulating Testings. Pharmazie, **49**, 605 (1994).
63. KRASNOFF, S.B., and S. GUPTA: Identification of the Antibiotic Phomalactone from the Entomopathogenic Fungus *Hirsutella thompsonii* var. *synnematosa*. J. Chem. Ecol., **20**, 293 (1994).
64. HONDA, T., T. KAMETANI, K. KANAI, Y. TATSUZAKI, and T. TSUBUKI: Enantioselective Syntheses of (+)-Acetylphomalactone and (6R)-(+)-Goniothalamin. J. Chem. Soc. (London), Perkin Trans. 1, 1733 (1990).
65. YANG, Z.C., X.B. JIANG, Z.M. WANG, and W.S. ZHOU: Total Synthesis of (+)-Asperlin, (+)-Acetylphomalactone and (5*S*,6*S*,7*R*,8*S*)-Asperlin Based on the Kinetic Resolution of 2-Furylmethanols. J. Chem. Soc (London) Perkin Trans. 1, 317 (1997).
66. SHING, T.K.M., and M. ALOUI: The Stereochemistry of the Epoxypropyl Side-chain of Asperlin. Chem. Commun., 1525 (1988).
67. SHING, T.K.M., and M. ALOUI: Enantiospecific Synthesis of the (6*R*,7*S*)-Diastereoisomer of Asperlin. Canad. J. Chem., **68**, 1035 (1990).
68. RAMESH, S., and R.W. FRANK: Total Synthesis of (+)-Asperlin. Tetrahedron Asymmetry, **1**, 137 (1990).
69. MASAKI, Y., T. IMAEDA, H. ODA, A. ITOH, and M. SHIRO: Total Synthesis of (+)-Asperlin Starting with (*S*,*S*)-Tartaric Acid. Chem. Letters, 1209 (1992).

70. Honda, T., N. Sano, and K. Kanai: Synthesis of (+)-Asperlin. Heterocycles, **41**, 425 (1995).
71. Grove, J.F.: Phomopsolide A and B, Tiglic Esters of Two 6-Substituted 5,6-Dihydro-5-hydroxy-pyran-2-ones. J. Chem. Soc. (London) Perkin Trans. 1, 865 (1985).
72. Noshita, T., T. Sugiyama, K. Yamashita, and T. Oritani: Total Synthesis of Natural (+)-Phomopsolide B, an Antifeedant Against Elm Bark Beetle. Biosci. Biotechnol. Biochem., **58**, 740 (1994).
73. O'Connor, B., and G. Just: Syntheses of Argentilactone and Goniothalamin. Tetrahedron Letters, **27**, 5201 (1986).
74. Carretero, J.J., and L. Ghosez: A Practical Route towards αβ-Unsaturated δ-Lactones based on a [3 + 3] Strategy. Synthesis of (−)-Argentilactone. Chem. Letters, **29**, 2059 (1988).
75. Rahman, S.S., B.J. Wakefield, S.M. Roberts, and M.D. Dowle: Intramolecular Nucleophilic Addition to Photochemically Generated Ketones as a Versatile Route to Lactones and Lactams: Synthesis of a Mosquito Pheromone, Goniothalamin, Argentilactone and the Streptomyces L-factor. Chem. Commun., 303 (1989).
76. Chidambaram, N., K. Satyanarayana, and S. Chandrasekaran: A General Approach to the Synthesis of 5,6-Dihydro-2(2H)pyranones: Simple Synthesis of (+)-Argentilactone and (±)-Goniothalamin. Tetrahedron Letters, **30**, 2429 (1989).
77. Tsubuki, M., K. Kanai, and T. Honda: Enantioselective Synthesis of 6-Substituted 5,6-Dihydro-α-pyranones, (+)-Goniothalamin and (−)-Argentilactone. Heterocycles, **35**, 281 (1993).
78. Matsuda, M., Y. Endo, S. Fushiya, T. Endo, and S. Nozoe: Cytotoxic 6-Substituted 5,6-Dihydro-2H-pyran-2-ones from a Brazilian Medicinal Plant, *Chorisia crispiflora.* Heterocycles, **38**, 1229 (1994).
79. Van Puyvelde, L., S. Dube, E. Uwimana, C. Uwera, R.A. Domisse, E. Lesmans, O. Van Schoor, and A.J. Vlietinch: New α-Pyrones from *Iboza riparia.* Phytochem., **18**, 1215 (1979).
80. Davies-Coleman, M.T., and D.E.A. Rivett: Structure of the 5,6-Dihydro-α-pyrone, Umaravumbolide. Phytochem., **38**, 791 (1995).
81. Pereda-Mirando, R., L. Hernandez, M.J. Villavicencio, M. Novello, P. Ibarra, H. Chai, and J.M. Pezzuto: Structure and Stereochemistry of Pectinolides A-C, Novel Antimicrobial and Cytotoxic 5,6-Dihydro-α-pyrones from *Hyptis pectinata.* J. Nat. Prod., **56**, 583 (1993).
82. De Vivar, A., P. Vidales, and A.L. Perez: An Aliphatic δ-Lactone from *Hyptis urticoides.* Phytochem., **30**, 2417 (1991).
83. Davies-Coleman, M.T., and D.E.A. Rivett: 5,6-Dihydro-α-pyrones from *Syncolostemon parviflorus.* Phytochem., **41**, 1085 (1996).
84. Pereda-Miranda, R., M. Garcia, and G. Delgado: Structure and Stereochemistry of Four α-Pyrones from *Hyptis oblongifolia.* Phytochem., **29**, 2971 (1990).
85. Aycard, J.P., F. Kini, B. Kam, E.M. Gaydou, and R. Faure: Isolation and Identification of Spicigera Lactone : Complete ^{1}H and ^{13}C Assignments using Two-dimensional NMR Experiments. J. Nat. Prod., **56**, 1171 (1993).
86. Almtorp, G.T., A.C. Hazell, and K.B.G. Torsell: A Lignan and Pyrone and Other Constituents from *Hyptis capitata.* Phytochem., **30**, 2753 (1991).
87. Lu, G.H., F.P. Wang, J.M. Pezzuto, T.C.M. Tamm, I.D. Williams, and C.T. Che: 10-Epiolguine from *Rabdosia ternifolia.* J. Nat. Prod., **60**, 425 (1997).
88. Collett, L.A.: MSc thesis, Rhodes University, Grahamstown, South Africa, 1997.
89. Drewes, S.E., M.M. Horn, and C.S. Wijewardene: α-Pyrone from *Cryptocarya latifolia* – A Structural Isomer of Umuravumbolide. Phytochem., **41**, 333 (1996).

90. YOSHIDA, T., K. KOIZUMI, Y. KAWAMURA, K. MATSUMOTO, and H. ITAZAKI: Lactone with Immunosuppresive Activity and its Manufacture with *Streptomyces prunicolor.* European Patent 560389 (1993).
91. KOBAYASHI, S., K. TSUCHIYA, T. HARADA, M. NISHIDE, T. KUROKAWA, T. NAKAGAWA, N. SHIMADA, and K. KOBAYASHI: Pironetin, a Novel Plant Growth Regulator Produced *by Streptomyces sp.* NK 10958. 1. Taxonomy, Production, Isolation and Preliminary Characterization. J. Antibiot., **47**, 697 (1994).
92. KOBAYASHI, S., K. TSUCHIYA, T. KUROKAWA, T. NAKAGAWA, and N. SHIMADA: Pironetin, a Novel Plant Growth Regulator Produced by *Streptomyces sp.* NK 10958. II. Structural, Elucidation. J. Antibiot., **47**, 703 (1994).
93. YASUI, K., Y. TAMURA, T. NAKATANI, K. KAWADA, and M. OHTANI: Total Synthesis of (−)-PA-48153C, a Novel Immunosuppressive 2-Pyranone Derivative. J. Organ. Chem. (USA), **60**, 7567 (1995).
94. NEEDHAM, J., R.J. ANDERSEN, and M.T. KELLY: Oncorhyncolide, A Novel Metabolite of a Bacterium Isolated from Seawater. Tetrahedron Letters, **32**, 315 (1991).
95. LICHTENTHALER, F.W., J. DINGES, and F. YOSHIMASA: ACRL Toxin I: Convergent Total Synthesis of its 3-Methyl Enol Ether from D-Glucose. Angew. Chem. Int. Ed. Engl., **30**, 1339 (1991).
96. AMEMIYA, M., T. SOMENO, R. SAWA, H. NAGANAWA, M. ISHIZUKA, and T. TAKEUCHI: Cytostatin, a Novel Inhibitor of Cell Adhesion to Components of Extracellular Matrix Produced by *Streptomyces sp.* MJ654-NF4. J. Antibiot., **47**, 541 (1994).
97. OHKUMA, H., N. NARUSE, Y. NISHIYAMA, T. TSUNO, Y. HOSHINO, Y. SAWADA, M. KONISHI, and T. OKI: Sultriecin, a New Antifungal and Antitumor Antibiotic from *Streptomyces roseiscleroticus.* Production, Isolation, Structure and Biological Activity. J. Antibiot., **45**, 1239 (1992).
98. HOSOKAWA, N., H. IINUMA, H. NAGANAWA, M. HAMADA, T. TAKEUCHI, T. ITOH, and M. HORI: A New Antibiotic, Structurally Related to Leptomycin A, Flattens the Morphology of V-ras^{ts} NRK Cells. J. Antibiot., **46**, 676 (1993).
99. ABE, K., M. YOSHIDA, H. NAOKI, S. HORINOUCHI, and T. BEPPU: Leptolstatin from *Streptomyces sp.* SAM1595, a New Gap Phase-specific Inhibitor of the Mammalian Cell Cycle. II. Physico-chemical Properties, and Structure. J. Antibiot., **46**, 735 (1993).
100. KOHAMA, T., T. NAKAMURA, T. KINOSHITA, I. KANEKO, and A. SHIRAISHI: Novel Microbial Metabolites of the Phoslactomycins Family Induce Production of Colony-stimulating Factors by Bone Marrow Stromal Cells. II. Isolation, Physicochemical Properties and Structure Determination. J. Antibiot., **46**, 1512 (1993).
101. GERTH, K., D. SCHUMMER, G. HOEFLE, H. IRSCHIK, and H. REICHENBACH: A New Antifungal Compound from *Sorangium cellulosum* (Myxobacteria). Production, Physico-Chemical and Biological Properties. J. Antibiot., **48**, 973 (1995).
102. SCHUMMER, D., K. GERTH, H. REICHENBACH, and G. HOEFLE: Ratjadone: A New Antifungal Metabolite from *Sorangium cellulosum.* Liebigs Ann. Chem. 685 (1995).
103. MERLIN, P., J.C. BRAEKMAN, D. DALOZE, J.M. PASTEELS, and A. DEJEAN: New C_{26} δ-Lactones from the Dufour's Gland of the Urticating Ant *Tetramorium aculeatum.* Experientia, **48**, 111 (1992).
104. WARNING, U., J. YAKUPOVIC, D. FRIEDRICH, V. CASTRO, and F. BOHLMANN: Further Seco-Labdanes from *Hebeclinum macrophyllum.* Phytochem., **26**, 2335 (1987).
105. ZDERO, C., F. BOHLMANN, and R.M. KING: Secolabdanes from *Tamaulipa azurea* and Constituents from other Eupatorieae. Phytochem., **31**, 155 (1992).
106. HASAM, C.M., M.A. HUSSAIN, M.Y. MIA, and M.A. RASHID: Goniothalamin from *Goniothalamus sesquipedalis.* Fitotherapia, **66**, 378 (1995).

107. Ahmad, F.B., W.A. Tukol, S. Omar, and A.M. Sharif: 5-Acetylgoniothalamin, a Styryldihydropyrone from *Goniothalamus uvaroides*. Phytochem., **30**, 2430 (1991).

108. Goh, S.H., G.C.L. Ee, C.H. Chuah, and T.C.W. Mak: 5β-Hydroxygoniothalamin, a Styrylpyrone Derivative from *Goniothalamus dolichocarpus (Annonaceae)*. Nat. Prod. Letters, **5**, 255 (1995).

109. Fang, X.P., J.E. Anderson, C.J. Chang, J.L. McLaughlin, and P.E. Fanwick: Two New Styryl Lactones, 9-Deoxygoniopyrone and 7-Epigoniofufurone, from *Goniothalamus giganteus*. J. Nat. Prod., **54**, 1034 (1991).

110. Wu, Y.C., C.Y. Duh, F.R. Chang, G.Y. Chang, S.K. Wang, J.J. Chang, D.R. McPhail, A.T. McPhail, and K.H. Lee: The Crystal Structure and Cytotoxicity of Goniodiol-7-monoacetate from *Goniothalamus amuyon*. J. Nat. Prod., **54**, 1077 (1991).

111. Wu, Y.C., F.R. Chang, C.Y. Duh, S.K. Wang, and T.S. Wu: Cytotoxic Styrylpyrones of *Goniothalamus amuyon*. Phytochem., **31**, 2851 (1992).

112. Alkofahi, A., W.W. Ma, A.T. McKenzie, S.R. Bryn, and J.L. McLaughlin: Goniotriol from *Goniothalamus. giganteus*. J. Nat. Prod., **52**, 1371 (1989).

113. Goh, S.H., G.C.L. Ee, C.H. Chuah, and C. Wei: Styrylpyrone Derivatives from *Goniothalamus dolichocarpus*. Austral. J. Chem., **48**, 199 (1995).

114. Hasan, C.M., M.Y. Mia, M.A. Rashid, and J.D. Connolly: 5-Acetoxyisogoniothalamin Oxide, an Epoxystyryl Lactone from *Goniothalamus sesquipedalis*. Phytochem., **37**, 1763 (1994).

115. Grazca, T., and V. Jager: Synthesis of Natural and Unnatural Enantiomers of Goniofufurone and its 7-Epimers from D-Glucose. Application of Palladium(II)-Catalyzed Oxycarbonylation of Unsaturated Polyols. Synthesis, 1359 (1994).

116. Mu, Q., C.M. Li, H.J. Zhang, Y. Wu, and H.D. Sun: A New Styryllactone Compound from *Goniothalamus leiocarpus*. Chinese Chemical Letters, **7**, 617 (1996).

117. Yang, Z.C., and W.S. Zhou: Total synthesis of Goniodiol 8-Monoacetate from Cinnamyl Alcohol. Chem. Commun., 743 (1995).

118. Surivet, J.P., J. Gore, and J.M. Vatele: Enantioselective Synthesis of (+)-Goniodiol and its Naturally Occurring Acetylated Analogs. Tetrahedron, **52**, 14877 (1996).

119. Yang, Z.C., and W.S. Zhou: Asymmetric Total Synthesis of (+)-Goniotriol and (+)-Goniofufurone. Tetrahedron, **51**, 1429 (1995).

120. Shing, T.K.M., and J.G. Gillhouley: Enantiospecific Synthesis of (+)-Altholactone and its Three Stereoisomers. Tetrahedron, **50**, 8685 (1994).

121. Somfai, P.: An Enantioselective Total Synthesis of (+)-Altholactone from Diethyl L-Tartrate. Tetrahedron, **50**, 11315 (1994).

122. Yang, Z.C., and W.S. Zhou: Asymmetric Total Synthesis of (+)-Goniopyrone and (+)-9-Deoxygoniopyrone. Chinese J. Chem, **14**, 152 (1996).

123. Friesen, R.W., and S. Bissada: Total Synthesis of (±)9-Deoxygoniopyrone. Tetrahedron Letters, **35**, 5615 (1994).

124. Colegate, M.C., L.B. Din, K.M. Salleh, M.W. Samsudin, B.W. Skelton, K. Tadano, A.H. White, and Z. Zakaria: (+)-Isoaltholactone: A Furanopyrone Isolated from *Goniothalamus* species. Phytochem., **29**, 1701 (1990).

125. Friesen, R.W., and C. Vanderwal: Total Synthesis of (±)-Dihydrokawain-5 ol. Regioselective Monoprotection of Vicinal *Syn*-Diols Derived from the Iodocyclofunctionalization of Allenic Alcohols. J. Organ. Chem. (USA), **61**, 9103 (1996).

126. Mori, Y., and H. Furukawa: Synthesis of Cryptocaryolactone, a 1,3-Polyol-derived Unsaturated Lactone. Chem. Pharm. Bull. (Japan), **42**, 2161 (1994).

127. Drewes, S.E., M.M. Horn, and R. Scott-Shaw: α-Pyrones and Their Derivatives from Two *Cryptocarya* species. Phytochem., **40**, 321 (1995).

128. SEHLAPELO, B.M., S.E. DREWES, and R. SCOTT-SHAW: A 6-Substituted 5,6-Dihydro-α-pyrone from Two Species of *Cryptocarya*. Phytochem., **37**, 847 (1994).

129. DREWES, S.E., M.M. HORN, and S. MAWI: *Cryptocarya liebertiana* and *Ocotea bullata* – Their Phytochemical Relationship. Phytochem., **44**, 437 (1997).

130. FU, X., T. SEVENET, A. HAMID, A. HADI, F. REMY, and M. PAIS: Kurzilactone from *Cryptocarya kurzii*. Phytochem., **33**, 1272 (1993).

131. ZDERO, C., F. BOHLMANN, and G.M. MUNGAI: Clerodanes, Secoclerodanes, Geranyl Geraniol Derivatives and Unusual Sesquiterpenes from *Conyza hypoleuca*. Phytochem., **30**, 575 (1991).

(Received July 23, 1997)

Author Index

Page numbers printed in *italics* refer to References

Subject Index

Springer-Verlag and the Environment

WE AT SPRINGER-VERLAG FIRMLY BELIEVE THAT AN international science publisher has a special obligation to the environment, and our corporate policies consistently reflect this conviction.

WE ALSO EXPECT OUR BUSINESS PARTNERS – PRINTERS, paper mills, packaging manufacturers, etc. – to commit themselves to using environmentally friendly materials and production processes.

THE PAPER IN THIS BOOK IS MADE FROM NO-CHLORINE pulp and is acid free, in conformance with international standards for paper permanency.

MIX
Papier aus verantwortungsvollen Quellen
Paper from responsible sources
FSC® C105338

If you have any concerns about our products,
you can contact us on
ProductSafety@springernature.com

In case Publisher is established outside the EU,
the EU authorized representative is:
Springer Nature Customer Service Center GmbH
Europaplatz 3, 69115 Heidelberg, Germany

Printed by Libri Plureos GmbH
in Hamburg, Germany